Aerothermodynamics of Gas Turbine and Rocket Propulsion

Third Edition

Gordon C. Oates
University of Washington
Seattle, Washington

EDUCATION SERIES
J. S. Przemieniecki
Series Editor-in-Chief
Air Force Institute of Technology
Wright–Patterson Air Force Base, Ohio

Published by
American Institute of Aeronautics and Astronautics, Inc.
1801 Alexander Bell Drive, Reston, VA 20191

American Institute of Aeronautics and Astronautics, Inc.,
Reston, Virginia

Library of Congress Cataloging-in-Publication Data

Oates, Gordon C.
    Aerothermodynamics of gas turbine and rocket propulsion / Gordon C. Oates.—3rd ed.
       p.   cm.—(AIAA education series)
    Includes bibliographical references and index.
    1. Aerothermodynamics.   2. Aircraft gas-turbines.   3. Rocket engines.
    I. Title. II. Series.
TL574.A45028 1998   629.134'353—dc21   97-51303
ISBN 1-56347-241-4

Eighth Printing

Foreword

The revised and enlarged text of *Aerothermodynamics of Gas Turbine and Rocket Propulsion* by the late Gordon C. Oates, published in 1988, continued to fulfill the need for a comprehensive, modern book on the principles of propulsion, both as a textbook for propulsion courses and as a reference for the practicing engineer. The original edition of this book was published in 1984 as the second volume of the then newly inaugurated AIAA Education Series. The Third Edition of this text adds now a companion software representing a set of programs for use with the problems and design analyses discussed in the book. The computer software has been prepared by Daniel H. Daley (U.S. Air Force, retired), Williams H. Heiser (formerly with the U.S. Air Force Academy), Jack D. Mattingly (Seattle University), and David T. Pratt (University of Washington).

The revised and enlarged edition contained major modifications to the original text, and some of the text was rearranged to improve the presentation. Chapter 5 included performance curves, design parameters values, and illustrations of several typical modern turbofan engines. Chapter 7 included a method of analysis to account for the effect of nonconstant specific heats in the cycle analysis equations, and in Chapter 8 a new section was added for an analysis of engine behavior during transient operation. For completeness, Appendices A and B were added: Standard Atmosphere and SAE (Society of Automotive Engineers) Gas Turbine Engine Notation. The Third Edition now has Appendix C, which gives an overview of the companion software.

The AIAA Education Series of textbooks and monographs embraces a broad spectrum of theory and application of different disciplines in aeronautics and astronautics, including aerospace design practice. The series includes texts on defense science, engineering, and management. The complete list of textbooks published in the series (over 50 titles) can be found following page 456. A typical book in the series presents subject material tutorially, discussing the fundamental principles and concepts, and additionally gives perspective on the state of the art. Thus the series serves as teaching texts as well as reference materials for practicing engineers, scientists, and managers.

J. S. PRZEMIENIECKI
Editor-in-Chief
AIAA Education Series

Acknowledgments

This volume has been adapted and extended from a portion of an Air Force report, *The Aerothermodynamics of Aircraft Gas Turbine Engines*, AFAPL-TR-78-52, Wright-Patterson Air Force Base, published in 1978. The author was editor of this report, as well as author of several chapters; thus, it is appropriate to once again acknowledge the support accorded the original report, as well as the present volume.

Much of my personal motivation in writing this volume and the AFAPL report arose from stimulating encounters with Dr. W. H. Heiser in the mid-1970s. At that time we thought that the propulsion industry could well use an up-to-date reference volume on airbreathing engines. In subsequent years Dr. Heiser's support and encouragement in the production of both the present volume and the former report have been greatly appreciated.

My thanks also to Lt. Col. Robert C. Smith of the Air Force Office of Scientific Research (AFOSR) for his encouragement and assistance in the production of the original report and to Dr. Francis R. Ostdiek of the Air Force Aeropropulsion Laboratory (AFAPL), whose support has never wavered from the conception of the original report through production of the present volume.

Many people have affected my career and interest in the aircraft gas turbine field. Most importantly in this respect, I must acknowledge Professor Frank E. Marble of the California Institute of Technology, whose influence will be most evident in this writing. Thanks are also due to my colleagues at Pratt & Whitney who presented me with so many challenging problems during my years of association with them, as well as to the several members of the Department of Aeronautics at the U.S. Air Force Academy whose encouragement and helpful comments greatly assisted me in the writing of the present text.

The writing of the present volume has been made easier by the encouragement and active criticism of Dr. J. S. Przemieniecki, Education Series editor, who himself has undertaken a most demanding task.

Finally, my heartfelt thanks to my wife, Joan, for her enthusiastic encouragement and endless patience.

GORDON C. OATES
University of Washington
Seattle, Washington

Table of Contents

Preface

This book was written with the intent of providing a text suitable for use in both graduate and undergraduate courses on propulsion. The format is such that some overlap will occur when the book is thus used, but the author has found that the diversified background typically found in most graduate classes is such that some repetition of undergraduate material is appropriate.

At the University of Washington, we have used this text for both graduate and undergraduate propulsion courses in two quarter sequences. Typical subject lists considered in the sequences are:

Undergraduate

• The introduction (Chapter 1), which could be considered "propulsion without equations," is discussed and assigned as outside reading.

• Thermodynamics and quasi-one-dimensional flows are reviewed. Because of the frequent use of the results in off-design performance analysis, the expressions for mass flow behavior are emphasized.

• The thermodynamics and fluid dynamics are first applied in the prediction of rocket nozzle behavior. Chemical thermodynamics is then reviewed so that rocket chamber conditions can be estimated.

• Usually, consideration of solid-propellant and nonchemical rockets is delayed until the second quarter. The extent of the consideration depends on the relative emphasis of aeronautical vs astronautical subjects desired.

• Airbreathing engines are introduced with the concepts of ideal cycle analysis. Simple design trends become evident and the simplicity of the equations helps to make the various optimal solutions somewhat transparent, as well as allowing time for the student to construct his own computer programs. Usually, time limitations do not allow considerations of the mixed-flow turbofan.

• Real engine effects are introduced through definition of the component measures. The relationship of the additive drag to the inlet lip suction is stressed.

• Selected examples of nonideal cycles are considered in detail and the student asked to "design" an engine. It is at this point that the student should realize that such a design cannot be determined properly without detailed information regarding the mission and the related aircraft configuration.

• The design concepts are extended to off-design estimation and the restrictive effects of fixed-geometry engines are revealed.

•The course concludes (about two-thirds of a quarter) with consideration of the elementary aerodynamics of rotating machinery. Three-dimensional effects are introduced via the free vortex theory (and its limitations) and through simple radial equilibrium concepts and examples.

Graduate

•Chapters 1 and 2 are briefly reviewed and given as a reading assignment.

•Rockets are not considered in the graduate course; rather the subject proceeds directly to ideal cycle analysis. The optimal solution techniques are emphasized and the mixed-flow turbofan is studied in detail.

•Component performance measures are reviewed, with emphasis placed upon the determination of appropriate average quantities. Supersonic inlet performance estimation is studied in detail.

•Detailed studies of both design and off-design examples of several engine types, including component losses, are considered.

•Blade aerodynamics is considered for both the turbine and compressor, including throughflow theory and cascade theory.

•The course concludes with topics of current interest such as engine poststall behavior, the effects of inlet distortion, etc.

An effort has been made throughout the text to develop the material to the point where computational examples may be easily obtained. Development of the required equations is often algebraically complex and somewhat tedious, but the ease of computation of the resulting equation sets through the use of modern calculators or small computers certainly justifies the effort required. In this respect, problem sets are provided in Chapters 2–11, and the student is urged to attempt as many problems as possible to develop both his problem-solving technique and his understanding of the engine and component behaviors predicted by the related analyses.

GORDON C. OATES
University of Washington
Seattle, Washington

1. INTRODUCTION

1.1 Purpose

The propulsion provided by airbreathing and rocket engines is basically similar in that thrust is obtained by generating rearward momentum in one or more streams of gas. In the case of a rocket the propulsive gas originates onboard the vehicle, whereas in the airbreathing engine most of the propellant gas originates from the free air surrounding the vehicle. This volume presents and explains the aerothermodynamics of rockets and airbreathing engines, detailing the mechanisms of the fluid and thermodynamic behavior in the engine components and revealing the overall behavior of engines and their interactions with the flight vehicles they power.

The interaction of the various components of aircraft and rocket engines, as well as the interactive nature of the entire engine with the flight vehicle, necessitates the extensive use of simplified physical models to provide analytical estimates of performance levels. As a result, the detailed calculations, although straightforward conceptually, can often be quite complex algebraically. For this reason, this introduction will outline many of the aspects of rocket and airbreathing engines in purely descriptive terms. The required analytical methods to support the stated behaviors are developed in subsequent chapters.

1.2 Chemical Rockets

Rockets are generally classified as either "chemical" or "nonchemical," depending upon whether the energy that eventually appears in the propellant stream arises from the release of internal chemical energy via a chemical reaction or is supplied to the propellant from an external source. Chemical rockets are further subdivided into the classes of solid-propellant and liquid-propellant rockets.

Liquid-Propellant Rockets

To date, the most frequently utilized rocket in large boosters has been of liquid-propellant design. Liquid-propellant rockets have several advantages for use as boosters, principal among which is that the most highly energetic propellants (in terms of enthalpy per mass) have been found to be liquid fuels and oxidizers. In addition, the separate fuel and oxidizer can be carried in low-pressure (and hence lightweight) tanks because the very

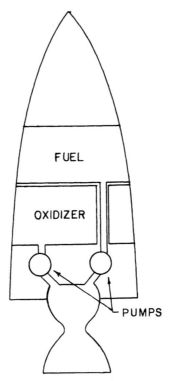

Fig. 1.1   Liquid-propellant rocket.

high pressure of the combustion chamber—so beneficial to efficient propulsion—need be contained only downstream of the fuel and oxidizer pumps. (See Fig. 1.1.)

Further advantages of liquid-fuel rockets can be exploited for use in the upper stages of large rockets. Thus, if maneuvering is required, it can be of benefit to have a variable thrust level capability: liquid propellants lend themselves to "throttling" much more easily than do solid propellants. Advantage can also be taken of the very energetic H_2-O_2 reaction to achieve very high rocket exhaust velocities. It is to be noted that the hydrogen-oxygen rocket is not as attractive for first-stage booster use, because the very low density of molecular hydrogen leads to a requirement for high-volume tankage. In a first stage, such a high-volume requirement has both large structural and large drag penalties due to the large vehicle cross section required within low-altitude, high-density air.

A disadvantage of a liquid-propellant rocket, as compared to a solid-propellant rocket, is that in order to generate large thrust levels, the fuel and oxidizer pumps and all associated piping must be increased in size, with a consequent increase in the overall mass of the vehicle. Very large booster rockets operate with surprisingly low thrust levels, typical values being

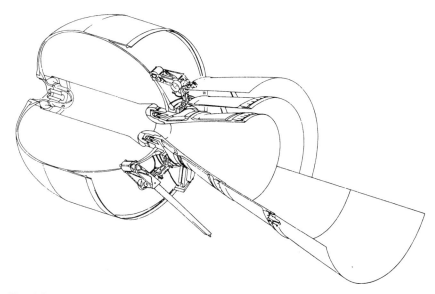

Fig. 1.2   Extendable exit cone (courtesy of United Technologies Chemical Systems Division).

about 1.2 times the rocket's initial weight. Such low thrust levels are utilized both because of the difficulty of providing pumps of sufficient size and because of the desire to restrict the "g loadings" just prior to stage burnout, to acceptable levels.

Design problems encountered in producing a successful liquid-propellant booster rocket include the provision of suitably matched pumps to supply the necessary fuel/oxidizer ratio to give maximum exhaust velocity, and to do so with such accuracy that the fuel and oxidizer tanks approach depletion at the same time. It is usual to maintain an almost constant combustion chamber pressure throughout a rocket firing; as a result, if the rocket climbs through a large altitude variation, a corresponding large variation in nozzle pressure ratio will occur. This variation in nozzle pressure ratio itself implies the use of a variable exit area nozzle if the maximum possible thrust for each altitude is to be approached. (Note that the maximum possible thrust occurs when the nozzle exhaust pressure is very near the ambient pressure.) It is a difficult task to provide a reliable, lightweight nozzle with variable geometry and an associated control system capable of adjusting appropriately for a given ambient pressure. Fortunately, however, several successful developments have occurred, giving the designer of modern rockets the possibility of exploiting rocket nozzles with more than one "design" altitude. Figure 1.2 shows an example of a recently developed rocket with an "extendable exit cone" (EEC) that allows exit pressure matching at three separate altitudes.

Perhaps the most persistent problem area encountered by the designer of any system utilizing very-high-energy sources is that of instabilities. There are several classes of instabilities to be found by an unfortunate designer of rocket engines. An example is that wherein a longitudinal disturbance of the rocket leads to a variation in the pumping rate of the fuel and oxidizer pumps (because of the associated pump inlet pressure fluctuation). The variation in pumping rate in turn leads to a variation in thrust level, which itself leads to a further variation in the pumping rate. Because many rockets are long and slender, and hence very flexible, such disturbances can couple ("feedback") in a way that leads to very large accelerative loads being transmitted to the payload. This class of instability, for rather obvious reasons termed the "pogo" instability, can force unpleasant design requirements, such as extra stiffening, upon the rocket designer.

Two classes of combustion instabilities have been found in the practice of rocket engine design. "Chugging," a relatively low-frequency oscillation, occurs when combustion chamber pressure variations couple with the liquid-fuel and oxidizer supply system. It can happen that, when the combustion chamber pressure momentarily exceeds the time-averaged chamber pressure, the fuel and oxidizer flow rates will decrease because of the decreased pressure drop across the injectors. As a result, the chamber pressure may drop, leading to an increased fuel and oxidizer flow with a subsequent pressure increase, etc. Chugging is usually eliminated by raising the fuel and oxidizer supply pressures so that the injector pressure drop will be so substantial that the chamber pressure fluctuations will not cause significant input flow rate fluctuations. Such a "cure" leads to the requirement for heavy piping and pump equipment.

"Screaming" combustion instability is an acoustic instability identified with the increase in the thermal output of the fuel-oxidizer reaction found with the increases in pressure and temperature identified with an acoustic disturbance. Such disturbances can reflect from the chamber walls, leading to continued amplification of the waves to extreme levels. It appears that the primary source of energy for such disturbances exists in the two-phase region close to the injector heads, so careful development of the injector flow geometry is required to prevent the onset of screaming combustion. Screaming is further reduced by providing the chamber walls with "acoustic tiling" that greatly reduces the intensity of waves reflected from the walls.

Solid-Propellant Rockets

Several advantages of liquid-propellant rockets, as compared to solid-propellant rockets, were discussed above. It is to be noted, however, that there are many missions for which the solid-propellant rocket is the most logical choice. Thus, the relative simplicity of a solid-propellant rocket encourages its use for such purposes as weapons and "strap-on" booster rockets to very large orbiting rockets. The relatively low exhaust velocity provided by solid rocket propellants does not create as great a penalty in the overall rocket mass needed for missions requiring relatively small vehicle velocity changes as it does for missions requiring large velocity changes. (This is because the liquid rocket pumping equipment becomes a larger fraction of the overall mass as the required vehicle velocity change is reduced.) Even though the entire solid-propellant rocket is exposed to the high pressure of the "combustion chamber," the required structural weight is no longer extreme because of the development of the enormously structurally efficient fillament-wound rocket case.

An area of great advantage for the solid-propellant rocket is that of propellant density. With the development of heavily aluminized solid propellants, the propellant density has been greatly increased, leading to the production of rockets with very small cross sections and hence much reduced drag. Such an advantage is particularly pertinent for low-altitude weapons use. Recently, also, propellants with a high surface burning rate have been developed, with the result that it is relatively easy to design solid-propellant rockets with enormous thrust-to-weight ratios.

Development efforts continue along the lines of developing high-energy, high-density, high-burning-rate propellants. In addition, methods of thrust level variation and rapid thrust termination continue to be investigated.

As with liquid rockets, screaming instabilities continue to be of development concern. Methods to reduce such instabilities, or their effects, include use of resilient propellant material and propellant grain cross-sectional shapes that reduce wave reflection.

1.3 Nonchemical Rockets

When " very-high-energy" missions are contemplated (missions for which the required change in vehicle velocity is very large), it is found that even with the use of the most energetic of chemical propellants, the required fraction of propellant mass to overall vehicle mass becomes excessive. Elementary considerations reveal that the rocket "mass ratio" (initial mass divided by final mass) is very sensitive to the ratio of required vehicle velocity change to rocket exhaust velocity. In order to reduce the mass ratios required, alternative schemes are investigated that allow the addition of energy to the propellant from sources other than the chemical energy of the propellant itself.

Once the possibility of an external energy source is considered, the problem of the energy supply becomes separate from the problem of choosing the most suitable propellant. Thus, the energy could be supplied to a propellant directly by thermally heating the propellant, the thermal energy itself being supplied by a nuclear reactor, a solar concentrator, radiative energy supplied from a remote energy source, or any other of a wide variety of schemes.

When very-high-energy levels are desired, a variety of electrically powered devices deserve consideration, two examples of which are briefly described in the following. The electrical power for the electrically driven rocket might be supplied by a nuclear-powered motor-generator set or possibly by a solar-powered motor-generator set. Provision of space power at manageable power-to-mass ratios remains one of the most perplexing problems in the next generation of spacecraft. It is to be noted that systems delivering power for such high-energy levels of propellant must be equipped with "waste heat" radiators. Such radiators must be extremely large or must operate at very high temperatures with a consequent penalty in the cycle thermodynamic efficiency (hence requiring a massive "engine"!).

Nuclear-Heated Rocket

A conceptually simple idea, the nuclear-heated rocket operates by having the propellant pass through heat-exchange passages within a nuclear reactor and then through a propelling nozzle. Conventional nuclear reactors must operate with a limit upon the maximum solid-surface temperature found within the reactor, in order to ensure the reactor's structural integrity. Thus, quite unlike the conditions found in chemical rockets where the energy release is within the propellant, the propellant temperature in nuclear reactors is restricted to being less than the wall temperatures and hence substantially less than that found within chemical rocket propellants.

The advantage of a nuclear-heated rocket arises because of the freedom in the choice of propellants. The most desireable propellant for such a system is that which gives the maximum possible specific enthalpy for the given limiting temperature. The specific enthalpy of a perfect gas is (nearly) inversely proportional to the molecular weight, so the logical choice of propellant for a nuclear-heated rocket is evidently molecular hydrogen (molecular weight of two).

The Rover and Nerva programs successfully demonstrated that nuclear-heated rockets utilizing molecular hydrogen as a propellant could achieve exhaust velocities almost twice those of the best chemical rockets. The related mass ratio for a very-high-energy mission could be less than one-third that for a chemical rocket!

It is unfortunate, however, that even such an enormous decrease in the mass ratio (or an equivalent increase in the payload) is such that even nuclear-heated propulsion gives an insufficient exhaust velocity for use in manned planetary missions. To date, it has also been found that the additional mass of the reactor and its shielding, as well as the enormity of the development problems expected, have precluded the use of nuclear-heated propulsion for lunar or near-Earth use. It is possible, however, that the future may see the use of "nuclear tugboats" for reusable lunar transport and synchronous orbit transport applications.

Electrical Rockets

At the very-high-energy end of the propulsion spectrum, so much energy must be added to the propellant that "self-cooling" schemes (such as the nuclear rocket) cannot provide sufficient energy; thus, systems that provide energy through use of a motor-generator configuration and its required radiator become mandatory. Relatively straightforward analysis shows that for such cases the optimum choice of exhaust velocity is not a limitingly large value. This is because the mass of the power supply and radiator increases as the propellant stream energy increases, so that the combination of propellant mass and power supply mass passes through a minimum at an intermediate value of exhaust velocity. Detailed studies of possible manned solar missions indicate the optimal exhaust velocities to be in the range of 30,000–50,000 ms^{-1}.

Electrothermal thrustors (arcjets).   Electrothermal thrustors are conceptually simple devices that operate by passing an electrical current directly through the propellant so that the electrical energy is deposited as thermal energy within the propellant. The high-enthalpy propellant is then expanded through a conventional nozzle.

Electrothermal thrustors are limited in performance by the onset of high ionization (or dissociation) losses, as well as high thermal losses to the containing walls. At present, attainable exhaust velocities are limited to about 17,000 ms^{-1}, so that the devices are inappropriate for planetary missions. Their relative simplicity makes them viable candidates for use in orbit perturbation and stationkeeping.

Electrostatic rockets (ion rockets).   When very-high-energy exhaust streams are considered, the particle energies are many times larger than typical ionization energies, so the loss (for propulsive purposes) of the ionization energy can be considered of small import. If the exhaust stream is fully ionized, however, the exhaust stream can be contained and directed through the use of electric (and possibly magnetic) fields alone. As a result, "viscous containment" by solid boundaries is not required and the problem of solid-surface erosion is vastly reduced.

Electrostatic thrustors operate by accelerating a stream of ions in an electrostatic field and subsequently neutralizing the exhaust stream by the injection of electrons. With such very-high-energy devices, a performance limitation occurs because of the difficulty of creating sufficient thrust per area. The thrust limitation occurs because the beam flow rate is restricted due to the proximity of the departing ions to the ion emitter surface, which much reduces the ion departure rate. (The beam becomes "space charge limited.") Straightforward analysis shows that the beam thrust is proportional to the square of the mass/charge ratio, to the fourth power of the exhaust velocity, and to the inverse square of the anode-cathode spacing. As a result, very small spacings and propellants with very high mass/charge ratios (cesium or mercury) are used.

To date, electrostatic thrustors have demonstrated successful performance in the range of exhaust velocities in excess of 50,000 ms^{-1}. The great remaining problem for future electrical rocket development is the generation of the required power at acceptable power-to-mass ratios.

1.4   Airbreathing Engines

Performance Measures and Engine Selection Considerations

The two most commonly used performance measures for airbreathing engines are specific thrust (the thrust force divided by the total mass flow rate of air through the engine) and specific fuel consumption (the mass flow rate of the fuel divided by the thrust force of the engine). These performance measures are themselves related to the more fundamental efficiency measures, the engine thermal efficiency and the propulsive efficiency.

It is to be noted that the (useful) mechanical output of the engine appears entirely as the rate of the generation of kinetic energy in the exhaust stream or streams. (Note that to a thermodynamicist kinetic energy is entirely equivalent to work.) It is fortunate for the aircraft engine designer that the kinetic energy of the exhaust stream is already in a form appropriate for the purpose of providing thrust. This is in contrast to (for example) a ground-based gas turbine engine that would have to be designed with subsequent turbine stages to remove the exhaust stream kinetic energy and convert it to shaft power. Such subsequent stages would have further component losses, leading to an engine thermal efficiency substantially less than that of an equivalent aircraft engine. Because of the equivalency of kinetic energy and work, the thermal efficiency can be obtained as the ratio of the rate of kinetic energy generation to the rate of thermal (chemical) energy input.

The propulsive efficiency gives a measure of how well the energy output of the engine is utilized in transmitting useful energy to the flight vehicle. It is defined as the ratio of the power transmitted to the flight vehicle and the rate of kinetic energy generation.

Elementary manipulations show that a propulsive efficiency increase will be accompanied by a specific thrust decrease, a situation that adds to the designer's dilemma. The specific fuel consumption is inversely proportional to the product of the thermal and propulsive efficiencies (as well as being proportional to the flight velocity). Hence, it is obvious that it would be desirable to increase the propulsive efficiency in order to reduce the specific fuel consumption. Inevitably, however, the amount of air handled by the engine would have to be increased (to maintain the same level of thrust) because of the related decrease in specific thrust. The requirement to increase the quantity of air handled by the engine can lead to difficult engine installation problems. The use of a very-large-diameter fan with a large bypass ratio, for example, might require an inordinately long landing gear as well as, perhaps, a gearbox to better match the fan tip speed with the tip speed of the turbine driving the fan.

It is evident that the optimum choice will depend much on the "mission." Thus, because for long-range transport aircraft fuel consumption is of dominant concern, the optimal design favors use of an engine with a high bypass ratio and a low fan pressure ratio. Figure 1.3 shows the PW2037 engine recently introduced into commercial service. This engine has a 5.8 bypass ratio and 1.4 fan pressure ratio (giving a high propulsive efficiency and hence a low specific fuel consumption, but also a low specific thrust). The engine has a high compressor pressure ratio (≈ 32), which helps to give a very high thermal efficiency, but also somewhat further contributes to the low specific thrust.

The extreme performance demands of the military environment lead to the selection of quite different design choices. Thus, high specific thrust is required for flight at high Mach numbers or for maneuvering flight at transonic Mach numbers. As a result, lower fan bypass ratios and higher fan pressure ratios are found to be suitable. Even then, such aircraft must have

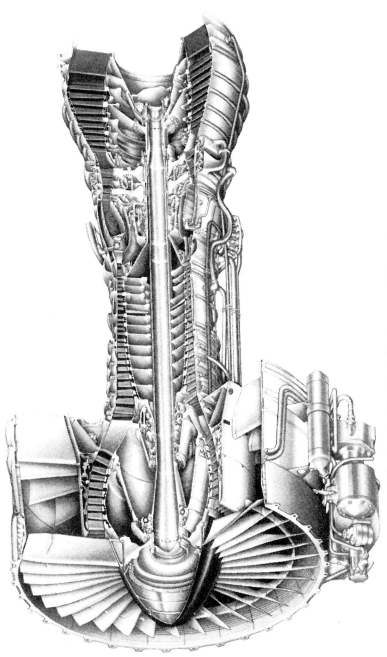

Fig. 1.3 PW2037 turbofan (courtesy of Pratt & Whitney).

acceptable subsonic cruise capability; thus, a compromise between high specific thrust and low specific fuel consumption is inevitable. The demand for compromise of such multimission aircraft is somewhat eased through incorporation of afterburning, which greatly increases the specific thrust at the high-performance condition and hence allows use of a more fuel-efficient system for subsonic cruise. Figure 1.4 shows the Pratt & Whitney F100 afterburning turbofan engine. It is to be noted that this engine has a three-stage fan with a pressure ratio of ≈ 3 and bypass ratio of 0.78. The compressor pressure ratio is 25, which is relatively high for this class of

Fig. 1.4   F100 afterburning turbofan (courtesy of Pratt & Whitney).

engine and is clearly incorporated to aid the subsonic fuel efficiency.

The choice of the appropriate engine is also much affected by the possibly conflicting requirements of takeoff thrust and cruise thrust. It is evident that for zero flight velocity the power required to supply a given thrust level is proportional to the exhaust velocity (inversely proportional to the mass flow). At high flight velocities, large powers are required; as a result engines with low specific thrusts sized for takeoff thrust requirements are found to be underpowered at high forward speeds (as is the case for conventional turboprops). Conversely, engines with very high specific thrust (turbojets) must be made oversize to satisfy the takeoff requirement and hence tend to be too powerful for cruise flight at subsonic speeds. This latter condition results in throttled-back operation at cruise with a consequent loss in thermal efficiency because of the related reduction in compressor pressure ratio.

It is of interest to note the recent development of very-high-power turboprops designed to fly at Mach numbers up to 0.8 (Fig. 1.5). Such engines are so powerful that in the takeoff condition they are operated at part throttle so as to allow the use of lighter gearboxes and to prevent propeller stalling.

It is also worthy of note that the turbofan, so popular in commercial airline use, provides an excellent balance between the takeoff thrust and cruise thrust requirements.

Engine Components

The major components of an aircraft gas turbine engine are the inlet, compressor (and fan), combustor, turbine, and nozzle. In this section the

Fig. 1.5   Test model of high-disk-loading propfan in wind tunnel.

principles of operation of each component and the design limitations and problem areas still found in practice will be briefly described.

Inlets. The design characteristics of an inlet very much depend on whether the inlet is to be flown at subsonic or supersonic speed. In either case, the requirement of the inlet is to provide the incoming air to the compressor (or fan) face at as high a (stagnation) pressure as possible and with the minimum possible variation in both stagnation pressure and temperature. For both supersonic and subsonic flight, modern design practice dictates that the inlet should deliver the air to the fan or compressor face at a Mach number of approximately 0.45. As a result, even for flight in the (high) subsonic regime, the inlet must provide substantial retardation (diffusion) of the air.

The design of subsonic inlets is dominated by the requirements to retard separation at extreme angle of attack and high air demand (as would occur in a two-engine aircraft with engine failure at takeoff) and to retard the onset of both internal and external shock waves in transonic flight. These two requirements tend to be in conflict, because a somewhat "fat" lower inlet lip best suits the high angle-of-attack requirement, whereas a thin inlet lip best suits the high Mach number requirement. Modern development of the best compromise design is greatly aided by the advent of high-speed electronic computation, which allows analytical estimation of the complex flowfields and related losses.

Estimation of the losses within supersonic inlets is an easier task than for subsonic inlets for the simple reason that the major losses occur across the shock waves, and hence may be estimated using the relatively simple shock wave formulas. More exacting estimates require estimation of the boundary-layer and separation losses.

There is a wide variety of design possibilities for supersonic inlets, ranging from the simple normal shock inlet (which has a single normal shock wave located in the flowfield ahead of the inlet lip) to the internal, external, or mixed compression inlets depicted in Fig. 1.6.

The design of an inlet and its related control system is a demanding task, particularly for an aircraft with very high Mach number capability. Optimal performance at a given Mach number requires exacting definition of the inlet geometry. [This is so that the shock wave strengths as well as wall impingement locations (in the neighborhood of suction slots) can be accurately determined.] When such inlets are flown at speeds other than the design Mach number, complex geometrical variation must occur if the inlet performance is not to deteriorate excessively.

It is to be noted that the great difficulty of providing acceptable inlet performance over a wide range interacts with the proper determination of an aircraft's flight envelope. The necessary variable geometry and actuation equipment can so increase the vehicle weight that insistence upon a high Mach number capability can greatly compromise the aircraft performance at lower Mach numbers. This situation is particularly true for military aircraft, where it is found in combat that the aircraft energy degradation in

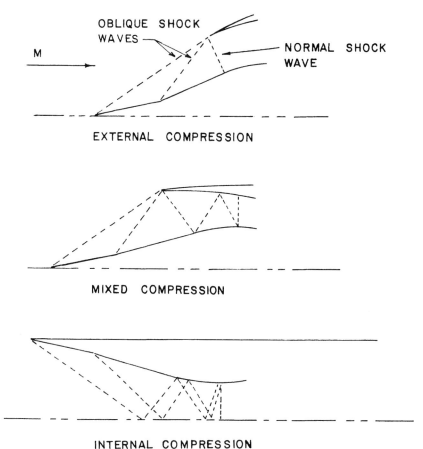

Fig. 1.6 Supersonic inlets.

severe maneuvering is so extreme that most actual combat occurs in the neighborhood of a unit flight Mach number! Clearly the F-16 aircraft, which has a simple normal shock inlet, has been designed to optimize its performance in this lower Mach number regime.

Compressors and fans. There are two major classes of compressors used in aircraft gas turbines, the centrifugal and the axial. In the centrifugal compressor, air is taken into the compressor near the axis and "centrifuged" to the outer radius. Subsequently, the swirl of the outlet air is removed and the air diffused prior to entry into another compressor stage or into the combustor. Centrifugal compressors have the advantage in that they are rugged and deliver a high-pressure ratio per stage. In addition, they are easily made in relatively small sizes. The disadvantages of the centrifugal compressor are that it is generally less efficient than an axial compressor and it has a large cross section compared to the cross section of the inlet flow. In modern usage, centrifugal compressors are used with relatively small engines or as a final stage (following an axial compressor) in larger engines.

Axial compressors are used in the majority of the larger gas turbine engines. In such compressors, enthalpy addition occurs in the rotating rows (rotors) in which, usually, both the kinetic energy and static pressure are increased. The stator rows remove some of the swirl velocity, thereby decreasing the kinetic energy and consequently increasing the static pressure. The limiting pressure rise through an axial compressor row occurs when the adverse pressure gradient on the blade suction surface becomes so severe that flow separation occurs. When substantial separation occurs, the entire compressor may surge (that is, massive flow reversal will occur) or rotating stall may result. Rotating stall is the condition where the flow in several blades stalls (becomes almost stagnant) and the "package" of stalled fluid then rotates around the blade row. The rotating stall condition is particularly dangerous, because very large vibratory stresses can occur as the blades enter and depart the stall.

In order to achieve a high limiting pressure rise per stage, it is beneficial to design the stage so that the static pressure rise in each row is almost the same (so that one row will not stall prematurely). The degree of reaction $°R$ is defined as the ratio of the static pressure rise in the rotor divided by the static pressure rise across the stage and provides a measure of how well balanced the blade row loadings are. When detailed designs are investigated, however, it is found that, inevitably, the degree of reaction increases with increase in the radius. A related result is that stator blades are limited in their performance at the hub, whereas rotor blades are usually limited at the tip. Further, the effect of the variation in $°R$ with the radius results in rows with large tip-to-hub ratios being more limited in attainable pressure rise than rows with small tip-to-hub ratios. This result in itself provides the designer with yet another compromise, in that a compressor with a small tip-to-hub ratio would require fewer stages to attain a given pressure ratio than would a compressor with a large tip-to-hub ratio, but would also require a greater outer diameter in order to handle the same quantity of air.

By and large, fuel efficiency increases with an increase in the compressor pressure ratio. The optimal pressure ratio is, however, constrained by several design limitations and tradeoffs. A compressor with a very-high-pressure ratio could require an excessively heavy casing if the compressor was to be used to its maximum capability in low-altitude (high-ambient-pressure) conditions. In addition, the high pressure tends to increase the casing expansion and distortion. The effects of such expansion appear in increased losses due to the flow around the blade tips. This situation is even further aggravated for very-high-pressure-ratio compressors, because the high-pressure blades in even large engines are very small and the tip leakage affects a proportionately larger portion of the flowfield.

High-pressure ratios also greatly compromise the off-design performance. It is evident that the overall contraction of the compressor annulus area will be chosen so as to provide the correct axial velocity throughout the compressor for the design condition. Thus, at off-design operation the axial velocity distribution will not be appropriate for the then present blade speeds. Consider, for example, the conditions existing at very low blade speed as would be found during starting. Under such conditions, the increase in pressure, and hence density, across each stage is far below that to be found when the compressor is at design speed. As a result, the axial speed of the flow must increase greatly as the air proceeds rearward into the contracted annular cross section. This effect can be so extreme that the flow can approach "choking" (approach Mach 1). Under these conditions, the flow tends to drive the rearward blades ("windmilling"), whereas the resultant back pressure slows the incoming flow and causes the frontward blades to stall.

The demands of these off-design considerations have lead to several ingenious "fixes." Thus, modern high-pressure-ratio compressors utilize "bleed valves" that release a portion of the air from the intermediate blade rows so as to reduce the axial velocity in subsequent stages. Several of the early stages of the compressor are equipped with variable stators so that the flow can be directed in the direction of the rotation of the rotor and so reduce the angle of attack and hence the tendency to stall. Finally, modern compressors are equipped with "multiple spools" such that portions of the compressors are driven by their own separate portions of the turbine. By so doing, each portion of the compressor (and its related turbine!) tends to adjust its speed better to the then present axial velocity.

Problems of scale are to be found when larger engines are scaled down for use in smaller aircraft. Tip clearance problems will obviously become greater, and the high-pressure blading can become extremely small. For this reason, it is often advantageous to employ a centrifugal compressor as a final stage, rather than the equivalent several stages of an axial compressor. A further problem of considerable consequence arises in the design of the first rows of a small-scale compressor or fan. All aircraft compressors must have sufficient tolerance to withstand bird strikes, and it is a considerably more demanding task to provide the required structural integrity—while retaining aerodynamic performance—in a small-scale engine than in a large-scale one.

Combustors. Combustors operate by having fuel sprayed into a central "flame-stabilized" region where the droplets evaporate and the fuel ignites. The fuel-rich gas of the combustion region is mixed with cooling air passed through holes in the combustion liner. Good combustor design is directed toward achieving complete burning of the fuel with minimal pressure loss. Sufficient mixing must be introduced to reduce the presence of "hot spots" as much as possible, provided that the pressure drop is not excessive.

Present development efforts are directed toward the reduction of pollutant emissions, operation with alternative fuels, and the achievement of stable and efficient operation in off-design operation.

Turbines. Virtually all turbines used in aircraft gas turbine engines are of the axial flow type and hence are superficially similar to an axial compressor operating in reverse. The engineering limitations on the performance of a turbine stage are, however, very different than the engineering limitations for a compressor stage. The large decrease in pressure found in turbines much reduces the tendency of the suction surface flow to separate, so turbine stages can be designed with very large pressure ratios. The gas entering the turbine is at a very high temperature, however, and hence the initial turbine stages must be cooled by passing air from the compressor outlet through the turbine blades.

Turbine cooling proves to have fairly severe performance penalties, so there is a premium on the development of very-high-temperature materials that will allow the use of high turbine inlet temperatures with only minimal cooling provided. Such material must exist in an extremely demanding environment, for not only are high temperatures encountered, but both the temperatures and centrifugal stresses are frequently cycled. Dimensional stability must be high, because if excessive creep occurs (brought about by the high thermal and centrifugal stresses), excessive rubbing of the blade tips on the outer annulus could occur. The problems of tip rubbing are so severe that in recent years "active clearance control" has been introduced. Active clearance control is achieved by actively cooling the turbine annulus wall to achieve the appropriate tip clearance.

Nozzles. The final component of the aircraft gas turbine engine, the nozzle, accelerates the high-pressure exhaust gas to close to the ambient pressure. The primary design difficulties arise with nozzles intended for use in aircraft with wide Mach number capability. Flight over a wide Mach number range introduces a wide range of ram pressure ratios, with a consequent wide range of nozzle pressure ratios. Optimum nozzle performance occurs when the nozzle exit pressure is not far from ambient; thus, for nozzles with a large operating pressure ratio range, substantial geometrical variation must be possible.

As a result of the geometrical restraints required for good matching with the external flowfield, the major effects of nozzle performance tend to be identified with the effect of exit pressure mismatch and installation effects on the installed thrust through "boat-tail" drag or exhaust plume back pressuring.

Because of the relative ease of geometrical variation, two-dimensional nozzles are presently under consideration for use on missions with large area variation requirements or for missions utilizing thrust vectoring. An additional possible benefit of the geometric flexibility of two-dimensional nozzles arises through the possibility of utilizing such flexibility to shield the internal hot surfaces from heat-seeking weapons.

1.5 Summary

In the foregoing, the principles of operation, design considerations, and present status of some of the aspects of rocket and airbreathing engines have been reviewed in purely descriptive terms. In the chapters to follow, the basic thermodynamics and fluid mechanics necessary to allow a quantitative estimate of many of the behaviors described in this chapter will be reviewed. Subsequent chapters introduce many simplified models of the processes found within the engines and their components that lead to analytical estimates of the engine performances. The underlying methodology of the modeling techniques has far greater applicability than the limited number of examples presented and the reader is urged to ponder the solution methodology itself, as well as the implications of the analytical results.

2. THERMODYNAMICS
AND QUASI-ONE-DIMENSIONAL FLUID FLOWS

2.1 Introduction

This chapter will be limited to a very brief review of the concepts and laws of thermodynamics and to the description of quasi-one-dimensional flows. It is important to understand that the subject of thermodynamics itself is restricted to the study of substances in equilibrium, including thermal, mechanical, and chemical equilibrium. Hence, thermodynamics is more nearly analogous to statics than to dynamics. This limitation might seem to be hopelessly restrictive to an engineer because he is most often concerned with flow processes, and the substances involved in flow processes are not, strictly speaking, in equilibrium. In fact, however, in most cases of interest to the engineer, such substances may be considered to be in "quasiequilibrium" such that local values of the thermodynamic properties may be meaningfully defined.

Flow processes have losses associated with them that can be identified with the lack of equilibrium, and it should be realized that the quantitative prediction of such losses is beyond the scope of thermodynamics. The "theory of transport phenomena" must be applied in order to quantitatively estimate such losses, and the prediction of the various "transport coefficients" must rely upon the techniques of kinetic theory.

The complicated transport mechanism known as turbulence is an essentially macroscopic phenomenon. The accurate description of losses in turbomachines relies very heavily upon the accurate description of turbulent processes because the turbulent transport mechanisms contribute the dominant portion of the losses in virtually all turbomachine components.

2.2 Definitions

It is important to be precise in the definition of terms intended for use in the context of thermodynamics so that possible confusion with the colloquial usage of a term may be avoided. A very abbreviated list of definitions, as will be utilized herein, follows.

Property

A property is a characteristic (of a system) that can in principle be quantitatively evaluated. Properties are macroscopic quantities that involve

19

no special assumptions regarding the structure of matter (i.e., temperature, pressure, volume, entropy, internal energy, etc.).

Properties are grouped into two classes: (1) extensive properties that are proportional to the mass of the system, and (2) intensive properties that are independent of the mass of the system. Any extensive property can be made an intensive property simply by dividing by the mass of the system.

Thermodynamic State

The state of a system is its condition as described by a list of the values of its properties.

Thermodynamic Process

In the limiting case when a change in the properties of a thermodynamic system takes place very slowly, with the system at all times very close to equilibrium, the "in-between" states can be described in terms of properties. A change under such conditions is called a thermodynamic process.

Work

The concept of work is a familiar one from mechanics. Work is said to be done by a system when the boundary of the system undergoes a displacement under the action of a force. The amount of work is defined as the product of the force and the component of the displacement in the direction of the force. Work is so defined as to be positive when the system does work on its surroundings. It should be noted that work can by no means be considered a property, but rather is identified with the transitory process. In order to avoid possible confusion, the term "work interaction" will often be used when a system is undergoing a "work process." Thus, a positive work interaction occurs when the system does work on its surroundings and a negative work interaction occurs when the surroundings do work on the system.

Heat

In analogy to the work interaction defined above, a heat interaction can be defined. Thus, when a hot body is brought into contact with a cold body, the temperature of each changes. It is said that the cold body experiences a positive heat interaction. Similarly, the hot body simultaneously experiences a negative heat interaction. In order to define the heat interaction of a body quantitatively, the change in one or more properties (usually the temperature) of a standard system is measured when the standard system and the body reach equilibrium after being placed in contact. Like the work interaction, the heat interaction is used only in connection with the transitory process.

In order to further emphasize that the work interaction and heat interaction are defined only in connection with the transitory process, when infinitesimal increments of work and heat interactions are considered, the special symbols $d'W$ and $d'Q$ will be introduced. The prime in these symbols is to remind the reader that "W" and "Q" are not properties, and hence the infinitesimal increments cannot be integrated to give the change in W and Q.

2.3 The Laws of Thermodynamics

In the following sections, the first three laws of thermodynamics will be discussed. Because thermodynamics is primarily concerned with heat and work interactions, the experiments leading to the formulation of the laws are in all cases considered to deal with macroscopically stationary materials. That is, although the material boundaries may be movable, there is no contribution to the interchanges of energy, etc., due to a change in potential energy or kinetic energy of the macroscopic sample. It is to be understood that when such contributions are of importance in an interaction (as they obviously are in most processes in turbomachinery), they may be included later in a straightforward manner by applying the laws of mechanics. The interaction of thermodynamic and overall mechanical energy effects is considered in Secs. 2.15–2.17.

2.4 The Zeroth Law of Thermodynamics

This law is so fundamental in classical thermodynamics that it was at first accepted as being self-evident and was not formally denoted a law until after the "first" and "second" laws had become established. However, it is now recognized that it is of fundamental importance to the foundation of classical thermodynamics.

Experience has shown that if a hot body is brought into contact with a cold body, changes take place until eventually the hot body stops getting colder and the cold body stops getting hotter. At this point the bodies are in thermal equilibrium. The zeroth law states:

> If two bodies are separately in thermal equilibrium with a third body, they are in thermal equilibrium with each other.

It is evident that bodies in thermal equilibrium have some property in common and this property is the temperature. Thus, if desired, any reference temperature scale (a mercury thermometer, for example) can be used to determine the temperature of an object; but it can be shown that the second law allows the definition of a temperature scale independent of the properties of the reference substance.

Thus, the zeroth law allows definition of the property temperature, although it does not lead to the definition of any particular reference scale for temperature. The restriction of thermodynamics to the study of equilibrium conditions is very evident here, because temperature could be defined as that property the bodies had in common only if the bodies were allowed to reach equilibrium.

2.5   The First Law of Thermodynamics

In 1840, Joule conducted his famous experiment to establish the equivalence of a heat interaction and a work interaction. His result, available in many texts, allowed the definition of a new property, the energy E. Thus, denoting $d'Q$ as an incremental heat interaction and $d'W$ as an incremental work interaction results in

$$E - E_1 = \int_1 (d'Q - d'W) \qquad (2.1)$$

where E_1 is the energy in the reference state. (Recall the special notation $d'Q$ and $d'W$ introduced in Sec. 2.2)

Here it should be noted that (1) the energy can be defined in terms of the system properties only when the end states are equilibrium states, although the intervening states on the path need not be in equilibrium; and (2) the energy is given as a difference in magnitude between the two states and is not defined in absolute values.

The definition of a simple system states that such a system is completely defined in terms of any two intensive properties, and the energy in such a restrictive case is usually termed the internal energy and denoted by U. Usually, any two of the three properties—temperature (defined from the zeroth law), pressure (defined from mechanics), or volume per unit mass (defined from geometry)—are used as the independent properties. It is apparent also that the internal energy of a system can be "tapped" so that a net outflow of energy is obtained in the form of either a heat or a work interaction. Any observant person, however, can sense that there must be some restriction on the form in which this outflow of energy can occur because of the comparative ease of obtaining a negative heat interaction from a system as compared to that of obtaining a positive work interaction. This restriction is formalized in the statement of the second law of thermodynamics.

The differential form of the first law may be written

$$d'Q = dE + d'W \qquad (2.2)$$

2.6   The Reversible Process

A very useful reference process in thermodynamics is that of the reversible process. A thermodynamic process is defined as a process in which changes take place so slowly that the "in-between" states of the system are at all times close to equilibrium so that the intermediate states can be described in terms of properties. In addition, in the case where all external constraints vary only infinitesimally from equilibrium, the process is said to be "reversible." The terminology is obvious here in that if there is an infinitesimal change in the properties of the system or its surroundings, the process can be reversed and the system returned to its initial conditions.

Our interest in this study is in gases, which are very close to ideal "simple systems," in that their thermodynamic state is (very nearly) completely determined by any two thermodynamic properties. For such a substance the element of work done in a reversible process is simply $d'W = p\,dV$, in which, by the definition of a reversible process, the pressure must be defined at all times throughout the process. Thus, the first law for a gas undergoing a reversible process may be written

$$d'Q_r = dU + p\,dV \tag{2.3}$$

where U is the internal energy and V the volume of the gas.

2.7 Derived Properties: Enthalpy and Specific Heats

So far in this discussion of thermodynamics, only four properties have been defined and used—specific volume v, pressure p, temperature T, and internal energy U. These properties are defined very fundamentally, but there is great utility of notation allowed if certain properties derived from these fundamental properties are defined. It should be noted, of course, that any combination of properties is itself a property.

One group of properties that occurs frequently for gasdynamicists is $(u + pv)$, which is given the symbol h and the name specific enthalpy, i.e.,

$$h = u + pv \tag{2.4}$$

The first law may be written in terms of enthalpy for a gas undergoing a reversible process as

$$d'q_r = dh - v\,dp \tag{2.5}$$

where $d'q_r$ represents the heat increment per mass in a reversible process.

[Henceforth, for convenience we shall refer to (specific) quantities.]

Two further useful derived properties are the specific heat at constant pressure and the specific heat at constant volume. These specific heats (or specific heat capacities) are defined as the (differential) heat interaction (at constant pressure and volume) occurring in a reversible process, divided by the resultant (differential) temperature change. That is,

$$C_p = \frac{(\delta q_r)}{\delta T} \qquad p = \text{const} \tag{2.6a}$$

$$C_v = \frac{(\delta q_r)}{\delta T} \qquad v = \text{const} \tag{2.6b}$$

The two forms of the first law given above show that these terms are, in fact, properties and that they may be written

$$C_p = \left(\frac{\partial h}{\partial T} \right)_p \tag{2.7a}$$

$$C_v = \left(\frac{\partial u}{\partial T} \right)_v \tag{2.7b}$$

The ratio of the specific heats is also often used and is given the symbol γ, where

$$\gamma \equiv C_p/C_v \tag{2.8}$$

2.8   The Second Law of Thermodynamics

Joule's experiment leading to the establishment of the first law of thermodynamics involved a negative work interaction with a system and the consequent increase of the internal energy of the system, rather than the reverse process of a positive work interaction at the expense of the internal energy of the system. The engineer is usually concerned with the latter procedure. It is clear that a very desirable type of engine—one that would not violate the first law—would be one using a very large reservoir of internal energy (the ocean, for example) and converting the energy drawn from the reservoir entirely into a work interaction. Even though, wittingly or unwittingly, inventors still attempt to obtain patents for devices capable of performing in the manner described above, no working model of such a device has ever been constructed; and the very long history of failures to do so has long since led to the belief that it is impossible. This restriction on the first law has been formalized in the second law of thermodynamics, which may be stated as,

> It is impossible for any engine, working in a cyclic process, to draw heat from a single reservoir and convert it to work.

This statement (or any of its equivalent forms), when combined with the zeroth and first laws, allows many remarkable deductions to be drawn concerning the thermodynamic behavior of matter.

These deductions are usually presented as theorems and include among them the definition of a new intensive property, the entropy s. Thus,

$$s - s_1 = \int_1 \frac{\mathrm{d}'q_r}{T} \tag{2.9}$$

The differential form is

$$T\,\mathrm{d}s = \mathrm{d}'q_r \tag{2.10}$$

Note here:

(1) The entropy, like the internal energy, is given as a difference in magnitude between the two states and is not defined as an absolute value.

(2) It is very important to note that the definition of entropy in *no* way requires that the state to be described be reached reversibly from some reference state. The integral relation given above simply gives a procedure for calculating the entropy difference between the specified end states. The value of entropy itself, like temperature, pressure, or any other thermodynamic property, depends only on the (equilibrium) conditions at the specified state and in no way depends on the "history" of the processes leading to that state. As stated previously, the assignment of any two thermodynamic properties completely defines all further thermodynamic properties for a simple system. Hence, for example, if the pressure and temperature of the air in a given room are specified, so too is the entropy. Conversely, of course, if the temperature and entropy are specified, so too is the pressure.

A further theorem of enormous consequence is that the entropy of an isolated system cannot decrease. This theorem has great utility in investigating the possibility or impossibility of an assumed process.

2.9   The Gibbs Equation

An equation relating the five fundamental properties of thermodynamics —specific volume v (defined from geometry), pressure p (defined from mechanics), absolute temperature T (defined from the zeroth and second laws), internal energy u (defined from the first law), and entropy s (defined from the second law)—follows directly by combining Eqs. (2.3) and (2.10). Thus,

$$T\,ds = du + p\,dv \qquad (2.11)$$

This equation is known as the Gibbs equation and, as stated, relates the five fundamental properties of thermodynamics. Note that a similar equation is obtained in terms of the derived property, enthalpy, by combining Eqs. (2.5) and (2.10) to give

$$T\,ds = dh - v\,dp \qquad (2.12)$$

2.10   The Gibbs Function and the Helmholtz Function

Two further derived properties are defined by the relationships,

Gibbs function: $G = h - Ts$ (2.13)

Helmholtz function: $F = u - Ts$ (2.14)

In some applications, particularly those involving determination of chemical equilibrium, these newly defined properties have important physical interpretations. For the purposes here, however, note that expressions obtained for differential changes in G and F may be promptly utilized to obtain a very useful set of relationships known as Maxwell's relations.

2.11 Maxwell's Relations

By definition, when a simple thermodynamic system is considered, specification of any two (intensive) properties completely defines the thermodynamic state (and hence all properties) of the system. Thus, the differential change dz in a property z is given in the form

$$dz = M(x, y)\,dx + N(x, y)\,dy \tag{2.15}$$

In this expression

$$M = \left(\frac{\partial z}{\partial x}\right)_y \qquad \text{and} \qquad N = \left(\frac{\partial z}{\partial y}\right)_x \tag{2.16}$$

If now z is to be an exact differential (and hence a property), then the second derivative with x and y will be independent of the order of differentiation. That is,

$$\left[\frac{\partial}{\partial y}\left(\frac{\partial z}{\partial x}\right)_y\right]_x = \left[\frac{\partial}{\partial x}\left(\frac{\partial z}{\partial y}\right)_x\right]_y$$

or equivalently

$$\left(\frac{\partial M}{\partial y}\right)_x = \left(\frac{\partial N}{\partial x}\right)_y \tag{2.17}$$

Before utilizing these equations to generate Maxwell's relations, it is appropriate to reflect upon the necessity of utilizing the partial differential notation in which the variable being held constant is explicitly indicated. This notation is required in thermodynamics because, although (for a simple system) only two properties may be separately specified, there is a wide choice of which two properties may be selected. Thus, for example, the rate of change of pressure with density with the entropy held constant is not equal to the rate of change of pressure with density with the temperature held constant. Thus, a notation is needed that clarifies which partial derivative is intended.

Combining Eqs. (2.11–2.14) leads to

$$du = T\,ds - p\,dv \tag{2.18}$$

$$dh = T\,ds + v\,dp \tag{2.19}$$

$$dG = -s\,dT + v\,dp \tag{2.20}$$

$$dF = -s\,dT - p\,dv \tag{2.21}$$

Systematic application of Eqs. (2.16) and (2.17) then leads to:

$$\left(\frac{\partial u}{\partial s}\right)_v = T = \left(\frac{\partial h}{\partial s}\right)_p \tag{2.22}$$

$$\left(\frac{\partial u}{\partial v}\right)_s = -p = \left(\frac{\partial F}{\partial v}\right)_T \tag{2.23}$$

$$\left(\frac{\partial h}{\partial p}\right)_s = v = \left(\frac{\partial G}{\partial p}\right)_T \tag{2.24}$$

$$\left(\frac{\partial G}{\partial T}\right)_p = -s = \left(\frac{\partial F}{\partial T}\right)_v \tag{2.25}$$

$$\left(\frac{\partial T}{\partial v}\right)_s = -\left(\frac{\partial p}{\partial s}\right)_v \tag{2.26}$$

$$\left(\frac{\partial T}{\partial p}\right)_s = \left(\frac{\partial v}{\partial s}\right)_p \tag{2.27}$$

$$\left(\frac{\partial s}{\partial p}\right)_T = -\left(\frac{\partial v}{\partial T}\right)_p \tag{2.28}$$

$$\left(\frac{\partial s}{\partial v}\right)_T = \left(\frac{\partial p}{\partial T}\right)_v \tag{2.29}$$

This set of equations is known as Maxwell's relations. One of the prime utilizations of these relations is to obtain the behavior of certain properties in terms of the "properties of state," p, v, or T. Usually, a substance is described by its equation of state relating the three variables p, v, and T; and by appropriately manipulating the Maxwell's relations, the behavior of other properties may be deduced. Of course, the equation of state may not be available in an analytic form, but rather in the form of tables or graphs. However, turbomachinery problems most often involve gases that may be considered perfect.

2.12 General Relationships between Properties

An expression for a differential change in entropy, with the entropy considered to be a function of temperature and specific volume, is

$$ds = \left(\frac{\partial s}{\partial T}\right)_v dT + \left(\frac{\partial s}{\partial v}\right)_T dv = \frac{\left(\frac{\partial u}{\partial T}\right)_v}{\left(\frac{\partial u}{\partial s}\right)_v} dT + \left(\frac{\partial s}{\partial v}\right)_T dv$$

which, with Eqs. (2.7), (2.22), and (2.29), results in

$$T\,ds = C_v\,dT + T\left(\frac{\partial p}{\partial T}\right)_v dv \qquad (2.30)$$

Similarly, the entropy is considered to be a function of temperature and pressure to give

$$ds = \left(\frac{\partial s}{\partial T}\right)_p dT + \left(\frac{\partial s}{\partial p}\right)_T dp = \frac{\left(\frac{\partial h}{\partial T}\right)_p}{\left(\frac{\partial h}{\partial s}\right)_p} dT + \left(\frac{\partial s}{\partial p}\right)_T dp$$

which, with Eqs. (2.7), (2.22), and (2.28), results in

$$T\,ds = C_p\,dT - T\left(\frac{\partial v}{\partial T}\right)_p dp \qquad (2.31)$$

Applying the condition for exactness to the expressions for ds given by Eqs. (2.30) and (2.31) gives

$$\left(\frac{\partial C_v}{\partial v}\right)_T = T\left(\frac{\partial^2 p}{\partial T^2}\right)_v \qquad (2.32)$$

$$\left(\frac{\partial C_p}{\partial p}\right)_T = -T\left(\frac{\partial^2 v}{\partial T^2}\right)_p \qquad (2.33)$$

An expression for the difference of specific heats is obtained by subtracting Eq. (2.30) from Eq. (2.31) and in addition noting that the ratio dp/dT corresponds to $(\partial p/\partial T)_v$ for the case where $dv = 0$. Thus,

$$C_p - C_v = T\left(\frac{\partial p}{\partial T}\right)_v\left(\frac{\partial v}{\partial T}\right)_p \qquad (2.34)$$

Combination of Eqs. (2.11) and (2.30) gives an expression for the differential change in internal energy,

$$du = C_v\,dT + \left[T\left(\frac{\partial p}{\partial T}\right)_v - p\right]dv \qquad (2.35)$$

and combination of Eqs. (2.12) and (2.31) gives an expression for the differential change in enthalpy,

$$dh = C_p \, dT + \left[v - T\left(\frac{\partial v}{\partial T}\right)_p \right] dp \qquad (2.36)$$

A final example is an expression for the rate of change in pressure with density at constant entropy. This ratio is of particular importance in fluid mechanics because (as follows from momentum considerations) it is equal to the square of the speed of small disturbances relative to the local fluid velocity. First,

$$\gamma = \frac{C_p}{C_v} = \frac{\left(\dfrac{\partial h}{\partial T}\right)_p}{\left(\dfrac{\partial u}{\partial T}\right)_v} = \frac{\left(\dfrac{\partial h}{\partial s}\right)_p \left(\dfrac{\partial s}{\partial T}\right)_p}{\left(\dfrac{\partial u}{\partial s}\right)_v \left(\dfrac{\partial s}{\partial T}\right)_v}$$

Noting $(\partial s/\partial T)_v \equiv (\partial s/\partial T)_\rho$ and utilizing Eq. (2.22) and the chain rule of calculus, it follows that

$$\gamma = \frac{\left(\dfrac{\partial s}{\partial T}\right)_p}{\left(\dfrac{\partial s}{\partial T}\right)_\rho} = \frac{-\left(\dfrac{\partial T}{\partial \rho}\right)_s \left(\dfrac{\partial \rho}{\partial s}\right)_T}{-\left(\dfrac{\partial T}{\partial p}\right)_s \left(\dfrac{\partial p}{\partial s}\right)_T} = \frac{\left(\dfrac{\partial p}{\partial \rho}\right)_s}{\left(\dfrac{\partial p}{\partial \rho}\right)_T}$$

and hence

$$\left(\frac{\partial p}{\partial \rho}\right)_s = \gamma \left(\frac{\partial p}{\partial \rho}\right)_T \qquad (2.37)$$

It is important to note that in the development of Eq. (2.37) no assumption was made regarding the equation of state. The expression is thus valid for situations in which the ratio of specific heats may vary substantially.

2.13 The Perfect Gas

A perfect gas is defined as a substance with equation of state given by

$$pv = RT \qquad (2.38)$$

In this expression, R is a constant termed the gas constant.

If, in addition to satisfying Eq. (2.38), the gas has a constant ratio of specific heats, it is termed a "calorically perfect" gas.

It should be noted that the gas constant R is given in terms of the "universal gas constant" R_u by

$$R = R_u/M \qquad (2.39)$$

where $M \equiv$ molecular weight.

The value of R_u in several systems is

$$R_u = 1545\left(\frac{\text{lbf}}{\text{lbm}}\frac{\text{s}^2}{\text{ft}}\right)\frac{\text{ft}^2}{\text{s}^2}\frac{\text{lbm}}{\text{lbm-mole}\cdot{}^\circ\text{R}}$$

$$= 49,700\frac{\text{ft}^2}{\text{s}^2}\frac{\text{lbm}}{\text{lbm-mole}\cdot{}^\circ\text{R}}$$

$$= 8317\frac{\text{m}^2}{\text{s}^2}\frac{\text{kg}}{\text{kg-mole}\cdot\text{K}} \tag{2.40}$$

For air of molecular weight 29 (approximately), R is

$$R = 53.3\left(\frac{\text{lbf}\cdot\text{s}^2}{\text{lbm}\cdot\text{ft}}\right)\left(\frac{\text{ft}}{\text{s}}\right)^2\frac{1}{{}^\circ\text{R}}$$

$$= 1714\left(\frac{\text{ft}}{\text{s}}\right)^2\frac{1}{{}^\circ\text{R}}$$

$$= 286.8\left(\frac{\text{m}}{\text{s}}\right)^2\frac{1}{\text{K}} \tag{2.41}$$

It should be noted from Eq. (2.39) that when a gas is in a regime where dissociation is occurring, it is not, strictly speaking, a perfect gas. This is because dissociation changes the value of M and hence of R. However, if no dissociation occurs but substantial vibrational excitation occurs, then the gas would be perfect, but not calorically perfect.

The behavior of a perfect gas may be illuminated by applying the relationships of Sec. 2.12. Thus, with Eq. (2.38) and Eqs. (2.32) and (2.33),

$$\left(\frac{\partial C_v}{\partial v}\right)_T = 0 \tag{2.42}$$

hence, C_v is a function of temperature only; and

$$\left(\frac{\partial C_p}{\partial p}\right)_T = 0 \tag{2.43}$$

hence, C_p is a function of temperature only.
From Eqs. (2.35) and (2.36)

$$du = C_v\,dT \tag{2.44}$$

$$dh = C_p\,dT \tag{2.45}$$

Thus, Eqs. (2.42) and (2.43) show that both the internal energy and the enthalpy are functions of temperature only. From Eq. (2.34)

$$C_p - C_v = R \qquad (2.46)$$

and from Eq. (2.37)

$$\left(\frac{\partial p}{\partial \rho} \right)_s = \gamma R T \qquad (2.47)$$

Thus, the many familiar relationships peculiar to a perfect gas follow directly from the equation of state and Maxwell's relations. Note particularly that Eqs. (2.46) and (2.47) are valid whether or not the ratio of specific heats γ varies.

2.14   Quasi-One-Dimensional Fluid Flows

A quasi-one-dimensional flow is defined as a flow in which the fluid properties can be described in terms of a single spatial coordinate (the axial dimension), the specified axial area variation of the containing tube or channel, and (if the flow is time dependent) the time. The simplicity introduced by utilizing such an approximate description of the flow in a channel is clearly a virtue, but it is equally clear that the regimes of validity of any analysis incorporating such an approximation should be thoroughly investigated.

Before investigating the expected regimes of validity of the quasi-one-dimensional approximation, note that the approximation is of particular utility in the study of aircraft gas turbine engines. In the first-order analysis of the various components of a turbomachine, it is customary to refer to the "inlet and outlet" conditions of each component. It is quite obvious that the conditions at the inlet and outlet of any component are not, in fact, uniform, and hence any single quantity chosen to represent a given property must be a properly chosen average quantity. When properly chosen, these average quantities may be utilized with ease in a cycle analysis to predict the overall performance of a given engine in terms of the performance of its individual components.

In many situations, however, it is the effect of the non-one-dimensionality that is at the core of the problem considered. Thus, for example, when describing the performance of ejectors, careful distinction must be made between the use of such averaging methods as area weighted, mass weighted, and mixed out. Another example of the importance of the non-one-dimensionality of a flow appears in the study of inlet distortion, which by its definition involves the study of the variation of the stagnation temperature and stagnation pressure about their appropriately chosen averages.

Considering a channel with solid walls in which a flow exists, it is clear that severe property variations will occur across the channel. To make this statement clear, consider the example of the behavior of the velocity that is known to have a value of zero on the wall and to be finite in the channel center. If the walls are heat conducting (or if the Mach number at the channel center is large), large variations in temperature across the channel would be expected. In such cases, the "temperature" or "velocity" appearing in the quasi-one-dimensional equations would in fact be an appropriately chosen average of the values found across the channel. The definition of this "average" would depend somewhat upon the particular problem to be investigated, but it can be seen that if the "shape" of the cross-channel property variation changes radically with the axial position, whatever averaging process is chosen will have to be modified with the axial position in order to obtain meaningful results from the analysis. Since the suitable averaging procedures would not be known a priori, such a procedure is tantamount to conducting a full two-dimensional investigation.

Conversely, however, there are two limits in which a quasi-one-dimensional assumption could be expected to be of use. The first is the so-called "fully developed flow" in which, in fact, the shape of the cross-channel property variations varies only slowly with the axial position. In this case, the quasi-one-dimensional assumption leads to an accurate description of the behavior of the average fluid properties. The other limit of channel flow behavior that allows an accurate analysis under the restrictions of the approximation is that limit where a dominant portion of the flow in the channel satisfies the assumption of very nearly uniform cross-channel conditions and only a small portion, "the boundary layer," has rapidly varying properties in the cross-channel direction. In this latter case, it is tacitly assumed that the boundary layer entrains such a small portion of the flow that the average quantities are not affected by excluding the effects of the boundary layer.

It would be well to mention here that a technique commonly used to extend the quasi-one-dimensional solution utilizes the results of the quasi-one-dimensional solution and applies these results as the boundary conditions ("freestream conditions") to be applied to the solution of the behavior of the (two-dimensional) boundary-layer equations.

The effect of the curvature of the channel must be considered carefully, because it is evident that channel curvature introduces cross-channel pressure gradients which, in turn, introduce cross-channel gradients in other fluid properties. In general, if the radius of curvature of the channel centerline is large compared to the channel dimension in the radial direction, the effects of the induced radial pressure fields will be small. This observation can be extended to indicate that the flow in individual stream tubes of a general three-dimensional flow can be expected to be described by the quasi-one-dimensional flow equations, although the determination of the behavior of the stream tube itself (through the pressure fields induced by other stream tubes) will rely upon solution of the complete equations.

In what follows the first law of thermodynamics is extended to include the effects of the kinetic energy and the potential energy of the flowing fluid in a form suitable to adapt for use in the quasi-one-dimensional equations. In this derivation, the very useful concept of the "control volume" will be introduced, which will be employed later in the derivation of the quasi-one-dimensional equations.

2.15 The First Law for a Flowing System—The Control Volume

An important application of the first law of thermodynamics is to be found in the analysis of a flowing system. A very convenient and often used concept is that of a "control volume," defined simply as a given region in space. This region in space may be moving, changing shape, etc., but most often in analyzing material behavior using the control volume technique, a simple control volume of fixed size, shape, and position is selected. The selection of the most suitable control volume will depend upon the problem at hand, but it can be stated in general that most often one is selected such that the material at the entrance and exit is in local thermodynamic equilibrium and in addition may have its behavior closely approximated by assuming that all properties at the entrance or exit are the average properties at those positions. This latter requirement is the equivalent to stating that conditions at the entrance and exit are "quasi-one-dimensional." As will be seen shortly, also required is some information concerning the heat and work interactions at the boundaries of the control volume, but it should be noted carefully that no requirement of reversibility or one-dimensionality for the processes within the control volume will be imposed.

An important point to recognize is that in the preceding sections on thermodynamics the first law was written as it pertained to a given mass of material (the thermodynamic system), and that now the focus is on a given volume in space through which the material passes. That portion of work, kinetic energy, or potential energy identified with the bulk motion of the thermodynamic system was not considered simply because such behavior was assumed to be already known from the laws of mechanics. In now considering a control volume through which the fluid enters and leaves, however, the energy identified with the bulk motion of the fluid must be included in order to account for any such bulk motion energy changes experienced by the fluid between the entrance and exit.

The method for extending the first law for a thermodynamic system to a control volume is to first write the first law for that *mass* contained within the control volume at time t. The mass is then followed until time $t + dt$ and the first law of thermodynamics is applied to the change experienced by the mass. These changes may then be related to the changes experienced by the material passing through the control volume.

Consider a control *mass* that at time t occupies the volume bounded by the two dotted lines 1 and 1' and the solid walls of the container shown in Fig. 2.1. At time $t + dt$ the control mass occupies the volume bounded by the two dotted lines 2 and 2' and the solid walls of the container. The control volume will be considered to be the volume bounded by the container and lines 2 and 1'.

The change in "total energy" (i.e., internal plus kinetic plus potential energy) of the control mass is given by

$$dE_{C.M.} = d'Q - d'W_T \tag{2.48}$$

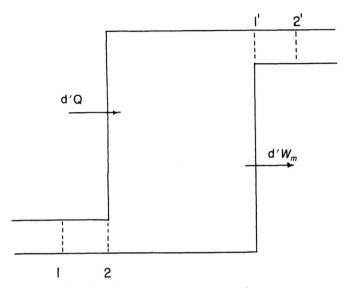

Fig. 2.1   Control volume and control mass.

where $d'W_T$ indicates the total work interaction of the control mass in the time dt. This work interaction can be considered to be of two parts:

(1) The work interaction with mechanical contrivances ($d'W_m$) such as shafts, etc.

(2) The net work interaction of the fluid in the control mass with the fluid external to the control mass by the moving interfaces at the entrance and exit. This form of work interaction, sometimes called the flow work, is equal to ($F_e V_e dt - F_i V_i dt$), in which F_i and F_e are the forces on the interfaces at the entrance and exit to the control volume, respectively. With the assumptions of local thermodynamic equilibrium and quasi-one-dimensionality, these forces can be written as the product of the pressure times the area so that the flow work may be written

$$p_e A_e V_e \, dt - p_i A_i V_i \, dt$$

The expression for the change in energy of the control mass may then be written

$$dE_{C.M.} = d'Q - d'W_m - \{ p_e A_e V_e - p_i A_i V_i \} \, dt \qquad (2.49)$$

The energy within the control volume is related to that within the control mass at times t and $t + dt$ by

$$E_{C.V.}(t) = E_{C.M.}(t) - (A_i V_i \, dt)\rho_i \left[u_i + (V_i^2/2) + \text{P.E.}_i \right]$$

$$E_{C.V.}(t + dt) = E_{C.M.}(t + dt) - (A_e V_e \, dt)\rho_e \left[u_e + (V_e^2/2) + \text{P.E.}_e \right]$$

Thus

$$d E_{C.V.} = d E_{C.M.} + A_i V_i \, dt \, \rho_i \left[u_i + \left(V_i^2/2 \right) + \text{P.E.}_i \right]$$

$$- A_e V_e \, dt \, \rho_e \left[u_e + \left(V_e^2/2 \right) + \text{P.E.}_e \right] \tag{2.50}$$

But $A \rho V = \dot{m}$, the mass transfer per second through the boundary. By equating the expressions for $d E_{C.M.}$, dividing by dt, and writing $u + p/\rho$ as the enthalpy h, there is obtained

$$\left(\frac{dE}{dt} \right)_{C.V.} = \left[\dot{m} \left(h + \frac{V^2}{2} + \text{P.E.} \right) \right]_i - \left[\dot{m} \left(h + \frac{V^2}{2} + \text{P.E.} \right) \right]_e$$

$$- \frac{d'W_m}{dt} + \frac{d'Q}{dt} \tag{2.51}$$

Example 2.1

As a very simple example, consider the adiabatic, steady flow of a fluid in a nozzle. The control volume is that volume bounded by the solid walls of the nozzle and the dotted lines shown in Fig. 2.2. The assumption of steady flow requires $(dE/dt)_{C.V.} = 0$ and $\dot{m}_i = \dot{m}_e$, the assumption of adiabatic flow requires $d'Q/dt = 0$, and, because the nozzle has no work interaction with mechanical contrivances, $d'W_m/dt = 0$. Assume very little, if any, change in potential energy across the nozzle so that the control volume form of the energy equation gives

$$\left[h + \left(V^2/2 \right) \right]_i = \left[h + \left(V^2/2 \right) \right]_e \tag{2.52}$$

If, as indicated in Fig. 2.2, the inlet surface is so chosen that the kinetic energy per unit mass of the fluid is very small compared to the enthalpy of the fluid, then

$$V_e = \sqrt{2 (h_i - h_e)} \tag{2.53}$$

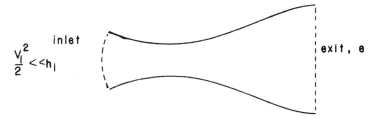

Fig. 2.2   Adiabatic nozzle.

This simple and convenient result can be interpreted in the following way. The function of the nozzle is to take the random thermal velocities of the fluid in the reservoir and to direct such velocities so as to give a directed kinetic energy to the expelled fluid. In addition, the pressure within the chamber also supplies "flow work" that adds further to the directed energy. Note that the flow work so supplied must come from a compressor or similar device if the reservoir conditions are not to change with time. Note also that without further specifying the process it cannot be stated what the conditions should be at a particular point in the nozzle. Thus, a very rough nozzle might cause the process to be highly irreversible, with the result that even if the pressure drops the same amount as it does in another smoother nozzle, the velocity at that point in the nozzle would be less than that in the smooth nozzle. The sum of the kinetic energy and enthalpy would be the same, however, indicating that the roughness has had the effect of slowing down the fluid and returning the directed velocity to the random thermal velocities identified with the temperature (and enthalpy).

Example 2.2

As a further example, consider the "blowdown" of a calorically perfect gas from a vessel through a nozzle. What is the relationship between the pressure, temperature, and density of the fluid in the container as mass is expelled? Again assume the container to be adiabatic. In this case, of course, the conditions in the control volume change with time and the control volume form of the energy equation hence gives

$$\left(\frac{dE}{dt}\right)_{\text{C.V.}} = -\dot{m}_e\left[h + \frac{V^2}{2}\right]_e \tag{2.54}$$

Now $\dot{m}_e$ is the mass flow *out* of the container, and may be written $-dm/dt$ where m is the mass within the container. For simplicity, the "exit portion" of the control volume is the dotted line shown in Fig. 2.3, which by assumption exists where $h = h_c \gg V^2/2$. In this case the energy E within the control volume consists of internal energy only, so that $E = mC_vT_c$ and

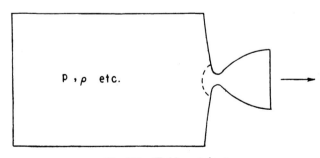

P , ρ etc.

Fig. 2.3   Fluid container.

there is obtained

$$C_v \frac{\mathrm{d}(m\,T_c)}{\mathrm{d}t} = \frac{\mathrm{d}m}{\mathrm{d}t} h_c = \frac{\mathrm{d}m}{\mathrm{d}t} C_p T_c$$

Expanding the derivative on the left side, cancelling the differential of time, and rearranging, it follows that

$$\frac{\mathrm{d}T_c}{T_c} = (\gamma - 1)\frac{\mathrm{d}m}{m}$$

This then integrates to give

$$\frac{T_{c_1}}{T_{c_2}} = \left(\frac{m_1}{m_2}\right)^{\gamma-1} = \left(\frac{\rho_{c_1}}{\rho_{c_2}}\right)^{\gamma-1} \tag{2.55}$$

It has been emphasized to this point that the assumptions leading to the control volume form of the first law do not require the presence of reversibility. If an adiabatic process is reversible, no entropy change occurs. Hence, calculate the entropy change occurring in this process. From the Gibbs' equation for a calorically perfect gas, it is found that

$$\mathrm{d}s = C_v \frac{\mathrm{d}T}{T} + R \frac{\mathrm{d}(1/\rho)}{1/\rho}$$

hence

$$s_{c_2} - s_{c_1} = C_v \ell n \frac{T_{c_2}}{T_{c_1}} + R \ell n \frac{\rho_{c_1}}{\rho_{c_2}} \tag{2.56}$$

Thus, in this case

$$s_{c_2} - s_{c_1} = -C_v(\gamma - 1)\ell n \frac{\rho_{c_1}}{\rho_{c_2}} + R \ell n \frac{\rho_{c_1}}{\rho_{c_2}} = 0$$

This seems to be a paradox, because no explicit statement has been made to introduce the assumption of reversibility, yet the result indicates that the process described must have been reversible. It is apparent that an assumption of reversibility must have been made implicitly. A little thought indicates that by assuming the enthalpy at the exit was that of the entire container, or more particularly by assuming the pressure at the exit was that of the entire chamber, in fact no "viscous drop" was assumed for the pressure and hence that the process within the chamber was reversible. If the expulsion of the fluid had been particularly rapid or the fluid particularly viscous, a pressure drop across the chamber would be expected and hence assuming the pressure at the exit and in the chamber to be identical would be in error.

Example 2.3

An example that will hopefully clarify the method and utility of utilizing the control volume or control mass method of "bookkeeping," the equation for the first law of thermodynamics follows by considering the combined behavior of Examples 2.1 and 2.2.

Consider the "blowdown" of a calorically perfect gas from a large container through a nozzle to an (approximately) zero exit pressure (Fig. 2.4). The behavior of the gas in the chamber and across the nozzle is that previously derived in Examples 2.1 and 2.2. Utilizing the control volume approach, find the directed kinetic energy per mass at the nozzle exit and integrate this energy over the entire mass outflow to determine the total directed kinetic energy in the departing fluid. This latter result will be checked by utilizing the control mass form of the first law.

First note from the Gibbs equation for a calorically perfect gas that

$$ds = C_p \frac{dT}{T} - R \frac{dp}{p}$$

so that

$$\frac{p_2}{p_1} = \left(\frac{T_2}{T_1} \right)^{\gamma/(\gamma-1)} e^{-[(s_2 - s_1)/R]} \tag{2.57}$$

Thus, for this case of expansion to zero exit pressure, the exit temperature and hence the exit internal energy and enthalpy are zero. Equation (2.53) thus gives

$$\text{Directed kinetic energy/mass} = V_e^2/2 = C_p T_c$$

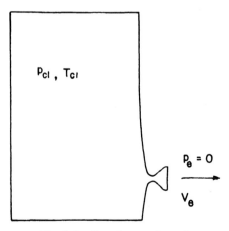

$$P_{cl} , T_{cl}$$

$$P_e = 0$$

$$V_e$$

Fig. 2.4   Container and nozzle.

At first this result might appear incongruous because the energy/mass escaping the container is larger (by a factor of γ) than the energy/mass within the container. Clearly, however, the additional energy is being transmitted to the escaping mass by the "flow work" of the internal fluid. This flow work is supplied by the expanding gas within the container that, as a result of its expansion (and hence work interaction), has its own internal energy and hence temperature reduced. The total outflow of energy from the nozzle is given by

$$\text{Total energy} = \int_0^{m_{c_1}} \frac{V_e^2}{2} \, dm = \int_0^{m_{c_1}} C_p T_c \, dm$$

$$= m_{c_1} T_{c_1} C_p \int_0^1 \frac{T_c}{T_{c_1}} \, d\left(\frac{m}{m_{c_1}}\right)$$

With Eq. (2.55), this becomes

$$\text{Total energy} = m_{c_1} T_{c_1} C_p \int_0^1 \left(\frac{m}{m_1}\right)^{\gamma-1} d\left(\frac{m}{m_{c_1}}\right)$$

$$= m_{c_1} T_{c_1} C_v$$

This latter result follows immediately when the control mass form of the first law is considered, because the control mass statement would simply be that all the energy originally contained in the vessel ($m_1 C_v T_{c_1}$) must be equal to all of the energy in the exhausted gas (namely, all of the directed kinetic energy).

2.16 The Channel Flow Equations

Consider now the steady flow of fluid in channels with rigid walls (Fig. 2.5). Quasi-one-dimensionality will be assumed, the effects of potential energy changes will be ignored, and, in addition, the wall slope and hence rate of change of the cross-sectional area will be assumed to be small. The effect of this latter assumption is that the cosine of the wall angle θ may be considered to be unity.

The conservation equations are developed by considering the conditions across a small axial segment of the duct, as indicated in Fig. 2.5. The conservation of mass gives immediately

$$\rho u A = \text{const}$$

or

$$\frac{d\rho}{\rho} + \frac{du}{u} + \frac{dA}{A} = 0 \qquad (2.58)$$

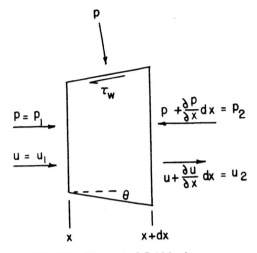

Fig. 2.5   Element of fluid in duct.

The first law of thermodynamics follows directly from Eq. (2.51) as applied across the element. Thus, noting that $\dot{m}_i = \dot{m}_e$ and that $(1/\dot{m})(d'Q/dt) = d'q$, the heat interaction per unit mass, the first law is obtained,

$$dh + u\,du = d'q \qquad (2.59)$$

The Gibbs equation may be applied directly,

$$T\,ds = dh - (1/\rho)\,dp \qquad (2.60)$$

and the first law and Gibbs equation may be combined to give

$$(1/\rho)\,dp + u\,du = d'q - T\,ds \qquad (2.61)$$

The momentum equation is found by writing Newton's law in a form appropriate for use with a control volume, namely

Sum of the forces = rate of production of momentum

or

Pressure forces + viscous forces = momentum convected out through surface 2 per second − momentum convected in through surface 1 per second

In symbols this is,

$$p_1 A_1 + \int_{\text{side}} p\,dA - p_2 A_2 - \int \tau c\,dx = \left(\rho u^2 A\right)_2 - \left(\rho u^2 A\right)_1$$

This expression introduces the shear stress τ and the circumference c. Retaining terms only to the first order and utilizing the continuity equation (2.58) results in

$$-\mathrm{d}(pA) + p\,\mathrm{d}A - \tau c\,\mathrm{d}x = \rho uA\,\mathrm{d}u$$

or the momentum equation

$$(1/\rho)\,\mathrm{d}p + u\,\mathrm{d}u = -(\tau c/\rho A)\,\mathrm{d}x \qquad (2.62)$$

It may be noted that this equation closely resembles the familiar Bernoulli equation of elementary ideal fluid mechanics. The only addition to the ideal form of the equation arises from the viscous shear stress contribution. An equation for the entropy variation follows by combining Eqs. (2.61) and (2.62) to give

$$T\,\mathrm{d}s = \mathrm{d}'q + (\tau c/\rho A)\,\mathrm{d}x \qquad (2.63)$$

It is apparent in this relationship that, for an ideal process in which the shear stress is zero, the entropy variation corresponds to that for reversible heat interaction, as already given in Eq. (2.10). The irreversibility of the viscous term is apparent in its contribution to the increase in entropy.

2.17 Stagnation Properties

A stagnation property is defined as that value of the property that would exist if the fluid were extracted and brought isentropically to rest. The process may be imagined to be that shown in Fig. 2.6 wherein the fluid flows isentropically through a duct from condition 1 to condition t_1.

Application of Eq. (2.59) gives an expression for the stagnation enthalpy. Thus, noting $\mathrm{d}'q = 0$ and integrating,

$$h_{t_1} = h_1 + u_1^2/2 \qquad (2.64)$$

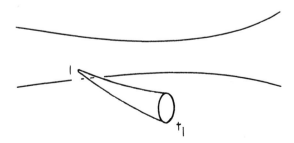

Fig. 2.6 Duct and imaginary duct for stagnation condition.

Obviously the location of point 1 in the duct in which the flow actually occurs is arbitrary, so in the following relationships between properties and their related stagnation values, the subscript 1 will be omitted.

The equations take on particularly simple forms when a calorically perfect gas is considered. Thus, Eq. (2.64) becomes

$$C_p T_t = C_p T + u^2/2 \tag{2.65}$$

Further manipulation leads to

$$\frac{T_t}{T} = 1 + \frac{\gamma R}{2C_p} \frac{u^2}{\gamma R T} = 1 + \frac{\gamma - 1}{2} M^2 \tag{2.66}$$

Here the Mach number is introduced, defined as $M = u/a$. The speed of sound a as stated in Sec. 2.12, is equal to $(\partial p/\partial \rho)^{\frac{1}{2}}$, which combined with Eq. (2.47) gives $a^2 = \gamma R T$. Equation (2.46) was utilized to give $\gamma R/C_p = \gamma - 1$.

Equation (2.57) may be applied directly (with $s_2 = s_1$) to give

$$\frac{p_t}{p} = \left(\frac{T_t}{T}\right)^{\gamma/(\gamma-1)} = \left(1 + \frac{\gamma - 1}{2} M^2\right)^{\gamma/(\gamma-1)} \tag{2.67}$$

Then, also

$$\frac{\rho_t}{\rho} = \frac{p_t}{p} \frac{T}{T_t} = \left(1 + \frac{\gamma - 1}{2} M^2\right)^{1/(\gamma-1)} \tag{2.68}$$

Some important behaviors concerning the variation of stagnation properties in ducts can be illuminated by applying Eq. (2.57) directly to the conditions at t_1 and t_2 (Fig. 2.7). Thus,

$$\frac{p_{t_2}}{p_{t_1}} = \left(\frac{T_{t_2}}{T_{t_1}}\right)^{\gamma/(\gamma-1)} e^{-[(s_2 - s_1)/R]} \tag{2.69}$$

where by definition $s_1 = s_{t_1}$ and $s_2 = s_{t_2}$.

In the special case of adiabatic flow in the duct, from Eqs. (2.59) and (2.65) $T_{t_2} = T_{t_1}$. Then

$$p_{t_2}/p_{t_1} = e^{-[(s_2 - s_1)/R]} \tag{2.70}$$

The second law, or equivalently Eq. (2.63), showed that when shear exists in an adiabatic flow, $s_2 > s_1$. Thus, for adiabatic flow

$$p_{t_2}/p_{t_1} = \rho_{t_2}/\rho_{t_1} < 1 \tag{2.71}$$

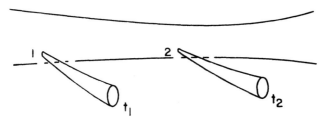

Fig. 2.7 Stagnation conditions at two axial locations.

This result has important consequences in turbomachinery, because the high performance of turbomachinery is dependent upon the efficient attainment of high stagnation pressures. It is clear from this application of the second law that nature naturally tends to erode the attained stagnation pressure.

2.18 Property Variations in Channels

Several important relationships may be obtained in a fairly general form by combining the conservation equations of Sec. 2.15. In so doing it will be appropriate to express the change in pressure in terms of corresponding changes in density and entropy. Thus Eq. (2.26) is used to obtain

$$d p = \left(\frac{\partial p}{\partial \rho} \right)_s d\rho + \left(\frac{\partial p}{\partial s} \right)_\rho ds$$

$$= \left(\frac{\partial p}{\partial \rho} \right)_s d\rho + \rho^2 \left(\frac{\partial T}{\partial \rho} \right)_s ds \qquad (2.72)$$

Then note that $dT = (\partial T/\partial \rho)_p d\rho + (\partial T/\partial p)_\rho dp$ gives

$$\left(\frac{\partial T}{\partial \rho} \right)_s = \left(\frac{\partial T}{\partial \rho} \right)_p + \left(\frac{\partial T}{\partial p} \right)_\rho \left(\frac{\partial p}{\partial \rho} \right)_s \qquad (2.73)$$

Thus, utilizing Eq. (2.37) and the relationship

$$\left(\frac{\partial T}{\partial p} \right)_\rho \left(\frac{\partial p}{\partial \rho} \right)_T = - \left(\frac{\partial T}{\partial \rho} \right)_p$$

Eqs. (2.72) and (2.73) may be combined to give

$$d p = \left(\frac{\partial p}{\partial \rho} \right)_s d\rho - \rho^2 (\gamma - 1) \left(\frac{\partial T}{\partial \rho} \right)_p ds \qquad (2.74)$$

This expression for the pressure increment may now be substituted into Eq. (2.61), whereupon by utilizing the continuity equation (2.58) together with $(\partial p/\partial\rho)_s = a^2$ and $M^2 = u^2/a^2$,

$$\frac{dA}{A} + (1 - M^2)\frac{du}{u} = \frac{1}{a^2}\left\{\left[T - (\gamma - 1)\rho\left(\frac{\partial T}{\partial\rho}\right)_p\right]ds - d'q\right\} \quad (2.75)$$

It will be recognized that this equation implies the requirement of the famous convergent-divergent duct shape if an adiabatic perfect flow ($d'q = ds = 0$) is to be accelerated from a Mach number less than unity to a Mach number greater than unity. When a perfect gas is considered, the equation reduces to

$$\frac{dA}{A} + (1 - M^2)\frac{du}{u} = \frac{1}{\gamma RT}(\gamma T ds - d'q) \quad (2.76)$$

Adiabatic Flow of a Perfect Gas

When no heat interaction occurs, Eq. (2.76) further reduces to

$$\frac{dA}{A} + (1 - M^2)\frac{du}{u} = \frac{ds}{R} \quad (2.77)$$

(appropriate for a perfect gas with adiabatic flow).

In an accelerating flow (i.e., in a nozzle), du is positive. In an adiabatic (real) flow, ds is positive. Thus, at the throat of a nozzle ($dA = 0$), M^2 must be less than unity. A method of estimating how much less than unity M^2 is at a nozzle throat will be developed shortly, but it is of interest at this point to relate this quasi-one-dimensional description to the two-dimensional description of a freestream interacting with a (growing) boundary layer. In the latter case, it is imagined that the growing boundary layer causes the effective throat of the nozzle (as "seen by the freestream") to shift slightly downstream, so that the Mach number at the geometrical throat remains less than unity. It can be seen that the two descriptions are not in opposition to each other.

To estimate the effect of shear upon the Mach number at the throat in an adiabatic nozzle ($dA = 0$, $d'q = 0$), Eqs. (2.62), (2.63), and (2.77) are combined to give

$$M^2_{\text{throat}} = 1 + \frac{\gamma(\tau c\, dx/A\, dp)}{1 - (\gamma - 1)(\tau c\, dx/A\, dp)} \quad (2.78)$$

It can be seen that the group ($\tau c\, dx/A\, dp$) is just the ratio of the shear force acting on the edges of an elemental volume of length dx to the pressure forces acting upon the cross section of the same volume. In order to further estimate the magnitude of this number, the skin-friction coefficient

C_f is introduced,

$$C_f = \tau / \tfrac{1}{2}\rho u^2 \tag{2.79}$$

Thus for a circular throat of diameter D,

$$\frac{\tau c \, dx}{A \, dp} = \frac{2\gamma M^2 C_f}{(D/p)(dp/dx)} \tag{2.80}$$

The denominator of Eq. (2.80) may be roughly approximated as unity with the assumption that the very rapid pressure changes found at a throat correspond to the pressure changing on the order of magnitude of its own value within one nozzle diameter. Such a coarse approximation is not wildly distant from the truth for typical nozzles. A typical value for C_f would be approximately 0.005, so that Eq. (2.78) yields $M_{\text{throat}} \approx 0.995$. (Note that dp/dx is negative.) This somewhat justifies the almost universally used approximation that $M_{\text{throat}} = 1$.

Nonadiabatic Flow of a Perfect Gas

As will be evident in Chap. 5, the behavior (and preservation) of the stagnation pressure has a vital effect upon the performance of gas turbine engines. The effects upon stagnation pressure of heat transfer and shear may be obtained by combining Eqs. (2.59) and (2.65), together with Eqs. (2.63) and (2.57). These may be written

$$C_p dT_t = d'q$$

$$ds = \frac{d'q}{T} + \frac{\tau c}{\rho T A} \, dx$$

$$ds_t = ds = C_p \frac{dT_t}{T_t} - R \frac{dp_t}{p_t}$$

Combination of these three equations and Eq. (2.79) leads to

$$\frac{dp_t}{p_t} = -\frac{\gamma M^2}{2} \left(\frac{d'q}{C_p T_t} + C_f \frac{c}{A} \, dx \right) \tag{2.81}$$

This expression makes it clear that both heat interaction and frictional effects cause a degradation in stagnation pressure when the Mach number is other than zero. The source of this degradation becomes clear when Eq. (2.69) is considered along with the above equations. Thus, from Eq. (2.69) it follows that, for a given stagnation temperature, the stagnation pressure decreases with increasing entropy. If a positive heat interaction is to occur and the entropy increase kept to a minimum, the (static) temperature at which the heat interaction takes place must be kept as high as possible. Clearly, when a flow of given stagnation temperature exists at finite Mach number, the static temperature is reduced, leading to a lower stagnation pressure than that attainable for zero Mach number heat addition.

This effect upon the stagnation pressure becomes of paramount importance in both the combustion chamber and afterburner of an aircraft gas turbine engine. In some modern, high-performance aircraft, the maximum engine cross section is determined by the requirement to keep the Mach number at the entrance to the afterburner at an acceptably low value.

It would appear at first glance that Eq. (2.81) implies that, in the case of a negative heat interaction, the stagnation pressure of the fluid could be increased. To investigate this concept, assume that the heat interaction may occur by either convective heat transfer $(d'q_T)$ or radiative heat transfer $(d'q_R)$.

Introducing the heat-transfer coefficient h results in (by definition of h)

$$\dot{m}\,d'q_T = h(T_w - T_t)c\,dx \tag{2.82}$$

(See Fig. 2.8.)

The Stanton number N_{St} is defined by

$$N_{St} = \frac{hA}{\dot{m}C_p} \tag{2.83}$$

Thus

$$\frac{d'q_T}{C_p T_t} = N_{St}\frac{T_w - T_t}{T_t}\frac{c}{A}\,dx \tag{2.84}$$

Hence Eq. (2.81) becomes

$$\frac{dp_t}{p_t} = -\frac{\gamma M^2}{2}\frac{c}{A}\,dx\left(\frac{T_w - T_t}{T_t}N_{St} + C_f\right) - \frac{\gamma M^2}{2}\frac{d'q_R}{C_p T_t} \tag{2.85}$$

A remarkable relationship termed the "Reynolds analogy" relates the skin-friction coefficient and Stanton number over a wide range of flow conditions,

$$N_{St} \approx C_f/2 \tag{2.86}$$

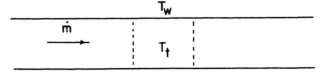

Fig. 2.8   Element of channel.

Incorporating this relationship in Eq. (2.85) yields

$$\frac{\mathrm{d}p_t}{p_t} \approx -\frac{\gamma M^2}{4}\frac{c}{A}C_f\mathrm{d}x\left(\frac{T_w + T_t}{T_t}\right) - \frac{\gamma M^2}{2}\frac{\mathrm{d}'q_R}{C_p T_t} \qquad (2.87)$$

This latter result indicates that (assuming the Reynolds analogy is approximately valid) no matter how far the wall temperature is reduced, the stagnation pressure cannot be increased through convective transfer effects. The possibility remains, however, that extreme radiative transfer (as might occur in a high-powered gas laser) might contribute to a stagnation pressure increase. Similarly, if evaporation occurs, the heat transfer is not limited by the Reynolds analogy and the stagnation pressure can increase.

Constant-Area Heat Interaction

Combination of the continuity equation (2.58) (with $\mathrm{d}A = 0$), the momentum equation (2.62), and Eq. (2.79) yields

$$\mathrm{d}(p + \rho u^2) = -\tfrac{1}{2}\rho u^2 C_f(c/A)\,\mathrm{d}x$$

Then, noting $M^2 = \rho u^2/\gamma p$, this expression may be manipulated to give

$$\frac{\mathrm{d}p}{p} + \frac{\gamma\,\mathrm{d}M^2}{1 + \gamma M^2} = -\frac{\gamma}{2}\frac{c}{A}C_f\frac{M^2}{1 + \gamma M^2}\,\mathrm{d}x \qquad (2.88)$$

The logarithmic derivative of Eq. (2.67) gives

$$\frac{\mathrm{d}p_t}{p_t} = \frac{\mathrm{d}p}{p} + \frac{(\gamma/2)\,\mathrm{d}M^2}{1 + [(\gamma - 1)/2]\,M^2} \qquad (2.89)$$

Thus, noting that $C_p\,\mathrm{d}T_t = \mathrm{d}'q$, Eqs. (2.81), (2.88), and (2.89) may be combined to yield

$$\frac{\mathrm{d}T_t}{T_t} = -C_f\frac{c}{A}\frac{\gamma M^2}{1 + \gamma M^2}\,\mathrm{d}x + \frac{2\,\mathrm{d}M^2}{M^2(1 + \gamma M^2)} - \frac{\mathrm{d}M^2}{M^2\left(1 + \dfrac{\gamma - 1}{2}M^2\right)}$$

$$(2.90)$$

If the heat interaction rate is known, this equation may be numerically integrated to give M vs x. Two special cases are considered in the following sections.

Ideal constant-area heat interaction—thermal choking. In the limit where the flow may be considered to be ideal ($C_f = 0$), Eq. (2.90) may be integrated directly to give

$$\frac{T_{t_2}}{T_{t_1}} = 1 + \frac{\Delta q_{1-2}}{C_p T_{t_1}} = \frac{f(M_2^2)}{f(M_1^2)} \qquad (2.91)$$

where

$$f(M^2) \equiv \frac{1 + [(\gamma - 1)/2] M^2}{(1 + \gamma M^2)^2} M^2 \qquad (2.92)$$

Clearly, when the stagnation temperature increases, $f(M_2^2)$ must be larger than $f(M_1^2)$. It is hence of interest to investigate the form of the function $f(M^2)$ to see if it is always possible to increase the stagnation temperature. Straightforward differentiation shows that $\partial f(M^2)/\partial M = 0$ at $M = 1$, so also noting that $f(0) = 0$, $f(1) = 1/2(\gamma + 1)$, and $f(\infty) = (\gamma - 1)/2\gamma^2$, $f(M^2)$ can be plotted vs M as indicated in Fig. 2.9.

This figure aids in the prediction of certain famous characteristics of heat addition at constant area. Thus, Eq. (2.91) shows that if the stagnation temperature increases, $f(M_2^2)$ must be larger than $f(M_1^2)$. Figure 2.9 indicates it must then be true that, whether M_1 is greater than or less than unity, M_2 must be closer to unity than M_1.

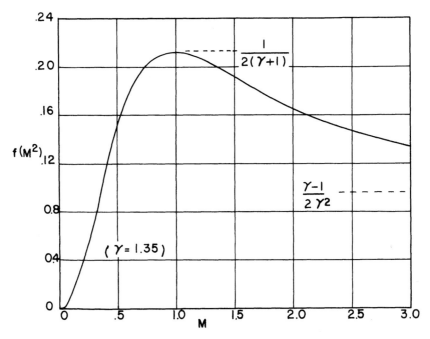

Fig. 2.9 $f(M^2)$ vs M, ideal constant-area heat interaction.

A further result, at first appearing to be a paradox, occurs when a heat interaction sufficient to require $f(M_2^2)$ to be greater than $1/2(\gamma + 1)$ is imposed. Since this requirement is impossible to satisfy, it appears that the analysis flies in the face of reality, because it is apparent that the amount of heat interaction that can be imposed is not limited experimentally. It must be recalled, however, that Δq_{1-2} is the heat interaction per unit mass, and the condition of $M_2 = 1$ represents that state when the maximum mass flow per unit area exists for the given p_{t_2} and T_{t_2}. Thus, in an experiment, if this condition (termed "thermal choking") is reached and it is attempted to cause a further heat interaction, the upstream conditions must change (usually the stagnation pressure must increase) if the mass flow is to be passed. It can be recognized that this phenomenon of thermal choking can be of vital importance in determining the maximum allowable heat interaction in a ramjet, an afterburner, or even a conventional combustor.

Equation (2.91) is a quadratic equation in M_2^2, which may be solved and manipulated to yield

$$M_2^2 = \frac{2f}{1 - 2\gamma f \pm \left[1 - 2(\gamma + 1)f\right]^{\frac{1}{2}}} \tag{2.93}$$

where

$$f = M_1^2 \frac{1 + \left[(\gamma - 1)/2\right] M_1^2}{\left(1 + \gamma M_1^2\right)^2} \left[1 + \frac{\Delta q_{1-2}}{C_p T_{t_1}}\right]$$

and where the $+$ sign corresponds to subsonic flow and the $-$ sign to supersonic flow.

The stagnation temperature follows immediately from Eq. (2.91), and combination of the continuity and momentum equations for constant-area flow gives

$$p_1 + \rho_1 u_1^2 = p_2 + \rho_2 u_2^2 \tag{2.94}$$

or

$$\frac{p_2}{p_1} = \frac{1 + \gamma M_1^2}{1 + \gamma M_2^2}$$

Then

$$\frac{p_{t_2}}{p_{t_1}} = \frac{1 + \gamma M_1^2}{1 + \gamma M_2^2} \left\{ \frac{1 + \left[(\gamma - 1)/2\right] M_2^2}{1 + \left[(\gamma - 1)/2\right] M_1^2} \right\}^{\gamma/(\gamma - 1)} \tag{2.95}$$

Example results are shown in Fig. 2.10.

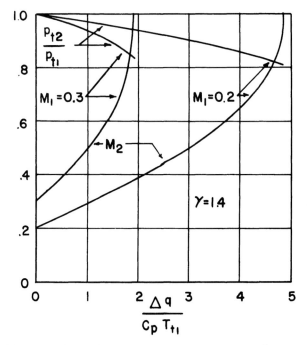

Fig. 2.10 Mach number and stagnation pressure ratio vs $\Delta q / C_p T_{t_1}$ for ideal constant-area heat interaction.

Adiabatic constant-area flow—viscous choking. When adiabatic flow is considered, no change in stagnation temperature occurs, so Eq. (2.90) yields

$$\gamma C_f \frac{c}{A}\,dx = \frac{1 - M^2}{M^4\left(1 + \dfrac{\gamma - 1}{2} M^2\right)}\,dM^2 \tag{2.96}$$

It is apparent that dM^2/dx is positive for $M < 1$ and negative for $M > 1$. Thus, as with thermal choking, the effect of viscosity is to drive the flow to Mach 1. This effect is termed viscous choking. As in the case of thermal choking, the condition represents that state where the maximum flow per unit area has been achieved for the given local values of stagnation pressure and temperature. If more flow is to be passed, the upstream conditions must change.

Equation (2.96) may be integrated in a straightforward manner to yield

$$\chi_2 - \chi_1 = f\left(M_2^2\right) - f\left(M_1^2\right) \tag{2.97}$$

where

$$f(M) = \frac{\gamma + 1}{2}\,\ell n\left\{\frac{1 + [(\gamma - 1)/2]\,M^2}{M^2}\right\} - \frac{1}{M^2} \tag{2.98}$$

and

$$\chi_2 - \chi_1 = \int_1^2 \gamma C_f \frac{c}{A}\, dx \equiv \int_1^2 d\chi \tag{2.99}$$

Combination of Eqs. (2.88) and (2.96) gives

$$\frac{dp}{p} = -\frac{1}{2}\left\{ \frac{1}{M^2} + \frac{(\gamma-1)/2}{1 + [(\gamma-1)/2]\,M^2} \right\} dM^2$$

from which

$$\frac{p_2}{p_1} = \frac{M_1}{M_2}\left\{ \frac{1 + [(\gamma-1)/2]\,M_1^2}{1 + [(\gamma-1)/2]\,M_2^2} \right\}^{\frac{1}{2}} \tag{2.100}$$

and

$$\frac{p_{t_2}}{p_{t_1}} = \frac{M_1}{M_2}\left\{ \frac{1 + [(\gamma-1)/2]\,M_2^2}{1 + [(\gamma-1)/2]\,M_1^2} \right\}^{(\gamma+1)/2(\gamma-1)} \tag{2.101}$$

Example results are indicated in Fig. 2.11.

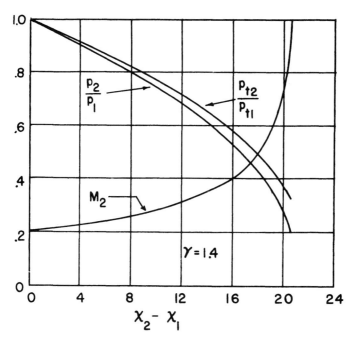

Fig. 2.11 Variation of static pressure, stagnation pressure, and Mach number with axial position, adiabatic constant-area flow.

2.19   The Nozzle Flow Equations

The flow of a calorically perfect gas in a channel of varying cross section is considered in this section. The mass flow rate $\dot{m}$ may be obtained in terms of the local area, Mach number, and stagnation properties directly from the continuity equation and Eqs. (2.67), (2.68), and $a^2 = \gamma RT$ as follows:

$$\dot{m} = \rho u A = \rho_t a_t A \frac{\rho}{\rho_t} \frac{a}{a_t} \frac{u}{a}$$

$$= \frac{A p_t}{\sqrt{T_t}} \sqrt{\frac{\gamma}{R}} \left(1 + \frac{\gamma - 1}{2} M^2\right)^{-[(\gamma + 1)/2(\gamma - 1)]} M \qquad (2.102)$$

Alternatively, this expression may be written in terms of the local static pressure by utilizing Eq. (2.67) to give

$$\dot{m} = \frac{A p_t}{\sqrt{T_t}} \left(\frac{p}{p_t}\right)^{1/\gamma} \left\{\frac{2}{R} \frac{\gamma}{\gamma - 1} \left[1 - \left(\frac{p}{p_t}\right)^{(\gamma - 1)/\gamma}\right]\right\}^{\frac{1}{2}} \qquad (2.103)$$

It is also common to reference conditions to conditions at the throat, which are here denoted by an asterisk. In Sec. 2.18 it was shown that the Mach number at the throat can be expected to be very close to unity and that it is usual to include this approximation. Thus,

$$\dot{m} = A^* p_t^* / C^* \qquad (2.104)$$

where by definition C^* is the characteristic velocity

$$C^* \equiv \left(\frac{\gamma + 1}{2}\right)^{(\gamma + 1)/2(\gamma - 1)} \sqrt{\frac{R T_t^*}{\gamma}}$$

It should be pointed out here that it is not inconsistent to consider the Mach number to be unity at the throat, but not to insist that $p_t = p_t^*$ throughout. Thus, $M^* = 1$ is an approximation that is numerically quite accurate, but it does not imply the assumption that the viscous effects are absent. It is possible to have the accumulated effects of viscosity upon the stagnation pressure be quite significant, but to still have the local viscous effect at the throat be very small.

Expressions for the area variation with Mach number and static pressure can be obtained directly by dividing Eq. (2.104) into Eqs. (2.102) and

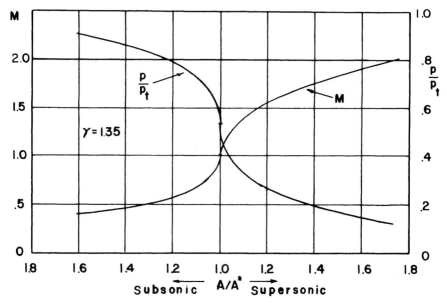

Fig. 2.12   Pressure ratio and Mach number vs area ratio for isentropic nozzle flow.

(2.103), respectively, to give

$$\frac{A}{A^*} = \frac{p_t^*}{p_t} \sqrt{\frac{T_t}{T_t^*}} \left[\frac{2}{\gamma+1} \left(1 + \frac{\gamma-1}{2} M^2 \right) \right]^{(\gamma+1)/2(\gamma-1)} \frac{1}{M} \qquad (2.105)$$

$$\frac{A}{A^*} = \sqrt{\frac{\gamma-1}{2}} \left(\frac{2}{\gamma+1} \right)^{(\gamma+1)/2(\gamma-1)} \frac{p_t^*}{p_t} \sqrt{\frac{T_t}{T_t^*}} \left(\frac{p_t}{p} \right)^{1/\gamma} \left[1 - \left(\frac{p}{p_t} \right)^{(\gamma-1)/\gamma} \right]^{\frac{1}{2}} \qquad (2.106)$$

Example results for isentropic flow ($p_t = p_t^*, T_t = T_t^*$) are shown in Fig. 2.12.

2.20   Numerical Solution of Equations

In several of the examples to follow (see also Problem 7.1), the desired variables appear in transcendental equations. Many iterative techniques are available for the solution of such equations. The numerical complications of such techniques have been greatly reduced with the advent of small computers with branching and looping capability, so that graphical techniques, etc., are no longer necessary. In the following, two well-known techniques are described.

Newtonian Iteration

Consider a transcendental equation of the form

$$F(x) = 0 \tag{2.107}$$

Now consider a jth estimate of x to be x_j and obtain a method of estimating a next (closer) quantity, x_{j+1}. The function $F(x_{j+1})$ is expanded in a Taylor's series to give

$$F(x_{j+1}) = F(x_j) + (x_{j+1} - x_j)F'(x_j) + \cdots \tag{2.108}$$

where $F'(x_j)$ is the derivative of the function $F(x)$ by x, evaluated at the value x_j. Now if x_{j+1} is to be close to the solution of $F(x) = 0$, $F(x_{j+1})$ may be approximated as zero. Also, if x_j is not far from x_{j+1}, the higher-order terms in the series may be ignored to give the Newtonian iteration,

$$x_{j+1} = x_j - \left[F(x_j)/F'(x_j) \right] \tag{2.109}$$

Equation (2.109) gives a method for obtaining the next estimate for the solution to the equation, x_{j+1}, in terms of the previous approximation. In practice, the process would be continued until $(x_{j+1} - x_j)$ was less than the desired accuracy.

It is sometimes convenient to approximate the derivative of the function by a finite difference form, say

$$F'(x_j) \approx \frac{F(x_j + \delta) - F(x_j - \delta)}{2\delta} \tag{2.110}$$

where δ would be a suitably small quantity. The advantage of utilizing such an approximation is that if a computer is to be programmed to utilize the Newtonian iteration, only a subroutine to calculate the function $F(x)$ itself need be supplied—it is not necessary to provide a separate subroutine to calculate the derivative.

Newtonian iteration is sometimes unstable, but it is fortunate in the examples to follow that, provided a suitable first guess for the desired variable is made, Newtonian iteration or the simpler functional iteration is stable in all of the examples considered herein.

Functional Iteration

A simple form of iteration related to the Newtonian is the functional iteration. Now assume a transcendental equation of the form

$$x = f(x) \tag{2.111}$$

This may be formally converted to a form suitable for solution by Newtonian iteration by defining

$$F(x) = x - f(x)$$

hence from Eq. (2.109)

$$x_{j+1} = x_j - \frac{F(x_j)}{F'(x_j)} = x_j - \frac{x_j - f(x_j)}{1 - f'(x_j)} \tag{2.112}$$

Now, if the function $f(x)$ is slowly varying in x, then $f'(x_j)$ may be ignored compared to unity. In that case, Eq. (2.112) reduces to a functional iteration

$$x_{j+1} = f(x_j) \tag{2.113}$$

This extremely simple form would then be iterated until $(x_{j+1} - x_j)$ is less than the desired accuracy. This form is very simple and convenient, but is suitable only when $|f'(x_j)| \ll 1$.

Reference

[1] Barclay, L. P., "Pressure Losses in Dump Combustors," AFAPL-TR-72-57, 1972.

Problems

2.1   Consider a perfect gas for which the specific heat at constant volume C_v can be approximated by

$$C_v = A + BT + CT^2 + \cdots \qquad (A, B, \text{etc.} = \text{const})$$

Show that for such a gas undergoing an isentropic process, the density is given in terms of the temperature by

$$\rho = k \left[T \exp\left(\frac{B}{A} T + \frac{CT^2}{2A} + \cdots \right) \right]^{A/R}$$

2.2   A Van der Waals fluid obeys the equation of state

$$\left[p + (a/v^2) \right](v - b) = RT \qquad (a, b, R = \text{const})$$

Show that for such a fluid

(a) $$\left(\frac{\partial C_v}{\partial v}\right)_T = 0$$

$$\left(\frac{\partial C_p}{\partial p}\right)_T = \frac{R^2 T (2a/v^3)[1-(3b/v)]}{[p-(a/v^2)+(2ab/v^3)]^3}$$

(b) $$du = C_v\,dT + (a/v^2)\,dv$$

$$dh = C_p\,dT + v\left[\frac{RTb-(2a/v^2)(v-b)^2}{RTv-(2a/v^2)(v-b)^2}\right]dp$$

2.3   The "Joule-Thomson coefficient" is defined as

$$\left(\frac{\partial T}{\partial p}\right)_h$$

(Note that a fluid that is adiabatically "throttled" through a porous plug undergoes a pressure change at constant enthalpy. The Joule-Thomson coefficient provides a measure of the expected temperature change.)

(a)   Show that for a Van der Waals fluid

$$\left(\frac{\partial T}{\partial p}\right)_h = -\frac{v}{C_p}\frac{RTb-(2a/v^2)(v-b)^2}{RTv-(2a/v^2)(v-b)^2}$$

(b)   Show that at the "inversion condition" [where $(\partial T/\partial p)_h = 0$]

$$p = (2a/bv) - (3a/v^2)$$

2.4   Show that for any fluid

$$\frac{C_v}{T}\left(\frac{\partial T}{\partial v}\right)_s = -\left(\frac{\partial p}{\partial T}\right)_v$$

Hence, for a Van der Waals fluid with C_v given by $C_v = A + BT$, show that the equation relating v and T for an isentropic process is

$$v - b = ke^{-BT/R}/T^{A/R}   (k = \text{const})$$

Show also that, if C_v were constant with temperature, the equation for an isentropic process could be written as

$$\left[p + \left(a/v^2 \right) \right] \left(v - b \right)^{1 + (R/C_v)} = \text{const}$$

2.5   Dieterici's equation of state is given by

$$p(v - b) = RTe^{-a/RTv} \qquad (a, b, R = \text{const})$$

Show that for such a fluid the Joule-Thomson coefficient is given by

$$\left(\frac{\partial T}{\partial p} \right)_h = \frac{1}{C_p} \left\{ \frac{(2a/vRT)(v - b) - b}{1 - \left[a(v - b)/v^2 TR \right]} \right\}$$

Show also that the difference in specific heats is given by

$$C_p - C_v = Re^{-a/RTv} \frac{(1 + a/RTv)^2}{1 - (a/v^2)[(v - b)/RT]}$$

2.6   Air is contained in a stepped cylinder fitted with a frictionless piston, as indicated in Fig. A. The air is cooled as a result of heat transfer to the surroundings.

(a)   What is the ratio of the temperature that would exist just as the piston reaches the step to the initial temperature T_i in terms of the lengths L_1 and L_2 and related cross-sectional areas A_1 and A_2?

(b)   If the air is further cooled to a final temperature of T_f, what is the ratio of final pressure to initial pressure in terms of L_1, L_2, A_1, A_2, and T_f/T_i?

2.7   A thin-walled metal can of volume V_c contains a calorically perfect gas at pressure p_c and temperature T_c. Connected to the can is a capillary tube and stopcock. The stopcock is opened slightly and the gas leaks slowly into a heat-conducting cylinder equipped with a frictionless piston. The surroundings are at pressure p_a and temperature T_c.

(a)   Show that, after as much gas as possible has leaked out, a work

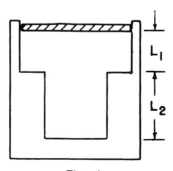

Fig.   A

interaction in the amount of $W = p_a(V_f - V_c)$ has occurred. Here, V_f is the final volume of the gas. Find V_f in terms of p_a, p_c, and V_c.

(b)   Show that the entropy gain of the system and surroundings is given by

$$\Delta s_{\text{tot}} = \frac{p_a V_f}{T_c}\left[\ell n \frac{V_f}{V_c} - \left(1 - \frac{V_c}{V_f}\right)\right]$$

2.8   A chamber contains a calorically perfect gas at pressure p_i and temperature T_i. It is connected through a valve with a vertical cylinder that is closed on top by a frictionless piston. The piston is loaded by a weight of such magnitude that a pressure of p_f is maintained within the cylinder. (See Fig. B.) Initially the piston is at the bottom of the cylinder; then, the valve is slightly opened to allow the pressures in chamber and cylinder to become equal. It may be assumed that the volume of the piping is negligible, that the expansion process in the chamber is reversible and adiabatic (no heat transfer back through the valve), and that there is no heat transfer to any of the walls.

Show that the final temperature in the cylinder is given by

$$T_{C_f} = \frac{1}{\gamma} T_i \left[\frac{1 - p_f/p_i}{1 - (p_f/p_i)^{1/\gamma}}\right]$$

2.9   A highly evacuated, thermally insulated flask is placed in a room with air temperature T_a. The outside air is then allowed to enter the flask through a slightly opened stopcock until the pressure inside equals the pressure outside, at which time the stopcock is closed.

Assuming that the air is calorically perfect, what would be the temperature of the air inside the flask after the process was completed?

2.10   Consider the frictionless flow of a calorically perfect gas in a channel with thermal interaction. For the case where the wall is shaped to

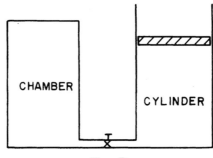

Fig. B

keep the static temperature constant:
  (a) Find an expression for the area ratio A/A_i in terms of γ, $(T_t - T_{t_i})/T$, and M_i (i refers to initial conditions).
  (b) Find an expression for the ratio p_t/p_{t_i} in terms of the same variables.

2.11   Consider the frictionless flow of a calorically perfect gas in a channel with thermal interaction. For the case where the wall is shaped to keep the Mach number constant:
  (a) Find an expression for the area ratio A/A_i in terms of γ, M, and T_t/T_{t_i} (i refers to initial conditions).
  (b) Find an expression for the ratio p_t/p_{t_i} in terms of the same variables.

2.12   Consider the frictionless flow of a calorically perfect gas in a channel with thermal interaction. The wall is shaped so as to keep the static pressure constant. The flow enters at condition 3 and departs at condition 4. Defined are $\tau_b = T_{t_4}/T_{t_3}$ and $\pi_b = p_{t_4}/p_{t_3}$.
  (a) Find an expression for M_4 in terms of γ, M_3, and τ_b.
  (b) Show that

$$\pi_b = \left[\frac{\tau_b}{\tau_b + [(\gamma - 1)/2]\, M_3^2(\tau_b - 1)} \right]^{\gamma/(\gamma-1)}$$

  (c) Show that

$$\frac{A_4}{A_3} = \tau_b + \frac{\gamma - 1}{2} M_3^2(\tau_b - 1)$$

2.13   Consider the frictionless flow of a calorically perfect gas in a constant-area channel with thermal interaction. The gas enters the duct at Mach number M_i and thermal energy addition occurs until the flow chokes ($M_f = 1$).
  (a) Obtain an expression for the related stagnation temperature ratio (T_{t_f}/T_{t_i}) in terms of γ and M_i.
  (b) Plot T_{t_f}/T_{t_i} vs M_i over the range $0 < M_i \leq 1$. Assume $\gamma = 1.29$.
  (c) Plot p_{t_f}/p_{t_i} over the same range.

2.14   Consider the adiabatic flow of a calorically perfect gas in a duct. The duct is shaped so that the velocity remains constant.
  (a) Assuming the duct is of circular cross section and that the skin-friction coefficient C_f may be approximated as being constant, show that the duct diameter D is given by

$$D = D_i + \gamma M^2 C_f x$$

where i refers to the initial condition and x_i is taken to be zero.

(b) Obtain an expression for p_t/p_{t_1} in terms of γ, M, C_f, and x/D_i.

2.15   A calorically perfect gas passing through a constant area duct enters a region where thermal addition occurs. The thermal addition is complete by station 2, after which the gas expands isentropically to station 3. See Fig. C. Skin friction may be ignored.

(a)   Assuming $\gamma = 1.32$, $M_1 = 0.3$, $T_{t_2}/T_{t_1} = 2$, and $A_3/A_2 = 0.9$, find M_3.

(b)   Find the value of T_{t_2}/T_{t_1} that will just cause the flow to choke.

(c)   Assuming that T_{t_1} does not change, if T_{t_2}/T_{t_1} is further increased by 10%, to what must M_1 be reduced to allow the process to continue?

(d)   If the mass flow is kept the same, what will be the required percentage increase in p_{t_1}?

2.16   A flow undergoes a sudden "dump" as indicated in Fig. D. A series of experiments has lead to the relationship of p_{t_2}/p_{t_1} in terms of the upstream conditions and the area ratio A_2/A_1. We are to determine the effective average (static) pressure on the "dome" p_D, where the appropriate average is considered to be that which would satisfy a momentum balance.

It is to be noted that such a pressure is indeed an average, because the Kutta condition at the pipe end would cause the local value of the wall pressure to be equal to the pressure in the stream at station 1.

It may be assumed that the sidewall friction is zero and that conditions are quasi-one-dimensional at station 2.

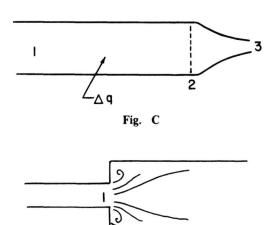

Fig. C

Fig. D

(a)   Show that

$$\frac{p_D}{p_1} = \frac{1}{\frac{A_2}{A_1} - 1} \left[\frac{M_1 \left(1 + \frac{\gamma - 1}{2} M_1^2\right)^{\frac{1}{2}}}{M_2 \left(1 + \frac{\gamma - 1}{2} M_2^2\right)^{\frac{1}{2}}} \left(1 + \gamma M_2^2\right) - \left(1 + \gamma M_1^2\right) \right]$$

and

$$M_2 = M_1 \frac{p_{t_1}}{p_{t_2}} \left(\frac{1 + \frac{\gamma - 1}{2} M_2^2}{1 + \frac{\gamma - 1}{2} M_1^2} \right)^{(\gamma+1)/2(\gamma-1)} \frac{A_1}{A_2}$$

(b)   It has been found[1] that in the range $0.2 < A_1/A_2 < 1$, the ratio p_{t_2}/p_{t_1} is given approximately by

$$\frac{p_{t_2}}{p_{t_1}} = \exp\left(-\overline{PLC}\frac{\gamma}{2} M_1^2\right)$$

where $\overline{PLC} \approx [1 - (A_1/A_2)]^2 + [1 - (A_1/A_2)]^6$.

  Calculate and plot p_D/p_1 vs M_1 over the range $0.1 \le M_1 \le 0.6$ for the case $\gamma = 1.30$, $A_1/A_2 = 0.5$.

2.17   An independent group of investigators approached the problem of the dump flow from another viewpoint. They heavily instrumented the "dome" of the pipe with static pressure instrumentation so that they could measure p_D. They then calculated p_{t_2}/p_{t_1} by assuming the sidewall friction to be zero.

  (a)   Assuming that A_1/A_2, p_D/p_1, M_1, and γ will be prescribed, obtain a series of relationships for M_2, p_2/p_1, and p_{t_2}/p_{t_1} in terms of the prescribed variables.

  (b)   For the case $p_D/p_1 = 1$, $M_1 = 0.5$, $\gamma = 1.3$, calculate and plot p_{t_2}/p_{t_1} for the range $0.2 < A_1/A_2 < 1$. Compare the ratio so obtained with that given by the formula of Problem 2.16(b).

2.18   The flow processes in a ramjet may be approximated as indicated in Fig. E. Thus, there is isentropic compression in the inlet from the freestream conditions (0) past the minimum area location (m) up to the upstream edge of a normal shock wave (su). Following passage through the shock the flow diffuses isentropically to station 3. Constant-area thermal addition occurs from stations 3 to 4, after which the flow expands isentropically through the nozzle.

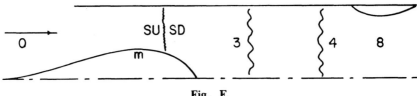

Fig. E

We are given

$$\frac{A_8}{A_4} = 0.8, \qquad \frac{T_{t_4}}{T_{t_3}} = 2.5, \qquad M_m = 1.3, \qquad M_8 = 1, \qquad \gamma = 1.35$$

The stagnation pressure ratio across a normal shock wave, p_{t_D}/p_{t_u}, is given in terms of the upstream Mach number by

$$\frac{p_{t_D}}{p_{t_u}} = \left(\frac{\frac{\gamma+1}{2}M_{su}^2}{1 + \frac{\gamma-1}{2}M_{su}^2} \right)^{\gamma/(\gamma-1)} \left[1 + \frac{2\gamma}{\gamma+1}(M_{su}^2 - 1) \right]^{-1/(\gamma-1)}$$

(a) What is M_3?
(b) If T_{t_4}/T_{t_3} is increased to 2.7, what is M_3?
(c) What is the required ratio of $(p_{t_3})2.7/(p_{t_3})2.5$ to bring about the change in M_3?
(d) If the shock was originally located where $M_{su} = 1.6$, what will the new value of M_{su} be?
(e) If the freestream Mach number is $M_0 = 2$, find the area ratios

$$(A_s/A_4)2.5, \qquad (A_s/A_4)2.7, \qquad (A_m/A_4), \qquad (A_0/A_4)$$

(f) What is the maximum value for T_{t_4}/T_{t_3} that could be achieved with the shock wave still contained within the ramjet?

3. CHEMICAL ROCKETS

3.1 Introduction

Methods for estimating the performance of both liquid- and solid-propellant rockets will be developed in this chapter. In order to make such estimates, it is necessary to predict the nozzle performance when given the thermodynamic conditions existing at the completion of combustion within the combustion chamber. Large booster rockets pass through very large altitude ranges with related large variations in ambient pressure (see App. A, Standard Atmosphere). The variation in ambient pressure has a significant effect upon the thrust level. Methods to predict the thrust level are presented.

In order to estimate conditions following combustion, it is necessary to apply concepts of equilibrium chemistry, and this subject is briefly reviewed. The designer can determine the mass flow rates of the fuel and oxidizer into the combustion chamber by correct pump design, and he can determine the chamber pressure by correct selection of the nozzle throat area. The chemical composition and temperature of the products of combustion can then be determined by applying the principles of equilibrium chemistry.

The temperature levels experienced in rocket combustion chambers are so extreme that a sizeable portion of the product gas remains dissociated. As a result, as the gas is accelerated through the nozzle (with a consequent decrease in static temperature), the chemical reactions continue, giving rise to further changes in the gas properties. The two limiting cases of nozzle flow, frozen and equilibrium flows, are illustrated in this chapter and may be used to estimate the possible ranges of the effects of the continued reaction within the nozzle.

Finally, simplified models are developed for the description of the processes within solid-propellant rockets. The models allow simple estimates of a solid-propellant rocket thrust history.

3.2 Expression for the Thrust

The thrust on a rocket can be expressed as the integral of the surface stresses over all of the rocket solid surfaces. Such an integral would include contributions over all the internal surfaces wetted by the fluid (chamber, pipes, pumps, etc.) and clearly would be most difficult to evaluate directly. Rather than attempting to do so, however, the internal force contributions are related to the fluid properties at the exit plane of the nozzle by use of the momentum equation.

Thus, the expression for the force on the thrust stand, depicted in Fig. 3.1, may be written

$$\mathbf{F} = -\iint_{(\Sigma_o + \Sigma_c)} (p - p_a)\, \mathbf{ds} + (\text{visc})_{\Sigma_c} + (\text{visc})_{\Sigma_o} \qquad (3.1)$$

In this expression $\mathbf{F}$ is the vector force transmitted by the rocket to the thrust stand, $\mathbf{ds}$ the outwardly directed vector area element, Σ_o and Σ_c the outer and inner (chamber) surface areas, respectively, and (visc) the viscous force over the given surface area. For later convenience, the pressure has been given relative to the ambient pressure p_a. It is to be noted that the area integral of p_a over the entire closed surface $(\Sigma_o + \Sigma_c)$ is zero.

Equation (3.1) is of little calculational use because of the complexity of the internal integrals. A more useful form follows by utilizing the momentum equation to relate the internal forces to conditions at the nozzle exit. The internal surface is $\Sigma_{c'}$. Note that the direction of the outward normal to the surface $\Sigma_{c'}$ is opposite to that of Σ_c (Fig. 3.2). Equating the internal forces to the rate at which momentum is convected through the surface results in

$$-\iint_{(\Sigma_{c'} + A_e)} (p - p_a)\, \mathbf{ds} + (\text{visc})_{\Sigma_{c'}} + (\text{visc})_{A_e} = \iint_{A_e} (\rho \mathbf{u})\mathbf{u} \cdot \mathbf{ds} \quad (3.2)$$

Noting that the force contributions from the surface $\Sigma_{c'}$ are just of opposite sign to the force contributions from Σ_c, and that the normal viscous force at exit, $(\text{visc})_{A_e}$, is negligible, combination of Eqs. (3.1) and (3.2) gives

$$\mathbf{F} = -\iint_{A_e} (\rho \mathbf{u})\mathbf{u} \cdot \mathbf{ds} - \iint_{A_e} (p - p_a)\, \mathbf{ds} + (\text{visc})_{\Sigma_o} - \iint_{\Sigma_o} (p - p_a)\, \mathbf{ds}$$

$$(3.3)$$

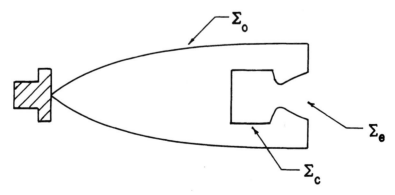

Fig. 3.1   Rocket on stand.

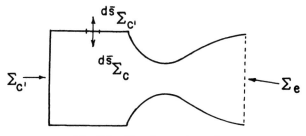

Fig. 3.2   Chamber and nozzle.

The third term in this expression represents the force of the external skin friction and is termed the skin drag. The fourth term represents the effect of the pressure imbalance on the external surface and is termed the form drag. The first two terms represent contributions to the thrust of the rocket, hence

$$\mathbf{T} = -\iint_{A_e} (\rho \mathbf{u}) \mathbf{u} \cdot \mathbf{ds} - \iint_{A_e} (p - p_a) \, \mathbf{ds} \tag{3.4}$$

If conditions at exit can be represented as one-dimensional, the magnitude of the thrust may be written

$$T = \dot{m} u_e + (p_e - p_a) A_e \tag{3.5}$$

Effective Exhaust Velocity

For convenience, the effective exhaust velocity C is defined by the relationship

$$T = \dot{m} C \tag{3.6}$$

hence

$$C = u_e + [(p_e - p_a)/\dot{m}] A_e \tag{3.7}$$

Note the alternative form

$$C = u_e \left[1 + \frac{(p_e - p_a) A_e}{u_e (\rho_e u_e A_e)} \right]$$

or

$$C = u_e \left[1 + \frac{1}{\gamma M_e^2} \left(1 - \frac{p_a}{p_e} \right) \right] \tag{3.8}$$

The Specific Impulse, I_{sp}

A commonly used measure of rocket performance is the specific impulse I_{sp}, defined as the ratio of thrust to the propellant weight flow per second. In the development leading to Eq. (3.5), it has been assumed that the units of the equation were in a "preferred" system, that is, one in which the units of force are defined to be those of mass times acceleration. In such a system weight is given in terms of mass by multiplying the mass by the magnitude of the standard acceleration of gravity, g_0 ($g_0 = 9.8067$ ms^{-2} or $g_0 = 32.174$ ft·s^{-2}). Note that although g_0 has the magnitude of a_0, the gravitational constant, the dimensions of the latter are kg/N·m/s^2 (or lbm/lbf·ft/s^2).

It follows from Eq. (3.6) that

$$I_{sp} = C/g_0 \qquad \text{(in seconds)} \qquad (3.9)$$

3.3 Acceleration of a Rocket

Trajectory analysis involves the analysis of a rocket flight path under the influence of the thrust, lift, drag, and gravitational force. However, rather than consider a complete trajectory analysis here, the simpler case of nonlifting motion with the thrust aligned in the direction of flight will be considered. The forces acting on the rocket (Fig. 3.3) may be resolved in the direction of flight to give

$$T - D - gm\cos\theta = m(\mathrm{d}v/\mathrm{d}t) \qquad (3.10)$$

Noting that the rate of mass flow through the nozzle is just equal to the negative of the rate of change of the vehicle mass, there is obtained

$$\mathrm{d}v = -C(\mathrm{d}m/m) - (D/m)\,\mathrm{d}t - g\cos\theta\,\mathrm{d}t \qquad (3.11)$$

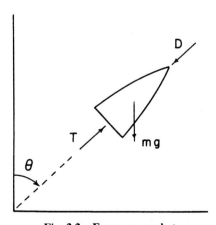

Fig. 3.3 Forces on rocket.

and

$$dh = v \cos \theta \, dt \qquad (3.12)$$

The drag is a complicated function of velocity and height: if the function is known analytically, this pair of equations may be numerically integrated in a straightforward manner. It is instructive to consider some simple special cases; if, for example, a suitable average value of the effective exhaust velocity can be obtained, the increment in vehicle velocity Δv found for a mass "burn" of m_p is seen to be given by

$$\frac{\Delta v}{C} = \ell n \left(\frac{m_0}{m_0 - m_p} \right) - \int_0^\tau \frac{D}{Cm} \, dt - \int_0^\tau \frac{g \cos \theta}{C} \, dt \qquad (3.13)$$

where m_0 is the initial mass and τ the burning time.

The first integral is termed the drag loss and the second integral the gravity loss. The latter term arises because during the finite firing time the mass of the still unburned propellant must be raised in the gravitational field. Obviously, the gravity loss term could be decreased by increasing the thrust level of the rocket (increasing $\dot{m}$ and thereby decreasing τ), but this in itself would introduce its own complications. The larger required pumps and related piping would increase the rocket mass, the maximum allowable acceleration could be exceeded, and the acceleration to high velocities in the lower altitudes would increase the overall drag loss. Good rocket design involves selection of the optimal balance between these competing tendencies.

In the very simple case of firing in "free space" where no drag or gravitational penalties exist, Eq. (3.13) reduces to

$$MR \equiv \frac{m_0}{m_0 - m_p} = e^{\Delta v / C} \qquad (3.14)$$

This expression emphasizes the sensitivity of the required mass ratio MR to the attainable effective exhaust velocity. This is particularly true for high-energy (large Δv) missions.

Multiple-Stage Rockets

Casual examination of Eqs. (3.13) or (3.14) indicates that if high-energy missions are to be contemplated utilizing the most energetic chemical propellants available to date ($C < 4500 \text{ ms}^{-1}$), distressingly large mass ratios will be required. It is to be remembered that the final mass consists of the "dead weight mass" m_d (mass of structure, engines, unused propellants, etc.), as well as the payload mass m_L itself, so that methods to increase the possible payload for given overall rocket mass are of great importance.

A fairly obvious method to accomplish an increase in payload capability is to stage the rocket so that unneeded mass can be discarded at opportune times in the flight. The expressions for the velocity increments given by Eqs. (3.13) and (3.14) remain valid, only now the total velocity increment Δv_{tot} would be obtained as the sum of the velocity increments of all stages. As a simple example of the effects of staging consider the "free space" case where, again, the drag loss and gravity loss terms do not contribute. Introducing the payload ratio λ and dead weight ratio δ,

$$\lambda = m_L/m_0 \qquad \text{and} \qquad \delta = m_d/m_0 \qquad (3.15)$$

note the relationships

$$\frac{m_0}{m_0 - m_p} = \frac{m_0}{m_d + m_L} = \frac{1}{\delta + \lambda} \qquad (3.16)$$

Thus

$$\Delta v_{tot} = \sum_{i=1}^{N} C_i \ell n \frac{1}{\delta_i + \lambda_i} \qquad (3.17)$$

where i refers to the ith of a total of N stages. Note that the payload of the ith stage is the sum of all succeeding stages, so that the overall payload ratio λ_0 is equal to the product of all λ_i. That is,

$$\lambda_0 = \prod_{i=1}^{N} \lambda_i \qquad (3.18)$$

Optimization of Multiple-Stage Rockets

The very simple formulas of Eqs. (3.17) and (3.18) allow simple determination of appropriate stage payload ratios to lead to a minimum overall mass for a given Δv_{tot} and payload mass. Normally, the engineer would be asked to consider the problem of designing the rocket for minimum overall mass given prescribed Δv_{tot} and payload mass. Mathematically, however, it is simpler to consider a given overall payload ratio and maximize the Δv_{tot} obtained. If the assumed δ_i and C_i are still found to be within reason after solution of the problem indicates the vehicle size, the λ_0 corresponding to the Δv_{tot} required, and hence the overall mass required, can be determined.

The mathematical problem is to maximize Δv_{tot} for given C_i and δ_i subject to the restriction

$$\lambda_0 = \prod_{i=1}^{N} \lambda_i \qquad (3.19)$$

Solution is facilitated by introduction of the Lagrange multiplier K and definition of the function F where

$$F = \sum_{i=1}^{N} C_i \ln \frac{1}{\delta_i + \lambda_i} + K \left[\ln \left(\prod_{i=1}^{N} \lambda_i \right) - \ln \lambda_0 \right] \tag{3.20}$$

Clearly F has a maximum at the same location as the maximum of Δv_{tot}. F can now be considered a function of all the λ_i, provided K is selected to ensure the restriction of Eq. (3.19). Taking the partial derivative, it follows that

$$\frac{\partial F}{\partial \lambda_i} = - \frac{C_i}{\delta_i + \lambda_i} + \frac{K}{\lambda_i} \tag{3.21}$$

Equating this expression to zero gives

$$\lambda_i = \frac{\delta_i}{[(C_i/K) - 1]} \tag{3.22}$$

A relationship for the unknown value of the Lagrange multiplier follows from Eqs. (3.19) and (3.22)

$$\prod_{i=1}^{N} \left(\frac{C_i}{K} - 1 \right) = \frac{\displaystyle\prod_{i=1}^{N} \delta_i}{\lambda_0} \tag{3.23}$$

This is an Nth-order equation for K. Following solution for K the desired optimal payload ratios are obtained from Eq. (3.22).

As a simple example consider the case where all the equivalent exhaust velocities are equal. It follows that

$$\lambda_i = \lambda_0^{1/N} \frac{\delta_i}{\left(\displaystyle\prod_{i=1}^{N} \delta_i \right)^{1/N}}$$

An even simpler example is that for which all the δ_i are equal, which gives $\lambda_i = \lambda_0^{1/N}$ and

$$\frac{\Delta v_{\text{tot}}}{C} = N \ln \frac{1}{\left(\lambda_0^{1/N} + \delta \right)}$$

from which

$$\lambda_0 = \left(e^{-[\Delta v_{\text{tot}}/NC]} - \delta \right)^N$$

Thus, for example, if $\Delta v_{tot}/C = 2$ and $\delta = 0.1$ are prescribed,

N	1	2	3
λ_0	0.035	0.072	0.071

It would appear that for the given velocity ratio and dead weight ratio, a two-stage rocket would be the optimal choice. (Why doesn't λ_0 keep going up with N?!)

As another example, consider a two-stage rocket. In this case, Eq. (3.23) is a quadratic equation for K, which upon solution yields

$$\lambda_1 = \frac{2\,\delta_1 C_2}{(C_1 - C_2) + \sqrt{(C_1 - C_2)^2 + 4C_1C_2\delta_1\delta_2/\lambda_0}}$$

$$\lambda_2 = \frac{2\,\delta_2 C_1}{(C_2 - C_1) + \sqrt{(C_2 - C_1)^2 + 4C_1C_2\delta_1\delta_2/\lambda_0}}$$

If Δv_{tot}, C_1, C_2, δ_1, and δ_2 were again prescribed, λ_0 would be obtained from Eq. (3.17).

3.4 Rocket Nozzle Performance

As stated in the introduction, the variation in ambient pressure with altitude causes significant variations in thrust level. An estimate of such variations can be obtained by utilizing the very simple approximation that the flow within the nozzle is isentropic and that the gas is calorically perfect. The validity of these assumptions is further investigated in Sec. 3.7.

Denoting by a subscript c the stagnation conditions within the rocket chamber, the first law of thermodynamics and the isentropic relationships give

$$\frac{u_e^2}{2} = h_c - h_e = \frac{\gamma}{\gamma - 1} RT_c \left[1 - \left(\frac{p_e}{p_c} \right)^{(\gamma-1)/\gamma} \right]$$

hence

$$u_e = \sqrt{\frac{2\gamma}{\gamma - 1}} \, \Gamma C^* \left[1 - \left(\frac{p_e}{p_c} \right)^{(\gamma-1)/\gamma} \right]^{\frac{1}{2}} \tag{3.24}$$

where C^* is the characteristic velocity defined in Eq. (2.104) and $\Gamma \equiv [2/(\gamma + 1)]^{(\gamma+1)/2(\gamma-1)}\sqrt{\gamma}$.

The thrust coefficient C_F is defined by

$$C_F = T/p_c A^* \tag{3.25}$$

With the relationship

$$M_e^2 = \frac{2}{\gamma - 1} \left[\left(\frac{p_c}{p_e} \right)^{(\gamma - 1)/\gamma} - 1 \right] \tag{3.26}$$

and Eqs. (2.104), (3.6), (3.8), (3.24), and (3.25), it follows that

$$C_F = \sqrt{\frac{2\gamma}{\gamma - 1}} \, \Gamma \left[1 - \left(\frac{p_e}{p_c} \right)^{(\gamma - 1)/\gamma} \right]^{\frac{1}{2}}$$

$$\times \left\{ 1 + \frac{\gamma - 1}{2\gamma} \left(\frac{p_c}{p_e} \right)^{1/\gamma} \frac{[(p_e/p_c) - (p_a/p_c)]}{\left[1 - (p_e/p_c)^{(\gamma - 1)/\gamma} \right]} \right\} \tag{3.27}$$

and also

$$C = C^* C_F \tag{3.28}$$

This is a convenient formulation for the effective exhaust velocity C, because C^* is primarily a function of the propellants (it is a weak function, also, of the chamber pressure through the chamber pressure effect on T_c and hence γ). The functional dependence of the thrust coefficient is of the form $C_F(\gamma, p_c/p_e, p_c/p_a)$, and hence C_F is a function of the design choice of p_c/p_e (determined by choice of area ratio A_e/A^*) and choice of combustion chamber pressure and altitude (p_c/p_a).

Nozzle Sizing

It is easily verified analytically (see Problem 3.5) that the thrust coefficient has a maximum value (for prescribed ambient pressure and chamber pressure) when the exit pressure equals the ambient pressure. Such a relationship is obvious from physical reasoning also, as can be seen by imagining, for example, that the nozzle was equipped with an additional length of exit cone to continue the gas expansion to lower pressures (and higher velocities) than the local external pressure. Because the area is expanding and the internal wall static pressure is less than the external static pressure, it is evident that the additional length of nozzle would have a rearward force acting upon it. Similarly, if the nozzle was of insufficient expansion to reduce p_e to p_a, the additional forward force that would result on the length of nozzle necessary to reduce p_e to p_a would not be available.

The ratio of nozzle exit area to throat area required to provide the desired design pressure ratio p_{ed}/p_c follows from Eq. (2.106) to give

$$\frac{A_e}{A^*} = \sqrt{\frac{\gamma - 1}{2}} \left(\frac{2}{\gamma + 1} \right)^{(\gamma + 1)/2(\gamma - 1)} \left(\frac{p_c}{p_{ed}} \right)^{1/\gamma} \left\{ 1 - \left(\frac{p_{ed}}{p_c} \right)^{(\gamma - 1)/\gamma} \right\}^{-\frac{1}{2}}$$

$$\tag{3.29}$$

It is to be noted that when a conventional rocket (fixed A_e/A^*) is flown at altitudes above the design altitude (where $p_{ed} = p_{ad}$), the exit pressure will be fixed at p_{ed} and is larger than the local ambient pressure p_a. Under these circumstances, the pressure of the exhaust stream adjusts to ambient pressure by passing through expansion fans attached to the nozzle lip.

When flight at altitudes below the design altitude occurs, the ambient pressure exceeds the nozzle exit pressure. Under these conditions oblique shock waves are created that bring the fluid pressure up to the ambient pressure. When conditions are such that the ambient pressure greatly exceeds the design exit pressure, the strong oblique shock waves that are formed at the exit due to the overpressure move into the nozzle, thereby changing the effective exit pressure [for use in Eq. (3.27)].

The behaviors described above are illustrated and summarized in Fig. 3.4. Note that in addition to the design altitude h_d the "separation altitude" h_{sep} has been introduced. This latter altitude is defined as the lowest altitude at which the oblique shock waves remain on the nozzle lip. It is apparent that above this altitude the exit pressure will be equal to the design pressure p_{ed}, whereas below this altitude the shocks move into the nozzle and the effective exit pressure becomes a function of altitude.

A very simple approximate method of estimating the resulting effective exit pressure was suggested by Summerfield.[1] He observed that the flow in the vicinity of the walls just following the location of the strong oblique shock waves was largely separated, and as a result the wall static pressure downstream of the shock waves was nearly equal to the ambient pressure. As a result, the effective exit pressure [for use in Eq. (3.27)] could be considered the pressure just preceding the shock wave. Summerfield further suggested the use of the very simple estimate of this "shock pressure" p_s given by

$$p_s/p_c = (1/K)(p_a/p_c) \qquad (3.30)$$

where K is a constant of value approximately 2.7–2.8.

The entire altitude performance of a given nozzle can now be calculated by utilizing Eq. (3.30) for altitudes below which the value of p_s given by Eq. (3.30) is larger than the design exit pressure p_{ed} and using $p_e = p_{ed}$ above this separation altitude h_{sep}.

Particularly simple results can be obtained if the pressure variation with altitude may be approximated as exponential. Thus

$$p_a/p_{S.L.} = e^{-h/H_{scl}} \qquad (3.31)$$

where $p_{S.L.}$ is the sea level pressure and H_{scl} the scale height. It follows directly (Problem 3.6) that

$$h_{sep} = h_d - H_{scl} \ln K \qquad (3.32)$$

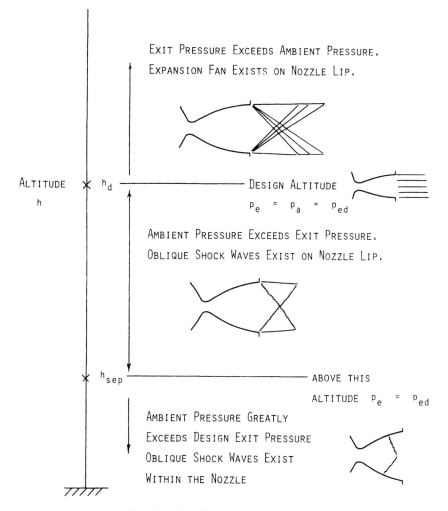

Fig. 3.4 Nozzle behavior with altitude.

It is apparent that an "ideal" rocket would have a completely variable nozzle area ratio, so that the exit area could be selected to give $p_e = p_a$ at all altitudes. It is hard to imagine a geometry that could give such capability, but a compromise concept is shown in Fig. 1.2 that illustrates a nozzle with three "design altitudes" capability. Optimal utilization of such a device would be obtained if the next larger area ratio skirt was translated at just the altitude where the thrust coefficients of the two nozzles are equal.

As an example, consider a rocket nozzle with a single translatable skirt and hence two design altitudes h_{d1} and h_{d2}. Assume that an exponential pressure variation is valid, and consider the case $h_{d1} = 30,000$ ft, $h_{d2} = 60,000$

ft, $K = 2.75$, $\gamma = 1.28$, $p_c = 40$ $p_{S.L.}$, and $H_{scl} = 23,000$ ft. It can be shown (Problem 3.7) that the correct altitude for skirt translation, h_{trans}, is given by

$$h_{trans} = H_{scl} \ln\left(\frac{p_{S.L.}}{p_c} \frac{B_1 - B_2}{A_1 - A_2}\right) \qquad (3.33)$$

where

$$A_i = \frac{1 - [(\gamma + 1)/2\gamma]\left(p_{e_i}/p_c\right)^{(\gamma-1)/\gamma}}{\left[1 - \left(p_{e_i}/p_c\right)^{(\gamma-1)/\gamma}\right]^{\frac{1}{2}}} \qquad (i = 1 \text{ or } 2)$$

$$B_i = \frac{[(\gamma - 1)/2\gamma]\left(p_{e_i}/p_c\right)^{-1/\gamma}}{\left[1 - \left(p_{e_i}/p_c\right)^{(\gamma-1)/\gamma}\right]^{\frac{1}{2}}}$$

$$\frac{p_{e_i}}{p_c} = \frac{p_{S.L.}}{p_c} e^{-h_{d_i}/H_{scl}}$$

For the given conditions it follows that $h_{sep} = 6733$ ft, $C_{Fsep} = 1.478$, $h_{trans} = 45,825$ ft, and $C_{Ftrans} = 1.6807$. The entire altitude performance for $0 \le h \le 100,000$ ft is shown in Fig. 3.5. The envelope of an ideally expanded nozzle has been included for comparison. Note the substantial improvement in performance evident for $h > h_{trans}$ because of the two-design altitude capability.

3.5 Elementary Chemistry

In the preceding sections, methods were presented that allowed estimation of rocket nozzle performance in terms of prescribed combustion chamber conditions. In fact, the designer has at his disposal the ability to prescribe the combustion chamber pressure (by matching pump capacity and nozzle throat size) and the fuel-to-oxidizer of the "reactants." The properties of the products can be estimated by using the methods of equilibrium chemistry.

Consider the flow of fuel and oxidizer (the reactants) into a duct wherein combustion occurs and products are formed. The process is approximated as adiabatic, it is assumed (for simplicity) that the flow Mach numbers are very low, and no work interaction occurs within the chamber. Under these circumstances the enthalpy of the product's h_p will equal the enthalpy of the reactants h_{RT_3}, so that

$$h_{RT_3} = h_{pT_4} \qquad (3.34)$$

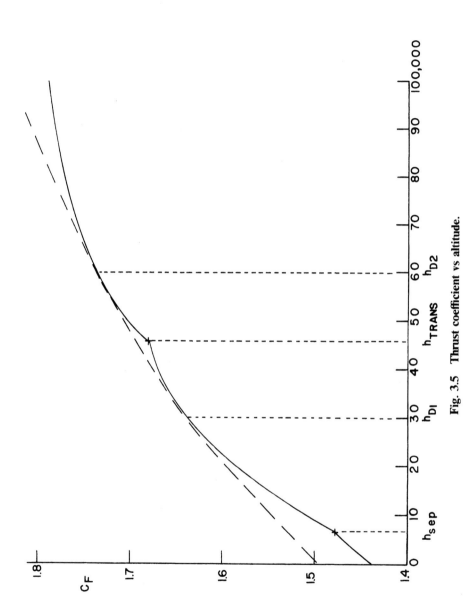

Fig. 3.5 Thrust coefficient vs altitude.

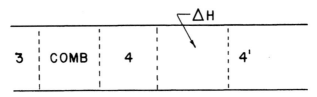

Fig. 3.6 Heat of reaction.

The reaction will bring about a change in temperature, and the "heat of reaction" ΔH is defined as the amount of (positive) heat interaction required to bring the products back to the same temperature as the reactants (at the same pressure). The actual process of combustion and the imagined additional process of heat interaction can be represented as illustrated in Fig. 3.6.

For the imagined process from point 4 to point 4', we have

$$h_{pT_{4'}} - h_{pT_4} = \Delta H_{T_{4'}} \tag{3.35}$$

By definition $T_{4'} = T_3$, so with Eq. (3.34),

$$h_{pT_3} - h_{RT_3} = \Delta H_{T_3} \tag{3.36}$$

and

$$h_{pT_4} = h_{pT_3} - \Delta H_{T_3} \tag{3.37}$$

Often, ΔH is available only at a reference value, say T_d. Thus,

$$\Delta H_{T_d} = h_{pT_d} - h_{RT_d}$$
$$= \left(h_{pT_3} - h_{RT_3} \right) - \left(h_{pT_3} - h_{pT_d} \right) + \left(h_{RT_3} - h_{RT_d} \right) \tag{3.38}$$

Combining Eqs. (3.37) and (3.38) and writing the general form T in place of T_3, it follows that

$$\Delta H_T = \Delta H_{T_d} + \left(h_{pT} - h_{pT_d} \right) - \left(h_{RT} - h_{RT_d} \right) \tag{3.39}$$

Note that combination of Eqs. (3.37) and (3.39) gives

$$h_{pT_4} = h_{pT_d} - \Delta H_{T_d} + \left(h_{RT_3} - h_{RT_d} \right) \tag{3.40}$$

The Heat of Formation, ΔH_f

The heat of formation is defined as the (positive) heat interaction required to form a compound from its elements at constant pressure and prescribed temperature. For perfect gases, the heat of solution is zero and, as a result, the heat of reaction may be obtained in terms of the heats of formation of the reactants and products. Thus,

$$\Delta H = \sum_p n_p \Delta H^\circ_{f_p} - \sum_R n_R \Delta H^\circ_{f_R} \tag{3.41}$$

where n_p and n_R are the number of moles of the product and reactant, respectively, and ΔH°_f is the heat of formation per mole.

The Law of Mass Action

It is usual to write the reaction equation for an example reaction in the form

$$\nu_1 A_1 + \nu_2 A_2 + \cdots \qquad \nu_\alpha A_\alpha = 0 \tag{3.42}$$

In this expression the ν_i are numbers (the stoichiometric coefficients) and the A_i units (usually moles) of the reactants and products. It is customary to write the products with positive stoichiometric coefficients and the reactants with negative coefficients. As an example consider the reaction

$$2H_2 + O_2 \rightarrow 2H_2O$$

With the suggested convention this equation would be written

$$H_2O - H_2 - \tfrac{1}{2}O_2 = 0$$

giving

$$\nu_1 = 1, \quad \nu_2 = -1, \quad \nu_3 = -\tfrac{1}{2}$$

The second law of thermodynamics states that the entropy of an isolated system cannot decrease. An extension to this result can be made in the statement, "the entropy of isolated systems tends to increase." Thus, the entropy of an isolated system will continue to increase until no further changes are possible. When the system reaches the state where no further increases in entropy are possible, it would hence have reached a state of equilibrium.

By applying this reasoning to a chemical mixture (making allowances for the possibility of heat and work interactions of the system with its surroundings), the condition for chemical equilibrium of a mixture can be deduced. The result is of a particularly simple form for perfect gases and is termed the law of mass action. Thus

$$p_1^{\nu_1} p_2^{\nu_2} p_3^{\nu_3} \cdots = K_p \tag{3.43}$$

Here K_p is the equilibrium constant. The detailed derivation of the expression reveals that K_p is a function of temperature only. Note that K_p is not dimensionless and that its value will depend on the units in which the partial pressures are expressed. An alternate form for the law of mass action is obtained in terms of the molal fractions of the species x_j. Thus, recalling that the molal fraction is equal to the ratio of the partial pressure to the total pressure, it follows that

$$p_j/p = n_i/\Sigma n_j = x_j \qquad (3.44)$$

and hence

$$x_1^{\nu_1} x_2^{\nu_2} \cdots x_j^{\nu_j} \cdots = p^{-(\nu_1 + \nu_2 + \cdots)} K_p \qquad (3.45)$$

Tables of values for the equilibrium constant are available in several references. Tables 3.1–3.3 have been taken from Ref. 2.

Symbols and Terminology for JANNAF Tables

The standard state is taken as the state at 1 atm pressure at the temperature under consideration for the solid, liquid, and ideal gas states. Only homogeneous substances are considered here.

The reference state applies to elements in their stable standard state. Consequently, the reference state tables presented here are either single-phase or polyphase tables; all other tables are single-phase.

A circular superscript ° indicates the thermodynamic standard state. The numerical subscript, as 298.15, denotes temperature in degrees Kelvin.

C_p° denotes the specific heat at the constant pressure of the substance in the thermodynamic standard state. S° represents the absolute entropy of the thermodynamic standard state at the absolute temperature T. $-(F^{\circ} - H_{298.15}^{\circ})/T$ denotes the free energy function in the standard state at temperature T and is defined as $S^{\circ} - (H^{\circ} - H_{298.15}^{\circ})/T$. $(H^{\circ} - H_{298.15}^{\circ})$ indicates the enthalpy (or heat content) in the standard state at the temperature T less the enthalpy in the standard state at 298.15 K. ΔH_f° represents the standard heat of formation, which is the increment in enthalpy associated with the reaction of forming the given compound from its elements, with each substance in its thermodynamic standard state at the given temperature.

When the reaction or process *evolves* heat, the sign of the heat term is arbitrarily taken to be negative. Conversely, when the reaction or process *absorbs* heat, the sign of the heat term is positive.

ΔF_f° denotes the standard free energy of formation, which is the increment in free energy associated with the reaction of forming the given compound from its elements, with each substance in its thermodynamic standard state at the given temperature.

Log K_p stands for the logarithm (to the base 10) of the equilibrium constant for the reaction forming the given compound from its elements, with each substance in its thermodynamic standard state at the given temperature.

Table 3.1 Equilibrium Constants for Hydrogen (H_2) (ideal gas, reference state, molecular weight 2.016, $H_2 \rightarrow H_2$, $K_{p_{H_2}} = p_{H_2}/p_{H_2} = 1$)

$T(K)$	C_p°	S°	$-\left(\dfrac{F^\circ - H_{298}^\circ}{T}\right)$	$H^\circ - H_{298}^\circ$	ΔH_f°	ΔF_f°	Log K_r
	(cal · mole^{-1} · deg^{-1})			(kcal · mole^{-1})			
0	0.000	0.000	Infinite	−2.024	0.000	0.000	0.000
100	5.393	24.387	37.035	−1.265	0.000	0.000	0.000
200	6.518	28.520	31.831	−0.662	0.000	0.000	0.000
298	6.892	31.208	31.208	0.000	0.000	0.000	0.000
300	6.894	31.251	31.208	0.013	0.000	0.000	0.000
400	6.975	33.247	31.480	0.707	0.000	0.000	0.000
500	6.993	34.806	31.995	1.406	0.000	0.000	0.000
600	7.009	36.082	32.573	2.106	0.000	0.000	0.000
700	7.036	37.165	33.153	2.808	0.000	0.000	0.000
800	7.087	38.107	33.715	3.514	0.000	0.000	0.000
900	7.148	38.946	34.250	4.226	0.000	0.000	0.000
1000	7.219	39.702	34.758	4.944	0.000	0.000	0.000
1100	7.300	40.394	35.240	5.670	0.000	0.000	0.000
1200	7.390	41.033	35.696	6.404	0.000	0.000	0.000
1300	7.490	41.628	36.130	7.148	0.000	0.000	0.000
1400	7.600	42.187	36.543	7.902	0.000	0.000	0.000
1500	7.720	42.716	36.937	8.668	0.000	0.000	0.000
1600	7.823	43.217	37.314	9.446	0.000	0.000	0.000
1700	7.921	43.695	37.675	10.233	0.000	0.000	0.000
1800	8.016	44.150	38.022	11.030	0.000	0.000	0.000
1900	8.108	44.586	38.356	11.836	0.000	0.000	0.000
2000	8.195	45.004	38.678	12.651	0.000	0.000	0.000
2100	8.279	45.406	38.989	13.475	0.000	0.000	0.000
2200	8.358	45.793	39.290	14.307	0.000	0.000	0.000
2300	8.434	46.166	39.581	15.146	0.000	0.000	0.000
2400	8.506	46.527	39.863	15.993	0.000	0.000	0.000
2500	8.575	46.875	40.136	16.848	0.000	0.000	0.000
2600	8.639	47.213	40.402	17.708	0.000	0.000	0.000
2700	8.700	47.540	40.660	18.575	0.000	0.000	0.000
2800	8.757	47.857	40.912	19.448	0.000	0.000	0.000
2900	8.810	48.166	41.157	20.326	0.000	0.000	0.000
3000	8.859	48.465	41.395	21.210	0.000	0.000	0.000
3100	8.911	48.756	41.628	22.098	0.000	0.000	0.000
3200	8.962	49.040	41.855	22.992	0.000	0.000	0.000
3300	9.012	49.317	42.077	23.891	0.000	0.000	0.000
3400	9.061	49.586	42.294	24.794	0.000	0.000	0.000
3500	9.110	49.850	42.506	25.703	0.000	0.000	0.000
3600	9.158	50.107	42.714	26.616	0.000	0.000	0.000
3700	9.205	50.359	42.917	27.535	0.000	0.000	0.000
3800	9.252	50.605	43.116	28.457	0.000	0.000	0.000
3900	9.297	50.846	43.311	29.385	0.000	0.000	0.000
4000	9.342	51.082	43.502	30.317	0.000	0.000	0.000
4100	9.386	51.313	43.690	31.253	0.000	0.000	0.000
4200	9.429	51.540	43.874	32.194	0.000	0.000	0.000
4300	9.472	51.762	44.055	33.139	0.000	0.000	0.000
4400	9.514	51.980	44.233	34.088	0.000	0.000	0.000
4500	9.555	52.194	44.407	35.042	0.000	0.000	0.000

Table 3.2   Equilibrium Constants for Diatomic Oxygen (O_2) (ideal gas, reference state, molecular weight 32.00, $O_2 \rightarrow O_2$, $K_{p_{O_2}} = p_{O_2}/p_{O_2} = 1$)

$T(K)$	C_p°	S°	$-\left(\dfrac{F^\circ - H^\circ_{298}}{T}\right)$	$H^\circ - H^\circ_{298}$	ΔH_f°	ΔF_f°	Log K_p
		(cal · mole^{-1} · deg^{-1})		(kcal · mole^{-1})			
0	0.000	0.000	Infinite	−2.075	0.000	0.000	0.000
100	6.958	41.395	55.205	−1.381	0.000	0.000	0.000
200	6.961	46.218	49.643	−0.685	0.000	0.000	0.000
298	7.020	49.004	49.004	0.000	0.000	0.000	0.000
300	7.023	49.047	49.004	0.013	0.000	0.000	0.000
400	7.196	51.091	49.282	0.724	0.000	0.000	0.000
500	7.431	52.722	49.812	1.455	0.000	0.000	0.000
600	7.670	54.098	50.414	2.210	0.000	0.000	0.000
700	7.883	55.297	51.028	2.988	0.000	0.000	0.000
800	8.063	56.361	51.629	3.786	0.000	0.000	0.000
900	8.212	57.320	52.209	4.600	0.000	0.000	0.000
1000	8.336	58.192	52.765	5.427	0.000	0.000	0.000
1100	8.439	58.991	53.295	6.266	0.000	0.000	0.000
1200	8.527	59.729	53.801	7.114	0.000	0.000	0.000
1300	8.604	60.415	54.283	7.971	0.000	0.000	0.000
1400	8.674	61.055	54.744	8.835	0.000	0.000	0.000
1500	8.738	61.656	55.185	9.706	0.000	0.000	0.000
1600	8.800	62.222	55.608	10.583	0.000	0.000	0.000
1700	8.858	62.757	56.013	11.465	0.000	0.000	0.000
1800	8.916	63.265	56.401	12.354	0.000	0.000	0.000
1900	8.973	63.749	56.776	13.249	0.000	0.000	0.000
2000	9.029	64.210	57.136	14.149	0.000	0.000	0.000
2100	9.084	64.652	57.483	15.054	0.000	0.000	0.000
2200	9.139	65.076	57.819	15.966	0.000	0.000	0.000
2300	9.194	65.483	58.143	16.882	0.000	0.000	0.000
2400	9.248	65.876	58.457	17.804	0.000	0.000	0.000
2500	9.301	66.254	58.762	18.732	0.000	0.000	0.000
2600	9.354	66.620	59.057	19.664	0.000	0.000	0.000
2700	9.405	66.974	59.344	20.602	0.000	0.000	0.000
2800	9.455	67.317	59.622	21.545	0.000	0.000	0.000
2900	9.503	67.650	59.893	22.493	0.000	0.000	0.000
3000	9.551	67.973	60.157	23.446	0.000	0.000	0.000
3100	9.596	68.287	60.415	24.403	0.000	0.000	0.000
3200	9.640	68.592	60.665	25.365	0.000	0.000	0.000
3300	9.682	68.889	60.910	26.331	0.000	0.000	0.000
3400	9.723	69.179	61.149	27.302	0.000	0.000	0.000
3500	9.762	69.461	61.383	28.276	0.000	0.000	0.000
3600	9.799	69.737	61.611	29.254	0.000	0.000	0.000
3700	9.835	70.006	61.834	30.236	0.000	0.000	0.000
3800	9.869	70.269	62.053	31.221	0.000	0.000	0.000
3900	9.901	70.525	62.267	32.209	0.000	0.000	0.000
4000	9.932	70.776	62.476	33.201	0.000	0.000	0.000
4100	9.961	71.022	62.682	34.196	0.000	0.000	0.000
4200	9.988	71.262	62.883	35.193	0.000	0.000	0.000
4300	10.015	71.498	63.081	36.193	0.000	0.000	0.000
4400	10.039	71.728	63.275	37.196	0.000	0.000	0.000
4500	10.062	71.954	63.465	38.201	0.000	0.000	0.000

Table 3.3 Equilibrium Constants for Water (H_2O) [ideal gas, molecular weight 18.016, $H_2 + \frac{1}{2}O_2 \rightarrow H_2O$, $K_{p\,H_2O} = p_{H_2O}/(p_{H_2} \cdot p_{O_2}^{\frac{1}{2}})$]

T(K)	C_p°	S°	$-\left(\dfrac{F^\circ - H_{298}}{T}\right)$	$H^\circ - H_{298}$	ΔH_f°	ΔF_f°	Log K_p
	(cal · mole^{-1} · deg^{-1})			(kcal · mole^{-1})			
0	0.000	0.000	Infinite	− 2.367	− 57.103	− 57.103	Infinite
100	7.961	36.396	52.202	− 1.581	− 57.433	− 56.557	123.600
200	7.969	41.916	45.837	− 0.784	− 57.579	− 55.635	60.792
298	8.025	45.106	45.106	0.000	− 57.798	− 54.636	40.048
300	8.027	45.155	45.106	0.015	− 57.803	− 54.617	39.786
400	8.186	47.484	45.422	0.825	− 58.042	− 53.519	29.240
500	8.415	49.334	46.026	1.654	− 58.277	− 52.361	22.886
600	8.676	50.891	46.710	2.509	− 58.500	− 51.156	18.633
700	8.954	52.249	47.406	3.390	− 58.710	− 49.915	15.583
800	9.246	53.464	48.089	4.300	− 58.905	− 48.646	13.289
900	9.547	54.570	48.749	5.240	− 59.084	− 47.352	11.498
1000	9.851	55.592	49.382	6.209	− 59.246	− 46.040	10.062
1100	10.152	56.545	49.991	7.210	− 59.391	− 44.712	8.883
1200	10.444	57.441	50.575	8.240	− 59.519	− 43.371	7.899
1300	10.723	58.288	51.136	9.298	− 59.634	− 42.022	7.064
1400	10.987	59.092	51.675	10.384	− 59.734	− 40.663	6.347
1500	11.233	59.859	52.196	11.495	− 59.824	− 39.297	5.725
1600	11.462	60.591	52.698	12.630	− 59.906	− 37.927	5.180
1700	11.674	61.293	53.183	13.787	− 59.977	− 36.549	4.699
1800	11.869	61.965	53.652	14.964	− 60.041	− 35.170	4.270
1900	12.048	62.612	54.107	16.160	− 60.099	− 33.786	3.886
2000	12.214	63.234	54.548	17.373	− 60.150	− 32.401	3.540
2100	12.366	63.834	54.976	18.602	− 60.198	− 31.012	3.227
2200	12.505	64.412	55.392	19.846	− 60.242	− 29.621	2.942
2300	12.634	64.971	55.796	21.103	− 60.282	− 28.229	2.682
2400	12.753	65.511	56.190	22.372	− 60.321	− 26.832	2.443
2500	12.863	66.034	56.573	23.653	− 60.359	− 25.439	2.224
2600	12.965	66.541	56.947	24.945	− 60.393	− 24.040	2.021
2700	13.059	67.032	57.311	26.246	− 60.428	− 22.641	1.833
2800	13.146	67.508	57.667	27.556	− 60.462	− 21.242	1.658
2900	13.228	67.971	58.014	28.875	− 60.496	− 19.838	1.495
3000	13.304	68.421	58.354	30.201	− 60.530	− 18.438	1.343
3100	13.374	68.858	58.685	31.535	− 60.562	− 17.034	1.201
3200	13.441	69.284	59.010	32.876	− 60.596	− 15.630	1.067
3300	13.503	69.698	59.328	34.223	− 60.631	− 14.223	0.942
3400	13.562	70.102	59.639	35.577	− 60.666	− 12.818	0.824
3500	13.617	70.496	59.943	36.936	− 60.703	− 11.409	0.712
3600	13.669	70.881	60.242	38.300	− 60.741	− 10.000	0.607
3700	13.718	71.256	60.534	39.669	− 60.782	− 8.589	0.507
3800	13.764	71.622	60.821	41.043	− 60.822	− 7.177	0.413
3900	13.808	71.980	61.103	42.422	− 60.865	− 5.766	0.323
4000	13.850	72.331	61.379	43.805	− 60.910	− 4.353	0.238
4100	13.890	72.673	61.651	45.192	− 60.957	− 2.938	0.157
4200	13.927	73.008	61.917	46.583	− 61.006	− 1.522	0.079
4300	13.963	73.336	62.179	47.977	− 61.056	− 0.105	0.005
4400	13.997	73.658	62.436	49.375	− 61.109	1.311	− 0.065
4500	14.030	73.973	62.689	50.777	− 61.164	2.729	− 0.133

Example Calculation—Hydrogen-Oxygen Reaction

As a relatively simple example of the calculational aspects of chemical equilibrium chemistry, the hydrogen-oxygen reaction is considered here. This reaction has special interest, in addition to its simplicity, in that it is one of the most energetic reactions available and hence is of particular use in propulsion. A very simple form of the reaction will be considered here in detail, but the procedure for a more exact (and complicated) form of the reaction will be outlined in the following.

Thus consider 1 mole of molecular hydrogen to react with ℓ moles of molecular oxygen as in the following equation:

$$H_2 + \ell O_2 \rightarrow m H_2O + n H_2 + q O_2 \qquad (3.46)$$

The mole balances give

$$H: m + n = 1$$

$$O: m + 2q = 2\ell$$

In anticipation of considering the fuel-rich case $\left(\ell < \frac{1}{2}\right)$ and hence expecting q to be small, solve for m and n in terms of the prescribed ℓ and q. Thus

$$m = 2(\ell - q), \qquad n = 1 - 2(\ell - q), \qquad n_T = m + n + q = 1 + q \quad (3.47)$$

The remaining equation for q is provided by the law of mass action [Eq. (3.45)], which gives $\left(\text{with } \nu_1 = 1, \nu_2 = -1, \nu_3 = -\frac{1}{2}\right)$

$$(n_T/q)^{\frac{1}{2}}(m/n) = p_c^{\frac{1}{2}} K_p \qquad (3.48)$$

Combination of Eqs. (3.47) and (3.48) then gives a cubic equation for q, which may be written in the form

$$F(q) = 0 = Aq^3 + Bq^2 + Cq + D \qquad (3.49)$$

where

$$A = 4\left[1 - \left(1/p_c K_p^2\right)\right]$$

$$B = 4(1 - 2\ell)\left[1 - \left(1/p_c K_p^2\right)\right]$$

$$C = \left[(1 - 2\ell)^2 + \left(4\ell/p_c K_p^2\right)(2 - \ell)\right]$$

$$D = -4\ell^2/p_c K_p^2$$

This equation can be solved numerically using the procedure of Newtonian iteration (see Sec. 7.2). This procedure gives an updated value for q, q_{j+i}, in terms of the previous value of q, q_j, by the formula

$$q_{j+1} = q_j - [F(q)/F'(q)]_j \qquad (3.50)$$

For the value of $F(q)$ given above,

$$F'_{(q)} = 3Aq^2 + 2Bq + C$$

and hence with Eq. (3.50)

$$q_{j+1} = \left(\frac{2Aq^3 + Bq^2 - D}{3Aq^2 + 2Bq + C} \right)_j \qquad (3.51)$$

An appropriate first guess for q is

$$q_0 = -D/C$$

As an example, assume $\ell = 0.4$, $p_c = 10$ atm, and $T_c = 4000$ K. K_p is obtained from Table 3.3 ($= 10^{0.238}$), and iteration of Eq. (3.51) gives (in four iterations, accurate to four significant figures) $q = 0.092770$. Then, $m = 2(\ell - q) = 0.6145$ and $n = 1 - m = 0.3855$.

More General Hydrogen-Oxygen Reaction

The reaction analyzed above is a very simple representation of the process actually experienced in a high-temperature reaction. A more general form can be written

$$H_2 + \ell O_2 = mH_2O + nH_2 + qO_2 + pH + rO + sOH$$

The equilibrium concentration of the products can be determined by introducing further equilibrium constants corresponding to the appearance of the given product. For example, equilibrium constants could be

$$K_{p_1} = \frac{p_O}{p_{O_2}^{\frac{1}{2}}}, \qquad K_{p_2} = \frac{p_H}{p_{H_2}^{\frac{1}{2}}}, \qquad K_{p_3} = \frac{p_{OH}}{p_{H_2}^{\frac{1}{2}} p_{O_2}^{\frac{1}{2}}}, \qquad K_{p_4} = \frac{p_{H_2O}}{p_{H_2} p_{O_2}^{\frac{1}{2}}}$$

These expressions give four additional equations to the two mole balance equations and hence provide six equations for the six unknowns $m \rightarrow s$. As can easily be imagined, the numerical complexity of these larger reaction equations can become severe. Reference 3 considers the numerical aspects of such complicated examples in detail.

3.6   Determination of Chamber Conditions

The concepts developed in the preceding section can now be applied to the determination of chamber conditions. The incoming temperature T_{in}, fuel-to-oxidizer ratio, and chamber pressure will be provided. It will be necessary to determine the outgoing chamber temperature T_c and other thermodynamic properties. The procedure is complicated somewhat by the fact that T_c is an unknown of the problem, and hence must be determined by iteration. Whether or not a given value for T_c is correct is determined by comparing the enthalpy required to raise the reactants from T_{in} to T_d and the products from T_d to T_c. This procedure is summarized in the following section.

Calculation Procedure—Summary

(1) Assume a value for T_c.

(2) Determine the equilibrium composition of the products (n_p) for the assumed value of T_c.

(3) Calculate the enthalpy released by the reaction $(-\Delta H)$ using the heats of formation at T_d:

$$-\Delta H = -\left(\sum_p n_p \Delta H^\circ_{f T_d} - \sum_R n_R \Delta H^\circ_{f T_d} \right)$$

(4) Calculate the enthalpy required to raise the reactants from T_{in} to T_d plus the enthalpy required to raise the products from T_d to T_c:

$$\Delta H_{req} = \sum_R n_R \left(H_{T_d} - H_{T_{in}} \right) + \sum_p n_p \left(H_{T_c} - H_{T_d} \right)$$

(5) Compare $-\Delta H$ to ΔH_{req}. If $-\Delta H$ is larger than ΔH_{req}, assume a larger value for T_c and repeat the process.

(6) Calculate the specific enthalpy, entropy, etc., from the known composition of the products and tabulated values of the molal quantities.

Example Calculation—Hydrogen-Oxygen Reaction

As an example calculation, consider the combustion of hydrogen and oxygen. Take $p_c = 15$ atm, $\ell = 0.35$, and assume for simplicity that the gases enter the chamber at $T_{in} = 298$ K and that the simplified reaction model [Eq. (3.46)] is appropriate. For these simplified conditions, the relationships of steps 3 and 4 above reduce to

$$-\Delta H = -m \, \Delta H^\circ_{f H_2O} = m(57.798)$$

$$\Delta H_{req} = m \left(H - H_{298} \right)_{H_2O, T_c} + n \left(H - H_{298} \right)_{H_2, T_c} + q \left(H - H_{298} \right)_{O_2, T_c}$$

As a first guess, assume $T_c = 3700$ K, giving $K_p = 10^{0.507}$, Eq. (3.49) gives

$$m = 0.6531, n = 0.3469, \text{ and } q = 0.02343$$

Thus

$$-\Delta H = 0.6531(57.798) = 37.751$$

$$\Delta H_{req} = m(39.669) + n(27.535) + q(30.236) = 36.169$$

The enthalpy required to raise the products to 3700 K is less than that provided by the reaction, so T_c must be higher than that assumed. Assuming $T_c = 3800$ K leads to

$$m = 0.6369, n = 0.3631, q = 0.3157, -\Delta H = 36.809, \text{ and } \Delta H_{req} = 37.458$$

It is apparent that T_c is between the values 3700 and 3800 K. Linear interpolation gives the estimate

$$T_c = 3700 + \frac{37.751 - 36.169}{(37.751 - 36.169) - (36.809 - 37.458)}(3800 - 3700)$$

or

$$T_c \approx 3770 \text{ K}$$

Interpolation of the other quantities gives

$$m = 0.6418, n = 0.3582, \text{ and } q = 0.02913$$

It is of interest, also, to determine the enthalpy per mass of the products h_m which may be written

$$h_m - h_{m298} = \frac{m\left(-\Delta H^\circ_{fH_2O}\right)_{298}}{2 + 32\ell} = 2.810 \text{ kcal/g}$$

Note that here the molecular weights of H_2 (2) and O_2 (32) are introduced.

3.7 Nozzle Flow of a Reacting Gas

As the gases flow through the nozzle, the pressure and temperature decrease. As a result the gases would tend to react further to equilibrium conditions appropriate for the local value of pressure and temperature. With conventional propellants, the effect of the decreasing temperature within the nozzle dominates the effect of the decreasing pressure and the reaction tends further to completion. Such further reaction is beneficial, because the continued reaction releases the chemical energy to the translational energy and hence leads to an increased exit velocity.

The extent to which reactions tend to go to completion within a nozzle is dependent on the relative times spent for a reaction process (that is, for the required collision processes to occur) and for the gas to traverse significant pressure and temperature changes within the nozzles. Thus, for example, a reaction that requires a three-body collision could occur quite rarely, and hence at long time intervals. In such a case the reaction could effectively "freeze" (discontinue) during the transit time of the gas in the nozzle. The detailed estimation of the extent of freezing requires kinetic theory, and will not be considered in this book. It is of interest, however, to investigate the two limits to flow within nozzles, "equilibrium" and "frozen" flows.

Equilibrium Flow

In that circumstance where all chemical reactions occur in times very short compared to the time of fluid passage through the nozzle, the fluid will be at all times (almost) in a state of chemical equilibrium. The reactions will occur continuously throughout the nozzle, leading to a continuous passage of energy from the chemical binding and excitation modes to the translational modes. Because the fluid is at all times in equilibrium, the equivalent temperatures of all such modes of energy storage are equal and as a result the total entropy of the fluid remains constant.

The total entropy may be obtained as the weighted sum of each constituent; thus

$$S = \sum_{j=1}^{n} n_j S_j \qquad (3.52)$$

From the Gibbs equation

$$S_j = \int_{T_0}^{T} C_{p_j} \frac{dT}{T} - R_u \ell n \frac{p_j}{p_0} + b_j \qquad (3.53)$$

where R_u is the universal gas constant, 0 refers to reference conditions ($p_0 = 1$ atm, $T_0 = 298$ K), and $b_j = S_j$ at p_0, T_0.

The symbol S° given in Tables 3.1–3.3 is related to these terms by

$$S_j^\circ = \int_{T_0}^{T} C_{p_j} \frac{dT}{T} + b_j \qquad (3.54)$$

So

$$S = \Sigma n_j S_j^\circ - R_u \Sigma n_j \ell n \frac{p_j}{p_0} \qquad (3.55)$$

Noting

$$\ell n \frac{p_j}{p_0} = \ell n \frac{p}{p_0} + \ell n \frac{p_j}{p} = \ell n \frac{p}{p_0} + \ell n \, x_j$$

Eq. (3.55) becomes

$$S = \Sigma n_j S_j^\circ - R_u n_T \ell n \frac{p}{p_0} - R_u \Sigma n_j \ell n \, x_j \qquad (3.56)$$

Note also that the further release of chemical energy will appear as an effective addition in enthalpy per mass. Thus, for the simplified H_2-O_2 reaction considered earlier,

$$\text{Additional enthalpy/mass} = \frac{\Delta(-\Delta H)}{\text{mass}} = \frac{m_e - m_c}{2 + 32\ell} \left(-\Delta H_{f \, H_2 O}^\circ \right)_{298}$$

$$(3.57)$$

The nozzle exit velocity may then be calculated using the following procedure. The chamber conditions would have been previously determined.

Calculation Procedure—Equilibrium Flow

(1) Assume a value for the nozzle exit temperature T_e.
(2) Calculate the composition for the resulting K_p and prescribed p_e. *Note:* In most cases of interest the exit temperature will be sufficiently low that the reaction may be assumed to have gone to completion.
(3) Calculate S_e and S_c using Eq. (3.56). Iterate with T_e until equality is obtained.
(4) Calculate h_e and $\Delta(-\Delta H)/\text{mass}$.
(5) Obtain $U_e = \{2[h_c + \Delta(-\Delta H)/\text{mass} - h_e]\}^{\frac{1}{2}}$

Frozen Flow

In the limiting circumstance where the chemical reaction rates are so slow that the fluid passes through the nozzle with no further reaction (following combustion), the flow is said to be frozen. In this case, the mole fractions of all the constituents remain those in the combustion chamber. Further, because no chemical reactions occur, the entropy remains constant for this case of frozen flow, also. Note, however, that when the fluid eventually does reach equilibrium (outside the rocket) entropy increases will occur.

It is to be noted that the condition for the equality of entropy $s_e = s_c$ leads to a somewhat simpler form in this case of frozen flow. Thus, with Eq. (3.56),

$$\Sigma x_j S_{je}^\circ = \Sigma x_j S_{jc}^\circ + R_u \ell n \frac{p_e}{p_c} \qquad (3.58)$$

Calculation Procedure—Frozen Flow

(1) Assume a value for the nozzle exit temperature T_e.

(2) Evaluate $\Sigma x_j S_{je}^\circ$ and compare to the known value of $\Sigma x_j S_{jc}^\circ + R_u \ell n \, p_e/p_c$. Continue until equality is obtained.

(3) Calculate h_e.

(4) Obtain $U_e = [2(h_c - h_e)]^{\frac{1}{2}}$.

Example calculations (Problems 3.14–3.16) show that specific impulse differences of several percent can occur between some examples of frozen and equilibrium flow; so when particularly accurate estimates are required, it can become necessary to include estimates of the extent to which a given flow freezes. Note that the concept of "sudden freezing" is sometimes employed, wherein the flow is considered to be in equilibrium to an intermediate location (say the throat), at which point it "suddenly freezes" and retains the same composition from that point on.

3.8 Solid-Propellant Rockets

Solid-propellant rockets can be broadly classified as one of two types, end burning or erosive burning. In the end-burning type (Fig. 3.7), the propellant burns only at the end, the sidewall propellant being inhibited to prevent the flame front from traveling into the propellant along the sidewall.

In the erosive-burning type (Fig. 3.8), the grain is inhibited on the ends and the propellant burns in a direction perpendicular to the gas flow. It is apparent that the erosive-burning type of rocket will usually be a higher thrust, shorter duration rocket because the large burning area leads to large mass flow rates.

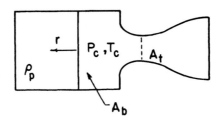

Fig. 3.7   End-burning solid-propellant rocket.

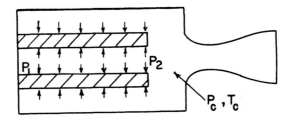

Fig. 3.8   Erosively burning solid-propellant rocket.

The combustion processes in the vicinity of the solid surface are extraordinarily complex, but it is fortunate that some purely empirical forms can be used to relate the local surface burning rate to the local fluid properties. Two often used forms are

$$r = ap^n \text{ (end-burning rocket)} \tag{3.59}$$

$$r = ap^n + k(\rho u) \tag{3.60}$$

where r is the surface burning rate, p, ρ, and u the fluid pressure, density, and velocity, respectively, and a, n, and k are empirically determined constants.

Calculation of the Chamber Pressure—End-Burning Grain

It is apparent that the chamber pressure will be determined by the requirement that the mass of gas produced by the surface must be sufficiently compressed to pass through the nozzle throat. Assume that the gas is calorically perfect, that the chamber Mach number is very low, and that the propellant density is very large compared to the gas density. Then, equating the rate of mass produced at the surface to that passing through the throat and employing Eqs. (2.104) and (3.59),

$$rA_b\rho_p = ap_c^n A_b\rho_p = p_c A_t/C^*$$

hence

$$p_c = \left[aC^*\rho_p(A_b/A_t)\right]^{1/(1-n)} \tag{3.61}$$

It is to be noted that the chamber pressure is a sensitive function of the exponent n. It is also clear that values of n in excess of (or equal to) unity will lead to unstable behavior. Thus, consider a rocket burning at "design" p_c, when a small increase in p_c occurs momentarily. The result would be that the surface burning rate would increase more than the flow rate through the throat. As a result, the propellant would accumulate in the chamber, leading to a further pressure rise with further increase in burning rate, etc. Practical propellants have values of n in the neighborhood of 0.75.

Calculation of the Chamber Pressure—Erosive-Burning Grain

It is evident that the determination of the chamber pressure will be substantially more complicated for an erosive-burning grain than for an end-burning grain, because determination of the mass flow rate from the surface will involve an integral over the entire surface in terms of the local fluid properties. In practice, the conditions within the rocket (at a given time) are determined by assuming a value for p_1 (see Fig. 3.8 for terminology) and then integrating the appropriate equations of motion along the grain to determine p_2. p_2 is then related to p_c (either by the assumption of isentropic flow or by use of a loss coefficient) and compared to the required

p_c to pass the flow through the throat. This process is repeated until convergence. It is then necessary to assume a small time step, with the burning rate, etc., as just calculated, to determine the amount of propellant consumed. The local conditions will then have changed because of the change in cross-sectional area, so the whole process is repeated for a new time step.

This procedure in fact requires relatively little time for calculation on a high-speed computer. However, an even simpler approximate form can be obtained if it is assumed that the ratio of throat area to flow cross-sectional area A_t/A is much less than 1.

Expression for the Downstream Pressure p_2

The mass flow at station 2 is equal to that through the throat, so

$$u_2 \rho_2 A_2 = \frac{p_c A_t}{C^*} = \frac{p_c A_t}{\sqrt{RT_c}} \Gamma \qquad (3.62)$$

then

$$u_2^2 = \left(\frac{p_c}{\rho_2}\right)^2 RT_c \left(\frac{A_t}{A_2}\right)^2 \Gamma^2$$

For simplicity the process from station 2 to station c is approximated as isentropic, so that

$$u_2^2 = \left(\frac{p_c}{p_2}\right)^{2/\gamma} RT_c \left(\frac{A_t}{A_2}\right)^2 \Gamma^2 = \frac{2\gamma}{\gamma-1} RT_c \left[1 - \left(\frac{p_2}{p_c}\right)^{(\gamma-1)/\gamma}\right] \qquad (3.63)$$

Now invoking the assumption that $A_t/A \ll 1$, we may assume that $p_2/p_c = 1 - \varepsilon$ where $\varepsilon \ll 1$, which after use of the binomial expansion in Eq. (3.63) gives

$$p_2/p_c = 1 - \tfrac{1}{2}[(A_t/A_2)\Gamma]^2 \quad (+ \text{ higher order terms}) \qquad (3.64)$$

Expression for the Local Pressure p

An approximation for the mass flow rate at location χ may be written

$$\rho u A \approx \bar{r}\rho_p A_b(\chi/L) \qquad (3.65)$$

where the average burning rate $\bar{r}$ is given by

$$\bar{r} = \frac{1}{L}\int_0^L r\,d\chi \qquad (3.66)$$

The momentum equation may be written

$$\rho u^2 = p_1 - p \qquad (3.67)$$

from which, with Eq. (3.65)

$$u = \frac{p_1 - p}{\bar{r}\rho_p A_b} \frac{L}{\chi} A \qquad (3.68)$$

The enthalpy equation coupled with Eq. (3.67) gives

$$C_p T_c = \frac{\gamma}{\gamma - 1} \frac{p}{\rho} + \frac{u^2}{2} = \left(\frac{\gamma}{\gamma - 1} \frac{p}{p_1 - p} + \frac{1}{2} \right) u^2 \qquad (3.69)$$

so that with Eqs. (3.68) and (3.69)

$$\left(\frac{\gamma}{\gamma - 1} \frac{p}{p_1 - p} + \frac{1}{2} \right) \left(\frac{p_1 - p}{\bar{r}\rho_p A_b} \frac{L}{\chi} A \right)^2 = C_p T_c \qquad (3.70)$$

The assumption $A_t/A \ll 1$ again allows an approximation $p/p_1 = 1 - \delta$, $\delta \ll 1$. Introducing this form into Eq. (3.70), noting that $p_c/p_1 = 1 +$ higher order terms, and noting

$$\dot{m} = \bar{r}\rho_p A_b = \frac{p_c A_t}{\sqrt{RT_c}} \Gamma \qquad (3.71)$$

it follows that

$$\frac{p}{p_1} = 1 - \left(\Gamma \frac{A_t}{A} \frac{\chi}{L} \right)^2 \quad (+ \text{ higher order terms}) \qquad (3.72)$$

It is consistent to assume $A_2/A \approx 1$, so that

$$\frac{p}{p_c} = \frac{p}{p_1} \frac{p_1}{p_2} \frac{p_2}{p_c} = \frac{1 - [\Gamma(A_t/A)(\chi/L)]^2}{1 - [\Gamma(A_t/A)]^2} \left[1 - \frac{1}{2} \left(\Gamma \frac{A_t}{A} \right)^2 \right] \qquad (3.73)$$

Continuing to expand the groups and retaining only terms to order $(A_t/A)^2$, there is finally obtained

$$\left(\frac{p}{p_c} \right)^n \approx 1 + n \left(\frac{A_t}{A} \Gamma \right)^2 \left[\frac{1}{2} - \left(\frac{\chi}{L} \right)^2 \right] \qquad (3.74)$$

The Burning Rate

Equations (3.60), (3.65), and (3.74) give

$$r = a p_c^n \left\{ 1 + n \left(\frac{A_t}{A} \Gamma \right)^2 \left[\frac{1}{2} - \left(\frac{\chi}{L} \right)^2 \right] \right\} + k\bar{r}\rho_p \frac{A_b}{A} \frac{\chi}{L} \qquad (3.75)$$

This expression can now be integrated as in Eq. (3.66) to give an expression for $\bar{r}$,

$$\bar{r} = ap_c^n\left[1 + \frac{n}{6}\left(\Gamma\frac{A_t}{A}\right)^2\right]\bigg/\left[1 - \frac{k}{2}\rho_p\frac{A_b}{A}\right] \tag{3.76}$$

Then with the continuity equation

$$\rho_p\bar{r}A_b = p_cA_t/C^*$$

it follows that

$$p_c = \left[aC^*\rho_p\frac{A_b}{A_t}\frac{1 + \frac{n}{6}\left(\Gamma\frac{A_t}{A}\right)^2}{1 - \frac{k}{2}\rho_p\frac{A_b}{A}}\right]^{1/(1-n)} \tag{3.77}$$

The thrust behavior of the rocket can now be calculated because, with p_c, C^*, and the exit pressure all known, the mass flow rate and exit velocity can be determined. In order to obtain the thrust behavior with time, it is necessary to calculate the variation of A_b and A. This requires numerical integration.

A simple and quite instructive example (Problem 3.17) is that of a grain so shaped that the burning area A_b remains constant in time. Such behavior can be approximated by employing a star grain (Fig. 3.9). In such a case the burning area would be equal to the cylindrical chamber area existing just at burnout, i.e.,

$$A_b = \pi DL \tag{3.78}$$

Also, the change in cross-sectional area in a small time interval δ_t would be given by

$$\delta A = A_{j+1} - A_j = \pi D\bar{r}\,\delta t \tag{3.79}$$

Numerical integration for this simple case is very straightforward. It is to be noted that, for this case of constant A_b, the thrust history will be "regressive," that is, it will decrease in time. This is apparent from Eqs.

Fig. 3.9   Star grain.

(3.76) and (3.77), which indicate that both $\bar{r}$ and p_c decrease as A increases. This result is simply a manifestation of the fact that the erosive contribution to the burning decreases as the cross-sectional area increases.

References

[1]Summerfield, M., Foster, C. R., and Swan, W. C., "Flow Separation in Over-Expanded Supersonic Exhaust Nozzles," *Jet Propulsion*, Vol. 24, Sept.–Oct. 1954, pp. 319–321.

[2]*JANNAF Thermochemical Tables*, 2nd ed., Office of Standard Reference Data, National Bureau of Standards, Washington, D.C., 1970.

[3]Penner, S. S., *Chemistry Problems in Jet Propulsion*, Pergamon Press, London, 1957.

Problems

3.1   An "optimal" two-stage rocket is to operate in a drag- and gravity-free environment. It is to provide a velocity increment Δv of 5000 ms^{-1} to a payload of 1000 kg. We are given (these figures are appropriate for a kerosene-oxygen first stage and hydrogen-oxygen second stage):

$$C_1 = 3000 \text{ ms}^{-1}, \qquad C_2 = 4000 \text{ ms}^{-1}, \qquad \delta_1 = 0.1, \qquad \delta_2 = 0.15$$

(a)   Find m_0, MR_1, MR_2, λ_1, and λ_2.
(b)   Find the same parameters if $C_1 = C_2 = 3000 \text{ ms}^{-1}$, and $\delta_1 = \delta_2 = 0.1$.

3.2   A rocket has identical stages in the sense that $C_i = C$, $\delta_i = \delta$, and $\lambda_i = \lambda = \lambda_0^{1/N}$. Gravitational and drag losses can be neglected. If $\lambda_0 = 0.05$ and $\delta_i = 0.1$, what is the optimum number of stages for the rocket?

3.3   Consider a rocket employing "continuous staging" in the sense that it discards all the dead weight (structural and engine weight) continuously at zero velocity relative to the rocket until only the payload is traveling at the final velocity.
(a)   Write the equation of motion for the rocket neglecting the drag and gravitational losses.
(b)   Integrate this equation to find Δv, assuming the rate of dead weight rejection $\dot{m}_d$, the propellant rejection rate m_p, and the exhaust velocity C are constant in terms of $\delta = m_d/m_0$ and $MR = 1/\lambda_0$.
(c)   What is the penalty paid in terms of percentage of ideal Δv (Δv achieved if no dead weight was present), if $\delta = 0.1$ and $\lambda_0 = 0.05$?
(d)   How much better (in terms of Δv achieved) is this continuously staged rocket than a three-stage rocket (optimized) with $\delta = 0.1$ and $\lambda_0 = 0.05$?

3.4 Consider a four-stage rocket for which the first two stages and the last two stages are similar. That is,

$$C_1 = C_2 = C, \qquad C_3 = C_4 = \beta C \quad \text{and} \quad \delta_1 = \delta_2 = \delta, \qquad \delta_3 = \delta_4 = \mu\delta$$

There are no drag or gravitational losses.

(a) Show that if the rocket is designed for a maximum Δv for a given λ_0:

$$\lambda_1 = \lambda_2 = \frac{2\,\delta\beta}{\left[(\beta-1)^2 + 4\mu\beta\,\delta^2/\sqrt{\lambda_0}\right]^{\frac{1}{2}} - (\beta-1)}$$

$$\lambda_3 = \lambda_4 = \sqrt{\lambda_0}/\lambda_1$$

(b) Calculate Δv_{tot} given that $C = 3000$ ms^{-1}, $\lambda_0 = 0.0016$, $\delta = 0.1$, $\mu = 1.3$, and $\beta = 1.2$.

3.5 Show that the derivative of the thrust coefficient is

$$\frac{\partial C_F}{\partial p_e/p_c} = -\sqrt{\frac{\gamma-1}{2\gamma}}\,\frac{\Gamma}{\gamma}\left[1 - \left(\frac{p_e}{p_c}\right)^{(\gamma-1)/\gamma}\right]^{-\frac{3}{2}}\left(\frac{p_e}{p_c}\right)^{-(\gamma+1)/\gamma}\left(\frac{p_e}{p_c} - \frac{p_a}{p_c}\right)$$

3.6 Derive Eq. (3.32).

3.7 Derive Eq. (3.33).

3.8 Consider a rocket with a translatable skirt, such that it has two design altitudes, 10,000 and 20,000 m. The ambient pressure vs altitude may be approximated by the formula

$$p_a = p_{S.L.}e^{-h/7000}$$

For this rocket the Summerfield criterion is

$$p_s = p_a/2.718$$

The ratio of specific heats is $\gamma = 1.22$ and the chamber pressure p_c is 50 atm. The skirt is translated at just the altitude where C_F for the skirt in the 10,000 m design condition equals C_F for the skirt in the 20,000 m design condition.

Calculate and carefully plot C_F over the range $0 \le h \le 40,000$ m. Carefully locate the altitudes h_{sep}, h_{trans}, h_{D1}, and h_{D2}. Include on the graph the envelope of a perfectly expanded rocket.

3.9 Molecular hydrogen (H_2) passes through heat-transfer passages and emerges at 2500 K. At this temperature significant dissociation can occur. What mass fraction of the hydrogen exists as atomic hydrogen (H) at each of the chamber pressures $p_c = 100$, 10^{-1}, and 10^{-4} atm?

At 2500 K the equilibrium constant K_p is approximately

$$K_p = p_H / (p_{H_2})^{\frac{1}{2}} = 10^{-2}$$

where p is in atmospheres.

3.10 Verify by direct calculation that Eq. (3.39) and the values tabulated in Tables 3.1–3.3 are in agreement for the case $T = 4000$ K.

3.11 Consider the reaction

$$H_2 + \ell O_2 \rightarrow m H_2 O + n H_2 + q O_2$$

(a) Obtain and plot n and q as a function of ℓ for the case where $T = 3500$ K. Plot in the range $0.2 \le \ell \le 0.5$ for the three pressures $p_c = 1, 10$, and 100 atm.

(b) Repeat part (a) for the case $T = 4000$ K.

3.12 ℓ moles of O_2 are mixed with 1 mole of H_2. The entering temperature of the O_2 is 100 K and of the H_2 is 200 K. Obtain and plot the enthalpy per mass of the products for the case where $p_c = 10$ atm. Calculate for the values $\ell = 0.2, 0.3, 0.4$, and 0.5.

3.13 Repeat Problem 3.12 for the case $p_c = 100$ atm.

3.14 The curves of enthalpy per mass vs ℓ obtained in Problems 3.12 and 3.13 should each contain a maximum. Select the nearest value of ℓ to the maximum that you calculated and for that case calculate the velocity at the exit of the rocket assuming frozen, isentropic flow to $p_e = 0.1$ atm for both chamber pressures.

3.15 For the conditions of Problem 3.14, calculate the exit velocity assuming isentropic equilibrium flow.

3.16 (a) For the situation where a nozzle has a large pressure ratio, what value of ℓ will lead to the maximum exit velocity if equilibrium flow occurs in the nozzle?

(b) Calculate the related exit velocities for the cases $p_c = 10$ and 100 atm with $p_e = 0.1$ atm.

3.17 Consider a solid-propellant rocket designed to produce a velocity increment Δv in a gravity- and drag-free environment. The empty mass of the rocket is to be m_f. The rocket may be considered to have a "star" grain, designed so that the burning area A_b remains constant with time and hence equal to the area of the cylinder of fuel existing just at burnout. The grain has a constant-area A gas passage and the length-to-diameter ratio L/D is prescribed, as are the initial value of $A/A_t = A_i/A_t$, and $p_c = p_{c_i}$. Other parameters to be provided are C_F (assumed constant), C^*, γ, ρ_p, n, k, and a.

Obtain analytical expressions for the following quantities in terms of the above prescribed parameters, or possibly in terms of parameters obtained in the following, preceding the parameter in question.

(a) The fuel volume V_f.
(b) The ratio of burning area to throat area A_b/A_t.
(c) The outside diameter D of the fuel charge.
(d) The burning area A_b.
(e) The throat area A_t and throat diameter D_t.
(f) Show that expressions for the chamber pressure and average burning rate may be obtained in the form

$$p_c = \left[\beta_1 \frac{1 + (\beta_2/A^2)}{1 - (\beta_3/A)} \right]^{1/(1-n)}$$

$$\bar{r} = a p_c^n \left[\frac{1 + (\beta_2/A^2)}{1 - (\beta_3/A)} \right]$$

and obtain β_1, β_2, and β_3 analytically.

(g) Summarize all of the above equations in a form that can be easily programmed for calculation, then noting the approximate form

$$A = A_i + \pi D \bar{r} \, \delta t$$

and the rough estimate for overall firing time τ given by

$$\tau \approx V_f / \bar{r}_i A_b$$

Obtain and plot a time history of A, p_c, $\bar{r}$, and thrust F from $t = 0$ to burnout, for the following input parameters:

$$\Delta v = 4500 \text{ ms}^{-1} \qquad p_{ci} = 10^7 \text{ N} \cdot \text{m}^{-2} \qquad \rho_p = 1750 \text{ kg} \cdot \text{m}^{-3}$$
$$m_f = 9000 \text{ kg} \qquad C_F = 1.90 \qquad a = 0.22 \times 10^{-6}$$
$$A_i/A_t = 2 \qquad C^* = 1700 \text{ ms}^{-1} \qquad n = 0.73$$
$$L/D = 1 \qquad \gamma = 1.25 \qquad k = 0.5 \times 10^{-5}$$

(h) Find the rocket acceleration at beginning and end of firing.

3.18   Consider an end-burning solid-propellant rocket with the same specifications as those listed in Problem 3.17. Find

(a) The throat diameter.
(b) The time of burning.
(c) The (constant) thrust.
(d) The initial and final accelerations.

4. NONCHEMICAL ROCKETS

4.1 Introduction

When considering space missions with very large velocity changes, it becomes apparent that huge mass ratios are required. Chemical rockets, which have the energy source coincident with the propellant, are fundamentally limited in their achievable specific impulse by the strength of the chemical bonds of the propellants (to about $I_{sp} = 450$ s). If higher specific impulses are to be obtained, an energy source other than, or in addition to, the propellant itself must be utilized.

Several methods of external energy addition suggest themselves. Examples are thermal addition of energy with the thermal energy provided by a nuclear reactor or electrical energy input to the propellant with the electrical energy provided by solar collectors, a nuclear-electric generator, a solar heater-electric generator combination, or any of several other competing concepts.

Before rational determination of the most promising concepts can be made, it is necessary to estimate the performance of the suggested systems. Simple performance models and example performance estimates are given in the following sections.

4.2 The Nuclear-Heated Rocket

A schematic diagram of a nuclear rocket is shown in Fig. 4.1. As illustrated, such a rocket operates by having the propellant pass through heat-transfer passages within the high-temperature core of the rocket where its enthalpy is raised by heat interaction with the walls of the passages. The energy necessary to maintain the core temperature is supplied by the nuclear reactions within the core material.

It can be noted here that an important engineering limitation is present, in that the temperatures within the (solid) core must be restricted to values that do not cause structural weakening of the rocket. This is in contrast to a conventional rocket where the highest temperatures within the system occur within the gas. The advantage of the nuclear rocket is not, then, that high temperatures are available, but rather that the choice of propellant is limited only by the requirement of chemical compatibility with the core surfaces.

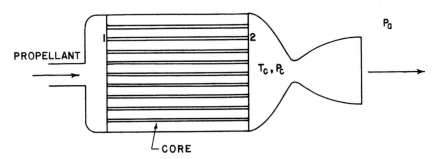

Fig. 4.1 Nuclear-heated rocket.

Choice of Propellant

The first priority in the selection of a propellant would obviously be identified with the propellant giving the highest exhaust velocity. Assuming for simplicity that the nozzle exit pressure is extremely low, Eq. (3.24) gives for the effective exhaust velocity

$$C \approx \sqrt{2C_p T_c} = \left(\frac{2\gamma}{\gamma - 1} \frac{R_u}{\mathcal{M}} T_c \right)^{\frac{1}{2}} \tag{4.1}$$

where T_c is the chamber stagnation temperature, R_u the universal gas constant, and $\mathcal{M}$ the molecular weight.

It is apparent that the largest exhaust velocity, for a given limited chamber stagnation temperature, will be found for hydrogen as propellant. Note that allowable temperatures are restricted to a range where virtually all of the hydrogen will be in molecular form. In this case, with $\gamma = 1.4$, $\mathcal{M} = 2$, and $R_u = 8320$ J (kg · mole)$^{-1}$ · K^{-1}, the maximum specific impulse is found to be

$$I_{sp} = \frac{C}{a_0} = \frac{1}{9.8} \left[\frac{2(1.4)}{0.4} \frac{8320}{2} T_c \right]^{\frac{1}{2}}$$

so $I_{sp} = 17.4 \sqrt{T_c}$.

The upper limit for the core temperature is approximately 2500 K, which gives for the related specific impulse

$$\left(I_{sp} \right)_{max} \approx 870 \text{ s}$$

This represents a substantial increase over the maximum specific impulse found in chemical rockets and justifies the considerable research and development directed toward the nuclear-heated rocket. It is clear that the advantages of a nuclear-heated rocket, as compared to a chemical rocket, will become more pronounced as more energetically demanding missions are contemplated. This is because as the mission demand increases, and hence propellant mass increases, the savings provided by the large specific impulse increase. In the case of low-energy missions, the propellant savings do not overcome the large mass penalty incurred by the reactor and all its related equipment.

With the success of the chemical rocket for manned missions to the moon, it appeared that the major hope for the nuclear rocket would be identified with manned planetary missions.

It happens, however, that for manned planetary missions, the energy requirements are so enormous that even the nuclear-heated rocket has too low a specific impulse, and the class of electrical rockets emerges as the most viable candidates for such missions.

With regard to the future, it is possible that nuclear rockets could be used on "space tugboats" between Earth orbit and the lunar surface or between vehicles in Earth orbit and geosynchronous orbit.

Approximate Performance Analysis

A relatively simple method for analyzing the performance of a nuclear-heated rocket was suggested in Ref. 1. An optimal design will include an appropriate choice of the length-to-diameter ratio of the heat-transfer tubes within the reactor core. It is evident that if tubes with very large length-to-diameter ratios are employed, then the propellant temperature will closely approach the allowable surface limit temperature T_s. The high propellant temperature would favor high specific impulse; but, if carried to extreme, the related stagnation pressure drop found in the long slender tubes would become excessive and lead to a reduced specific impulse.

In the following the temperature rise as a function of length-to-diameter ratio of the heat-transfer tubes is estimated. An approximation to the pressure loss is then obtained and the combined effect of temperature and pressure on the specific impulse estimated.

Heat Transfer and Power Balance

Consider a tube of length L and diameter D, as shown in Fig. 4.2. The thermal energy transferred into the flow in the elemental length element dx leads to a differential increase in stagnation temperature given by Eq. (2.84). This may be written in the form

$$dT_t = 4N_{st}(T_w - T_t)(dx/D) \qquad (4.2)$$

It will be convenient when searching for the optimal tube length to diameter to relate the stagnation temperature increase to the stagnation pressure loss. Hence, Reynolds' analogy [Eq. (2.86)] is assumed to be valid. Then, denoting the skin-friction coefficient by f, the expression for the increment in stagnation temperature becomes

$$dT_t = 2f(T_w - T_t)(dx/D) \qquad (4.3)$$

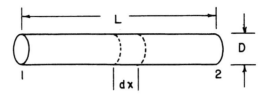

Fig. 4.2   Heat-transfer tube.

In order to obtain T_t as a function of x, it is necessary to know the variation of $f(T_w - T_t)$ with x. This variation is determined by the distribution of the power density within the reactor, because no matter what the local power density, the wall temperature (and hence T_t) must "float" to the value that equates the thermal transfer rate into the flow to the local generation of thermal energy by the nuclear reactions.

The power density distribution within a reactor is determined by the distribution of nuclear fuel as well as by the amount and location of shielding material. The power density distribution is somewhat amenable to design choice, so two example distributions are considered here. In each case, for simplicity, it is assumed that the skin-friction coefficient is constant.

Constant-power density.   A constant-power density insures that the rate of increase of stagnation temperature and the temperature increment $(T_w - T_t)$ are constant. The limiting wall temperature T_s will occur at the end of the tube, where

$$(T_w - T_t) = (T_s - T_{t_2})$$

Equation (4.3) can be immediately integrated to give

$$\frac{T_{t_2}}{T_s} = \frac{T_{t_1}/T_s + 2fL/D}{1 + 2fL/D} \tag{4.4}$$

Sine-power density.   This power distribution would (approximately) exist if the ends of the core were unshielded and the nuclear fuel distribution uniform. The power density is assumed to be proportional to $\sin \pi x/L$, so that the temperature increment is given by

$$\Delta T = \Delta T_m \sin(\pi x/L) \tag{4.5}$$

where $\Delta T \equiv T_w - T_t$ and $\Delta T_m =$ maximum increment (located at $x = L/2$).
With Eq. (4.5), Eq. (4.3) is easily integrated to give

$$T_t = T_{t_1} + (2fL/\pi D)\,\Delta T_m[1 - \cos(\pi x/L)] \tag{4.6}$$

thus

$$\Delta T_m = (\pi D/4fL)(T_{t_2} - T_{t_1}) \tag{4.7}$$

The limiting temperature T_s is located where the wall temperature T_w has a maximum. Thus, writing

$$T_w = T_t + \Delta T = T_{t_1} + \frac{2fL}{\pi D} \Delta T_m \left(1 - \cos \frac{\pi x}{L}\right) + \Delta T_m \sin \frac{\pi x}{L} \quad (4.8)$$

taking the derivative of this expression, and equating the result to zero, it follows that the location where $T_w = T_s$ is given by

$$\left(\frac{x}{L}\right)_{T_s} = \frac{1}{2} + \frac{1}{\pi} \arctan\left(\frac{2fL}{\pi D}\right) \quad (4.9)$$

At that location

$$\cos \frac{\pi x}{L} = \frac{-1}{\sqrt{1 + (\pi D/2fL)^2}} \quad \text{and} \quad \sin \frac{\pi x}{L} = \frac{\pi D/2fL}{\sqrt{1 + (\pi D/2fL)^2}}$$

$$(4.10)$$

Combination of Eqs. (4.8) and (4.10) then gives

$$T_s = T_{t_1} + \frac{2fL}{\pi D} \Delta T_m \left[1 + \sqrt{1 + \left(\frac{\pi D}{2fL}\right)^2}\right] \quad (4.11)$$

And finally, Eqs. (4.6–4.8) and (4.11) give

$$\frac{T_t}{T_s} = \frac{1}{1 + \sqrt{1 + (\pi D/2fL)^2}}$$

$$\times \left[1 + \frac{T_{t_1}}{T_s} \sqrt{1 + \left(\frac{\pi D}{2fL}\right)^2} - \left(1 - \frac{T_{t_1}}{T_s}\right) \cos \frac{\pi x}{L}\right] \quad (4.12)$$

$$\frac{T_{t_2}}{T_s} = \frac{1}{1 + \sqrt{1 + (\pi D/2fL)^2}} \left[2 + \frac{T_{t_1}}{T_s}\left(\sqrt{1 + \left(\frac{\pi D}{2fL}\right)^2} - 1\right)\right] \quad (4.13)$$

$$\frac{T_w}{T_s} = \frac{1}{1 + \sqrt{1 + (\pi D/2fL)^2}}$$

$$\times \left[1 + \frac{T_{t_1}}{T_s} \sqrt{1 + \left(\frac{\pi D}{2fL}\right)^2} + \left(1 - \frac{T_{t_1}}{T_s}\right)\left(\frac{\pi D}{2fL} \sin \frac{\pi x}{L} - \cos \frac{\pi x}{L}\right)\right]$$

$$(4.14)$$

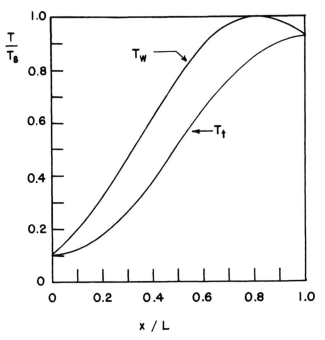

Fig. 4.3   Wall temperature and gas temperature vs length.

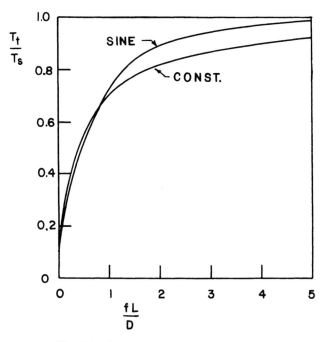

Fig. 4.4   Outlet gas temperature vs fL/D.

Figure 4.3 illustrates the temperature variation with axial position for the case $fL/D = 2.5$ and $T_{t_1}/T_s = 0.1$. The behavior of outlet stagnation temperature with fL/D for the constant-power density case and for the sine-power density case is shown in Fig. 4.4.

Core Pressure Drop

The preceding analysis led to simple expressions for the outlet stagnation temperature T_{t_2} in terms of T_s, T_{t_1}, and fL/D. As is obvious, if fL/D is increased, T_{t_2} will more nearly approach the maximum wall temperature T_s. However, an increase in fL/D causes some problems itself, in that the stagnation pressure drop will increase, the thrust coefficient will decrease, and the pressure drop across the core will increase, leading to possible structural problems.

To estimate the related performance penalties, first consider a very simplified form of the momentum equation. Thus, the shear force is assumed to be given in terms of the average of the upstream and downstream shear stresses,

$$\text{Shear force} = \tfrac{1}{2}\left(\tau_1 + \tau_2\right)\pi DL$$

$$= \left(f/4\right)\left(\rho_1 u_1^2 + \rho_2 u_2^2\right)\pi DL$$

The momentum equation thus becomes

$$\left(p_1 - p_2\right)\frac{\pi}{4}D^2 - \frac{f}{4}\left(\rho_1 u_1^2 + \rho_2 u_2^2\right)\pi DL = \left(\rho_2 u_2^2 - \rho_1 u_1^2\right)\frac{\pi}{4}D^2$$

hence

$$p_1\left[1 + \gamma M_1^2\left(1 - \frac{fL}{D}\right)\right] = p_2\left[1 + \gamma M_2^2\left(1 + \frac{fL}{D}\right)\right] \tag{4.15}$$

so

$$\frac{p_{t_2}}{p_{t_1}} = \frac{\left[1 + \gamma M_1^2\left(1 - \dfrac{fL}{D}\right)\right]}{\left[1 + \gamma M_2^2\left(1 + \dfrac{fL}{D}\right)\right]}\left[\frac{1 + \dfrac{\gamma - 1}{2}M_2^2}{1 + \dfrac{\gamma - 1}{2}M_1^2}\right]^{\gamma/(\gamma-1)} \tag{4.16}$$

An expression for the stagnation temperature ratio is obtained by first noting

$$p = R\rho T = R\frac{\dot{m}}{uA}T = \left(\sqrt{\frac{R}{\gamma}}\frac{\dot{m}}{A}\right)\frac{\sqrt{T_t}}{M}\frac{1}{\left(1 + \dfrac{\gamma - 1}{2}M^2\right)^{\frac{1}{2}}}$$

which, when substituted into Eq. (4.15) yields

$$\frac{T_{t_2}}{T_{t_1}} = \frac{M_2^2}{M_1^2} \frac{1 + \dfrac{\gamma - 1}{2} M_2^2}{1 + \dfrac{\gamma - 1}{2} M_1^2} \left[\frac{1 + \gamma M_1^2 \left(1 - \dfrac{fL}{D}\right)}{1 + \gamma M_2^2 \left(1 + \dfrac{fL}{D}\right)} \right]^2 \tag{4.17}$$

Note that if T_{t_1}, T_s, and fL/D were prescribed, Eqs. (4.4) or (4.13) would yield T_{t_2} and Eq. (4.17) would then relate M_2 and M_1. In fact, M_2 can be prescribed implicitly by the choice of the ratio of the nozzle throat area to core tube area, so M_1 could then be obtained from Eq. (4.17), which is a quadratic equation for M_1^2. The stagnation pressure ratio then follows from Eq. (4.16), and hence the performance variables can be estimated.

Performance Variables

It is assumed that no further decrease in stagnation pressure occurs after station 2, so the exit velocity may be written

$$u_e = \left\{ \frac{2\gamma}{\gamma - 1} R T_{t_2} \left[1 - \left(\frac{p_a}{p_{t_2}} \right)^{(\gamma-1)/\gamma} \right] \right\}^{\frac{1}{2}} \tag{4.18}$$

The maximum imaginable specific impulse I_m would be that for which the propellant reaches the maximum allowable wall temperature T_s and no pressure drop occurs ($p_{t_2} = p_{t_1}$). The ratio of actual I_{sp} to this maximum I_{sp} is hence

$$\frac{I}{I_m} = \left\{ \frac{T_{t_2} \left[1 - (p_a/p_{t_2})^{(\gamma-1)/\gamma} \right]}{T_s \left[1 - (p_a/p_{t_1})^{(\gamma-1)/\gamma} \right]} \right\}^{\frac{1}{2}} \tag{4.19}$$

The thrust coefficient based on flow cross-sectional area and conditions at station 1 may be written

$$F/Ap_{t_1} = \dot{m} u_e / Ap_{t_1} \tag{4.20}$$

Combining Eqs. (2.102), (4.18), and (4.20) gives

$$\frac{F}{Ap_{t_1}} = \gamma \sqrt{\frac{2}{\gamma - 1}} \left[1 - \left(\frac{p_a}{p_{t_2}} \right)^{(\gamma-1)/\gamma} \right]^{\frac{1}{2}} \frac{1 + \gamma M_1^2 \left(1 - \dfrac{fL}{D}\right)}{\left(1 + \dfrac{\gamma - 1}{2} M_1^2\right)^{\gamma/(\gamma-1)}} J_{(M_2)}$$

$$\tag{4.21}$$

where

$$J_{(M_2)} \equiv M_2 \frac{\left[1 + \frac{\gamma - 1}{2} M_2^2\right]^{\frac{1}{2}}}{1 + \gamma M_2^2 \left(1 + \frac{fL}{D}\right)} \qquad (4.22)$$

With T_{t_2} and p_{t_2} determined as suggested in the previous section, Eqs. (4.19) and (4.21) give the performance variables. An even simpler approximation was suggested by the author of Ref. 1, however, who noted that in the usual case where $T_{t_1}/T_s \ll 1$, the inlet Mach number would also be very small indeed. In such cases, approximate forms of Eqs. (4.16) and (4.21) may be used,

$$\frac{p_{t_2}}{p_{t_1}} \approx \frac{\left(1 + \frac{\gamma - 1}{2} M_2^2\right)^{\gamma/(\gamma - 1)}}{\left[1 + \gamma M_2^2 \left(1 + \frac{fL}{D}\right)\right]} \qquad (4.23)$$

$$\frac{F}{Ap_{t_1}} \approx \gamma \sqrt{\frac{2}{\gamma - 1}} \left[1 - \left(\frac{p_a}{p_{t_2}}\right)^{(\gamma - 1)/\gamma}\right]^{\frac{1}{2}} J_{(M_2)} \qquad (4.24)$$

Figure 4.5 shows the behavior of I/I_m and F/Ap_{t_1} vs M_2 for the case $T_{t_1}/T_s = 1/40$, $p_a/p_{t_1} = 1/80$, $fL/D = 1.5$, $\gamma = 1.4$, and power density constant. An erroneous drop in the thrust coefficient is predicted, which is introduced by the approximations leading to Eqs. (4.23) and (4.24). In any case, full calculation reveals that the thrust coefficient changes very slowly above the point that the false maximum is predicted to exist, whereas the specific impulse does continue to decrease.

A further simplification is then suggested, in that the Mach number at the exit M_2^* will be taken to be that at the maximum of the function $J_{(M_2)}$. This maximum (Problem 4.1) occurs at

$$M_2^* = \frac{1}{\sqrt{1 + \gamma fL/D}} \qquad (4.25)$$

The resulting equation set for this "optimum" choice is summarized as follows:

Inputs: $\gamma, fL/D, p_a/p_{t_1}, T_{t_1}/T_s$

Outputs: $M_2^*, I/I_m, F/Ap_{t_1}$

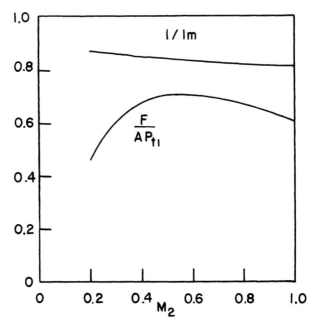

Fig. 4.5   Specific impulse and thrust coefficient vs M_2.

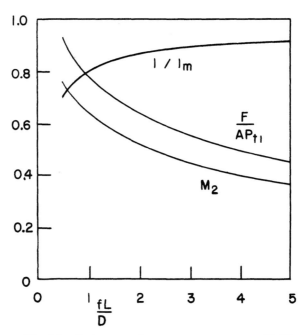

Fig. 4.6   Optimum performance parameters vs fL/D.

Equations:

$$M_2^* = \frac{1}{\sqrt{1 + \gamma f L / D}}$$

$$\frac{p_{t_2}}{p_{t_1}} = \frac{\left[1 + \frac{\gamma - 1}{2} M_2^2\right]^{\gamma/(\gamma-1)}}{1 + \gamma M_2^2 \left(1 + \frac{fL}{D}\right)}$$

$$\frac{T_{t_2}}{T_s} = f\left(\frac{fL}{D}, \frac{T_{t_1}}{T_s}\right) \qquad \begin{array}{l}[\text{Eq. (4.4), (4.13), or as appropriate} \\ \text{for chosen fuel loading}]\end{array}$$

$$\frac{p_a}{p_{t_2}} = \frac{p_a/p_{t_1}}{p_{t_2}/p_{t_1}}$$

$$\frac{F}{A p_{t_1}} = \gamma \sqrt{\frac{2}{\gamma - 1} \left[1 - \left(\frac{p_a}{p_{t_2}}\right)^{(\gamma-1)/\gamma}\right]^{\frac{1}{2}}} \frac{M_2 \left(1 + \frac{\gamma - 1}{2} M_2^2\right)^{\frac{1}{2}}}{\left[1 + \gamma M_2^2 \left(1 + \frac{fL}{D}\right)\right]}$$

$$\frac{I}{I_m} = \sqrt{\frac{T_{t_2}}{T_s} \left[\frac{1 - \left(p_a/p_{t_2}\right)^{(\gamma-1)/\gamma}}{1 - \left(p_a/p_{t_1}\right)^{(\gamma-1)/\gamma}}\right]^{\frac{1}{2}}}$$

An example calculation for the case $p_a/p_{t_1} = 1/80$, $T_{t_1}/T_s = 1/40$, $\gamma = 1.4$, and power density constant is shown in Fig. 4.6. It is to be noted that with typical values of the skin-friction coefficient of $f \approx 0.005$, the length-to-diameter ratios corresponding to the right portion of the graph can be very large. Note that $L/D = 600$ implies a tube diameter of 3.3 mm for a core length of 2 m. It is probable that fabrication limitations and core pressure drop problems will cause the selection to be more in the neighborhood of $fL/D = 1.5$ or lower.

4.3 Electrically Powered Rockets

When very-high-energy missions are to be considered, the specific impulse must be extremely high if the overall mass ratio of the rocket is not to become extreme. There are several concepts for providing electrical energy directly to the propellant, but an example configuration will be considered

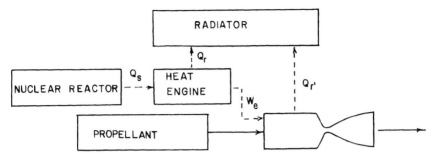

Fig. 4.7   Electric rocket schematic diagram.

first without regard to the details of the actual energy addition process. The major components of such a system are indicated schematically in Fig. 4.7.

The diagram of Fig. 4.7 indicates the primary work and heat interactions of a typical system. No heat exchanger is indicated between the propellant and heat engine, simply because in cases of very high specific impulse, the stagnation enthalpy of the propellant leaving the accelerator is so enormous that the savings of energy through the use of a heat exchanger is probably not worth the complexity of the required additional equipment.

It should be apparent that there will be some optimal choice of specific impulse for a given mission, because the required mass of the electrical supply equipment will increase with the increase in specific impulse, whereas the required mass of propellant decreases with the increase in specific impulse. In the following, a simple model is provided for estimating the optimum choice of specific impulse as a function of mission requirements and system parameters.

Selection of the Optimum Specific Impulse

A simple definition of the "optimum specific impulse" is that which leads to the minimum overall system mass. As a convenient approximation, it is assumed that the entire mass of the engine, radiator, heat exchanger, and accelerator (m_e) is proportional to the power delivered to the accelerator. Thus, the specific power α' (watt/kilogram) is defined by

$$\alpha' = W_e/m_e \tag{4.26}$$

The kinetic energy of the exhaust will be so enormous that to a good approximation the thermal energy remaining in the propellant may be ignored, so that

$$\eta_a W_e = \dot{m}\left(C^2/2\right) \tag{4.27}$$

where η_a is the accelerator efficiency. Hence,

$$m_e = \frac{\dot{m}}{\alpha'\eta_a}\frac{C^2}{2} \tag{4.28}$$

The remaining mass of the vehicle may be considered to consist of the propellant mass m_p and the payload mass m_L (which includes the structural mass, etc.). Thus,

$$m_0 = m_e + m_L + m_p \tag{4.29}$$

For a constant propellant flow rate, the firing time of the vehicle τ will be given by $\tau = m_p/\dot{m}$, so there is obtained

$$\frac{m_L}{m_0} = 1 - \frac{m_p}{m_0}\left[1 + \frac{C^2}{2\alpha'\eta_a\tau}\right] \tag{4.30}$$

Equation (3.14) gives

$$m_p/m_0 = 1 - e^{-\Delta v/c} \tag{4.31}$$

and combination of Eqs. (4.30) and (4.31) gives

$$\frac{m_L}{m_0} = 1 - (1 - e^{-\alpha})\left[1 + \left(\frac{\beta}{\alpha}\right)^2\right] \tag{4.32}$$

where $\alpha = \Delta v/c$ and $\beta = \Delta v/\sqrt{2\alpha'\eta_a\tau}$.

The optimum specific impulse (or C) occurs when the derivative of Eq. (4.32) with respect to α is zero. Hence, at the optimum condition

$$\left(\frac{\beta}{\alpha}\right)^2 = \frac{\alpha}{2(e^\alpha - 1) - \alpha} \tag{4.33}$$

and the equation for the optimal value of α for a prescribed payload ratio may thus be written

$$F(\alpha) = 1 - \frac{\alpha}{2} - e^{-\alpha} - \frac{m_L}{m_0}\left(e^\alpha - 1 - \frac{\alpha}{2}\right) = 0 \tag{4.34}$$

This equation is easily solved numerically using Newtonian iteration (see Sec. 7.2). Thus, with

$$F' = -\tfrac{1}{2} + e^{-\alpha} - (m_L/m_0)(e^\alpha - \tfrac{1}{2}) \tag{4.35}$$

it follows that

$$\alpha_{j+1} = (\alpha - F/F')_j \tag{4.36}$$

A suitable first guess for α, α_0, is

$$\alpha_0 = 1 - m_L/m_0 \tag{4.37}$$

Once α has been obtained, the related optimum value of β follows from Eq. (4.33), thence C and τ from the definitions of Eq. (4.32), and, finally, the initial acceleration a_i from

$$a_i = \frac{\dot{m}C}{m_0} = \frac{C}{\tau}\frac{m_p}{m_0} = \frac{C}{\tau}(1 - e^{-\alpha}) \tag{4.38}$$

These equations for electric rocket optimum specific impulse may be summarized as follows:

Input: m_L/m_0, Δv ms^{-1}, $\eta_a\alpha'$ W $\cdot$ kg^{-1}

Output: α, β, I_{opt} s, τ s, a_i ms^{-2}

Equations:

$$\alpha_0 = 1 - m_L/m_0$$

$$F = 1 - \frac{\alpha}{2} - e^{-\alpha} - \frac{m_L}{m_0}\left(e^\alpha - 1 - \frac{\alpha}{2}\right)$$

$$F' = -\frac{1}{2} + e^{-\alpha} - \frac{m_L}{m_0}\left(e^\alpha - \frac{1}{2}\right)$$

$$\alpha_{j+1} = (\alpha - F/F')_j$$

$$\beta = \frac{\alpha^{\frac{3}{2}}}{\sqrt{2(e^\alpha - 1) - \alpha}}$$

$$C = \Delta v/\alpha$$

$$I_{\text{opt}} = C/9.807$$

$$\tau = (1/2\alpha'\eta_a)(\Delta v/\beta)^2$$

$$a_i = (C/\tau)(1 - e^{-\alpha})$$

Example results for the case of $\Delta v = 20{,}000$ ms^{-1} and $\alpha'\eta_a = 100$ W $\cdot$ kg^{-1} are shown in Fig. 4.8 where the variation of I_{opt}, a_i, τ, and m_p/m_0 is shown plotted vs m_L/m_0. As is evident, the initial accelerations and hence firing times get very small and very large, respectively, for this very-high-energy mission when large mass ratios are desired. The related specific impulses also become large, clearly indicating why it is that electrical rockets must be employed.

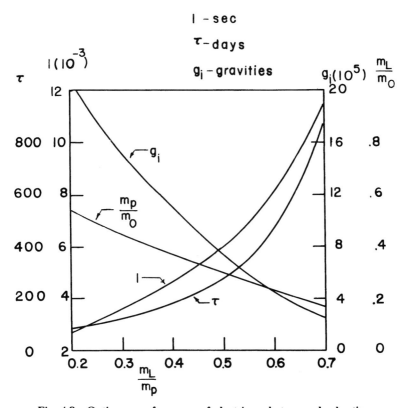

I - sec

τ - days

g_i - gravities

Fig. 4.8   Optimum performance of electric rocket vs payload ratio.

When more exacting mission analysis estimates are made, it is found that the optimum range of specific impulse for planetary missions is approximately $3000 < I < 5000$. It is a most unfortunate result that it is just this range of specific impulse that is the most difficult to achieve!

Classification of Electric Thrustors

Electric thrustors can be broadly classified in one of three categories, electrothermal, electromagnetic, or electrostatic. Each class has limitations that confine its use to a particular specific impulse range. In the following, the method of operation and operational limitations of each class will be briefly outlined.

Electrothermal thrustors.   In these devices the electrical energy provided to the thrustor is first converted to thermal energy of the propellant and hence to kinetic energy in the exhaust by expansion of the propellant through a conventional nozzle. A simple form of the electrothermal thrustor is the resisto jet, which operates by using the electrical energy to resistively heat filaments that in turn heat the propellant. Resisto jets are quite limited in attainable specific impulse ($I_{sp} \leq 800$ s) because of the thermal limitations of the filaments.

The most common form of the electrothermal thrustor is the arc jet, which operates by passing an electric current directly through the propellant itself, thereby depositing the electrical energy directly within the propellant. This "ohmic heating" appears as random (thermal) energy in the propellant. The arc jet, by depositing energy directly in the propellant, avoids the solid-surface temperature limitation of the resisto jet and nuclear-heated rocket. Engineering limitations do arise, however, in that dissociation (H_2) or ionization (H_2 or H_e) losses become substantial at high temperatures. Ionization losses are particularly aggravating because the electrical current tends to concentrate in filaments (much as do lightening bolts) where the ionization level becomes extremely high, far exceeding the level identified with the equilibrium temperature of the gas.

Successful designs have employed settling chambers to allow the electrons to recombine with the ions. The use of a settling chamber introduces problems of its own, in that substantial heat-transfer losses occur in the chamber. Other techniques to enhance arc jet performance include swirling the flows so as to increase the length of the current path from the anode to the cathode. Magnetic fields have also been applied to spin the electrical filaments, thereby not only increasing the filament length, but also reducing the probability of the arc "spotting" and damaging the electrode surfaces.

It appears, on the balance of performance to date, that arc jets show promise for specific impulses up to about 1500 s.

Electromagnetic thrustors.   Efforts to circumvent the ionization limit of the arc jet led to investigation of the electromagnetic thrustor. In this concept, it was hoped to utilize the Lorentz force (or $\mathbf{j} \times \mathbf{B}$ force) resulting from interaction of an electrical current with a magnetic field. By so doing, it was hoped to add a substantial portion of the electrical energy directly in the form of directed kinetic energy of the propellant. In bypassing the intermediate condition of very high static temperature of the propellant, the ionization losses could be substantially reduced.

Substantial investigations were conducted in the electromagnetic thrustor field, but were only moderately successful, primarily because two further engineering limitations on this class of device appeared. It was found that when the stagnation enthalpy of the propellant was increased (over that found in arc jets), the thermal transfer to the containing walls became so great that structural integrity could not be maintained.

Attempts to reduce the wall thermal transport by reducing the operating pressure of the thrustor introduced yet another engineering limitation. When the thrustors were run at the very low pressures necessary to prevent wall structural failure, the fluid density became so low that the ions (which are acted upon directly by the Lorentz force) tended to slip through the neutral particles. When the resulting "ion slip" becomes extreme, the jet exhaust tends to consist of rapidly moving electrons and ions and relatively slowly moving neutral particles. As a result, most of the electrical energy provided goes into accelerating the ions, but the neutral gas is accelerated much less, with a resulting inefficient and low specific impulse exhaust jet.

It appears at the present time that the electromagnetic thrustors do not hold the promise originally hoped.

Electrostatic thrustors.   The performance limitations brought about by the requirements of viscous containment led to the investigation of methods of propellant stream acceleration that could use purely electrical (and possibly magnetic) methods of containing the propellant. It is clear that any such method must utilize a virtually completely ionized stream, because any neutral particles would be unaffected by the imposed electrical fields. Once it has been decided to utilize a fully ionized stream, there is no benefit to be found in using low-molecular-weight propellants. Rather, as will be evident in the following analysis, propellants with very high molecular weights are found to be most suitable. It follows then that the energy levels of the exhausting propellants are extremely high (in the thousands of electron volts), so that the energy lost to the ionization process, or to the remaining thermal energy, is virtually negligible.

Figure 4.9 shows a schematic diagram of an electrostatic propulsion concept. As indicated, a source of ions is provided that is attracted to the highly negatively charged cathode. By properly shaping the anode and cathode, the beam can be held very nearly parallel. After passage through the cathode opening, the beam is immediately neutralized by electrons supplied by electron emitters. It is to be noted that the actual energy

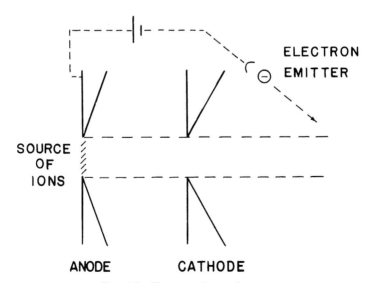

Fig. 4.9   Electrostatic accelerator.

provided goes into raising the electrons from the anode to cathode potential. Note also that if the electron beam was not provided, the entire spaceship would become highly negatively charged and would hence attract the departing ions back to the ship!

A highly simplified but revealing analysis can be constructed by assuming that conditions between the anode and cathode are one-dimensional and that only singly charged ions exist in the region. The geometry and related nomenclature are shown in Fig. 4.10. The voltage at location x is ϕ, the potential energy of a particle of charge q is $q\phi$, and the kinetic energy is $\frac{1}{2}mu^2$.

The conservation of total energy may be written

$$\tfrac{1}{2}mu^2 + q\phi = \text{const} = \tfrac{1}{2}mu_A^2 + q\phi_A \qquad (4.39)$$

hence

$$u^2 = 2(q/m)(\phi_A - \phi) + u_A^2 \qquad (4.40)$$

The electric current per area j may be written

$$j = un_i q \qquad (4.41)$$

where n_i is the number of ions/volume.

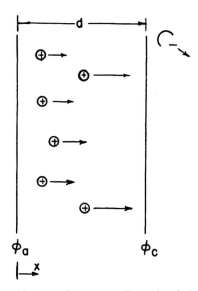

Fig. 4.10   Geometry and nomenclature, one-dimensional electrostatic accelerator.

An expression for the variation of voltage with position can be obtained by invoking the second Maxwell equation and by noting that the current density is constant with x. Thus with

$$\frac{d^2\phi}{dx^2} = -\frac{n_i q}{\varepsilon_0} \qquad (4.42)$$

it follows with Eqs. (4.40) and (4.41) that

$$\frac{d^2\phi}{dx^2} = -\frac{j}{\varepsilon_0} \frac{1}{\sqrt{2(q/m)(\phi_A - \phi) + u_A^2}} \qquad (4.43)$$

A first integral of this equation is obtained by noting

$$\frac{d^2\phi}{dx^2} = \frac{d\phi}{dx}\left(\frac{d}{d\phi}\frac{d\phi}{dx}\right) = \frac{1}{2}\frac{d}{d\phi}\left(\frac{d\phi}{dx}\right)^2 \qquad (4.44)$$

and hence from Eq. (4.43)

$$\left(\frac{d\phi}{dx}\right)^2 - \left(\frac{d\phi}{dx}\right)_A^2 = \frac{2jm}{\varepsilon_0 q}\left[\sqrt{2\frac{q}{m}(\phi_A - \phi) + u_A^2} - u_A\right] \qquad (4.45)$$

A second integral would now give the distribution of ϕ with x, and hence with Eq. (4.40) of u with x. It is of more interest, however, to use Eq. (4.45) to obtain an expression for the maximum current attainable for a given electrode spacing and cathode voltage. For simplicity, consider the case where the velocity at the anode is very small and note that the current will be a maximum (equal to j_{scl}) when $(d\phi/dx)_A = 0$. In this case

$$\frac{d\phi}{dx} = -(\phi_A - \phi)^{\frac{1}{4}}\left(\frac{4}{\varepsilon_0}\right)^{\frac{1}{2}}\left(\frac{m}{2q}\right)^{\frac{1}{4}} j_{scl}^{\frac{1}{2}} \qquad (4.46)$$

Integrating and rearranging there is obtained

$$j_{scl} = \frac{4\varepsilon_0}{9}\sqrt{2\frac{q}{m}}\frac{(\phi_A - \phi_c)^{\frac{3}{2}}}{d^2} \qquad (4.47)$$

This relationship is known as the Child-Langmuir law. The Child-Langmuir law indicates that even if ions are made available at a high rate at the anode surface, the "space charge" existing because of the departing ions will limit the rate at which ions can be attracted from the surface. This limiting current has an important influence on the performance of electrostatic thrustors through its related limitation on the thrust per area of the device.

Idealized Performance

The preceding equations may be arranged in order to give the performance of an ideal electrostatic thrustor as follows:

$$I_{sp} = \frac{u_c}{a_0} = \frac{1}{a_0} \sqrt{2\frac{q}{m}} \left(\phi_A - \phi_c\right)^{\frac{1}{2}}$$

$$\frac{F}{A} = \frac{\text{thrust}}{\text{area}} = \frac{m}{q} j \sqrt{2\frac{q}{m}} \left(\phi_A - \phi_c\right)^{\frac{1}{2}}$$

$$j_{scl} = \frac{4\varepsilon_0}{9} \sqrt{2\frac{q}{m}} \frac{\left(\phi_A - \phi_c\right)^{\frac{3}{2}}}{d^2}$$

$$\left(\frac{F}{A}\right)_{scl} = \frac{8\varepsilon_0}{9} \left(\frac{\phi_A - \phi_c}{d}\right)^2 = \left(\frac{2}{9}\varepsilon_0 a_0^4\right) \left(\frac{m}{qd}\right)^2 I_{sp}^4$$

As an example calculation consider an electrostatic accelerator utilizing cesium, with the following characteristics: $m/q = 1.38 \,(10^{-6})$ kg/C, $d = 0.01$ m, $(\phi_A - \phi_c) = 10,000$ V, and $\varepsilon_0 = 8.85 \,(10^{-12})$ F/m. Then

$$I_{sp} = \frac{1}{9.807} \sqrt{\frac{2}{1.38}} \,(10^3)(10^2) = 12,200 \text{ s}$$

$$j_{scl} = 47.2 \text{ A/m}^2$$

$$(F/A)_{scl} = 7.9 \text{ N/m}^2$$

These simple calculations indicate many of the engineering limitations of electrostatic thrusters. Thus, even for the very high specific impulse considered, the thrust per area is extremely low. When it is noted that the thrust per area goes as I_{sp}^4, it can be realized that operating at lower I_{sp} levels will cause even more unacceptable thrust levels to occur.

It would seem that this problem could be alleviated by decreasing the electrode spacing, but the assumed value of $d = 1$ cm would seem to be as small as reasonable to support such a huge voltage difference. Several efforts have been directed to increasing the mass-to-charge ratio m/q. (Note that cesium has an extremely high mass-to-charge ratio for an ionizable atom.)

One method suggested for increasing the mass-to-charge ratio is to attach charged particles to other tiny particles in the form of colloids. Although "colloid rockets" showed great promise, it was found to be very difficult to generate a uniform charge-to-mass ratio propellant. The resulting nonuniform propellant stream led to unacceptable beam efficiencies, as well as to unacceptable problems in beam focusing.

Another technique for increasing the thrust per area that has seen considerable success is to utilize an "accel-decel" system in which an intermediate electrode at very high voltage is utilized to increase the current density. Prior to departing the rocket, however, the ions are decelerated to a suitable low exhaust velocity. It is to be noted that the intermediate electrode draws only that power identified with "leakage" currents, the majority of the power requirement being identified only with the power required to supply the electron emitter at the outlet cathode voltage. Accel-decel systems have proved quite successful in practice, although care must be taken in the design to avoid beam instabilities in the decel portion.

The "space charge limitation" arises because the cloud of departing ions tends to reflect further ions. An attempt to surmount this difficulty that met with some success involved utilizing a (weak) magnetic field to contain electrons in the region between the cathode and anode. As a result the space charge was much reduced because, as they passed through the circling electrons, the ions would have their charge "cancelled" by a nearby electron.

In conclusion, it may be said that ion rockets presently give acceptable performances at specific impulses as low as about 7000 s. The great remaining problems of electric rockets remain not so much in the thrustors as in the power supplies. The future will see whether a mission arises sufficient to warrant further development in this fascinating field.

Reference

[1]Stenning, A. H., "Rapid Approximate Method for Analyzing Nuclear Rocket Performance," *ARS Journal*, Vol. 30, Feb. 1960, pp. 169–172.

Problems

4.1 Prove Eq. (4.25).

4.2 The nuclear fuel distribution in a nuclear reactor is chosen so that when in operation the wall temperature of the reactor is a constant equal to the maximum allowable temperature T_s.

(a) Find an expression for T_{t_2}/T_s in terms of T_{t_1}/T_s and fL/D.

(b) Show that if the rocket is to be designed to have the same T_{t_1}, T_{t_2}, and T_s as a nuclear-heated rocket with a constant-power density, then

$$G = \ell n(1 + F)$$

where $G = 2fL/D$, where the L/D is that for the case of constant wall temperature; $F = 2fL/D$, where the L/D is that for the case of constant-power density.

(c) Assuming that both rockets are designed for the "optimum case" $M_2 = M_2^*$ and that the incoming Mach number is very small, find an expression for p_{t_2}/p_{t_1} in terms of fL/D.

(d) Calculate p_{t_2}/p_{t_1} for both rockets for the case $F = 1$.

(e) Show that if the performance of the constant-wall-temperature rocket, or of the sine-power-density rocket is plotted as in Fig. 4.6, only the curve of I/I_m changes from that of Fig. 4.6.

4.3   Obtain the curve of F/Ap_{t_1} vs M_2 for the case illustrated in Fig. 4.5, but utilizing the full equations (that is, do not assume $M_1 \approx 0$).

4.4   Investigate the sensitivity of the "optimal" values of I/I_m and F/Ap_{t_1}, as shown in Fig. 4.6, to the assumed values of T_{t_1}/T_s and p_a/p_{t_1}.

4.5   An electrically powered rocket with $\eta_a \alpha' = 80$ W·kg^{-1} and a mission requirement of $\Delta v = 17,000$ ms^{-1} is designed to have a payload ratio of $m_L/m_0 = 0.4$.
       (a)   Assuming the optimal exhaust velocity C_{opt} is selected, find (i) C_{opt} ms^{-1}, (ii) τ s, (iii) a_i ms^{-2}, (iv) m_p/m_0 and m_e/m_0.
       (b)   If instead of the optimal exhaust velocity, $C = C_{\text{opt}}/1.1$ is selected, find τ, a_i, m_p/m_0, and m_e/m_0.

4.6   (a)   Consider an electric rocket with optimum specific impulse for the case $m_L/m_0 = 0.4$ and $\Delta v = 20,000$ ms^{-1}. Plot the variation of I_{opt}, τ, a_i, m_p/m_0, and m_e/m_0 vs $\alpha' \eta_a$ for the range $50 < \alpha' \eta_a < 200$.
       (b)   Consider the rocket of part (a) for the case $\alpha' \eta_a = 100$. If it is possible to attain $\alpha' \eta_a = 200$ and it is desired to keep the firing time the same, what will be the new payload ratio?

4.7   Show that the time of firing for an optimal electric rocket with given m_L/m_0 and $\alpha' \eta_a$ goes like $(\Delta v)^2$. Explain why this is so, rather than the time of firing being proportional to Δv.

4.8   (a)   Helium is heated by an electric current in an arc jet chamber to a stagnation temperature of 4000 K. It is then expanded (almost) isentropically through a nozzle to very low pressure. Estimate the specific impulse, assuming that the ionization effects can be ignored.
       (b)   Molecular hydrogen is used as a propellant in the arc jet of part (a). Estimate the specific impulse, assuming the approximations for the case of part (a) are valid and that in addition any dissociation effects can be ignored.
       (c)   Using approximations similar to the preceding parts, obtain an expression giving the "effective" chamber stagnation temperature T_c in terms of the specific impulse and other required material properties. Calculate T_c for the case with $I_{sp} = 1500$ s and propellant H_2.

4.9   (a)   It may be assumed that virtually all the energy acquired by the propellant in an electrostatic thrustor appears as the kinetic energy of the jet, $\frac{1}{2}mu_e^2$. If the particles are singly charged, this energy may be expressed in terms of the "electron volts," eV, of the exhaust beam. Show that the electron volts of the exhaust may be expressed in terms of the specific impulse and molecular weight of the propellant by

$$eV = K\mathcal{M}\left(I_{sp}/10^3\right)^2$$

where K is the constant of proportionality.

Noting that $\mathcal{M} = N_a m$, where N_a = Avogadro's number $\approx 6.03 \ (10^{26})$ and m = mass of particle, and also that the charge on an electron q_e = 1.6 (10^{-19}) C, find the numerical value of K.

(b) If cesium is the propellant ($\mathcal{M} = 133$, monatomic), find the voltage of the exhaust if $I_{sp} = 6000$ s.

(c) Using the results of Problem 4.8(c), calculate the effective reservoir temperature for the conditions of part (b).

(d) If the thrust level of this rocket is to be 100 N, what is the required power level in kilowatts?

(e) What is the mass flow rate in kg/s^{-1}?

4.10 Consider a colloid rocket with the properties $m/q = \beta (m/q)_R$ and $I_{sp} = \delta I_{sp_R}$. (Here, R refers to a reference electrostatic accelerator.)

(a) If the same electrode spacing is used as for the reference accelerator, find the ratio of the following terms to their reference values in terms of β and δ: j_{scl}, $(\phi_A - \phi_C)$, and $(F/A)_{scl}$.

(b) Find the same ratios as in part (a) for the case where the electrode spacing is chosen to keep the voltage gradient at the cathode of the accelerator the same as that of the reference accelerator.

4.11 Consider an "accel-decel" system consisting of an anode at voltage ϕ_A, cathode at voltage ϕ_C, and downstream cathode at ϕ_D. Consider the current in the first region ($A-C$) to be space charge limited.

(a) Show that the equation for the potential in the region downstream of C ($C - D$) may be written

$$\left(\frac{d\phi}{dx} \right)^2 - \left(\frac{d\phi}{dx} \right)_C^2 = \frac{4 j_{scl}}{\varepsilon} \sqrt{\frac{m}{2q}} \left(\sqrt{\phi_A - \phi} - \sqrt{\phi_A - \phi_C} \right)$$

(b) Introducing $\Phi = (\phi - \phi_C)/(\phi_A - \phi_C)$ and $\delta = x/d_{AC}$, where d_{AC} is the distance between anode and first cathode and $\delta = 0$ when $\Phi = 0$, show that the equation for δ in terms of Φ may be written

$$\delta = (1 + 2K)\sqrt{1 - K} - [\sqrt{1 - \Phi} + 2K]\sqrt{\sqrt{1 - \Phi} - K}$$

where

$$K \equiv 1 - \frac{9}{16} \left(\frac{d\Phi}{d\delta} \right)_C^2$$

4.12 A method of ionic propulsion has been suggested that utilizes a series of electrodes. If such a device is constructed so that the slope of the potential to the left of each electrode is to be the same as that preceding it and the slope to the right of each electrode is to be zero, show that the magnitude of the voltage on the nth electrode is given in terms of the voltage on the first electrode by

$$|\phi_n| = n^2 |\phi_1|$$

(The voltage of the "zeroth electrode" is taken to be zero.)

5. IDEAL CYCLE ANALYSIS

5.1 Introduction

In this chapter the systematic process termed "cycle analysis" will be applied to several different engine types. The object of cycle analysis is to obtain estimates of the performance parameters (primarily thrust and specific fuel consumption) in terms of design limitations (such as the maximum allowable turbine temperature), the flight conditions (the ambient pressure and temperature and the Mach number), and design choices (such as the compressor pressure ratio, fan pressure ratio, bypass ratio, etc.). In this chapter all components are considered to be ideal, with the result that the various algebraic manipulations will be quite simplified and, consequently, the methodology of the analysis comparatively transparent. The analytical results will, of course, be far more optimistic than would be the case if component losses were included, but many of the general trends will be valid. The effects of component losses will be thoroughly investigated in Chap. 7.

Gas turbine engine performance measures, selection considerations, and components are briefly discussed in Sec. 1.4. In order to further understand the engine components that are being idealized in the cycle analysis, and to be able to envision the individual engine parts and their relationships in engine configurations, it is useful at this point to examine illustrations, design parameter values, and performance curves of several typical modern turbofan engines. The design and performance data presented can serve as reference points for later numerical computations.

The Pratt and Whitney JT8D two-spool low-bypass ratio mixed flow (fan air and turbine exit gases mix and leave through the common exhaust nozzle) turbofan engine used in medium-range commercial aircraft is shown schematically in Fig. 5.1. The nominal gas property values listed in the figure are for an early member of the JT8D family at sea-level static takeoff thrust where the design values are 1.1 bypass ratio, 15.9 compressor (cycle) pressure ratio, and 1.9 fan pressure ratio. Estimated performance data for the JT8D-17 engine model are presented in Figs. 5.2 and 5.3.

The Pratt and Whitney JT9D two-spool high-bypass ratio separate flow turbofan engine shown in Fig. 5.4 is designed for long-range aircraft use. An early member of the JT9D family has a 5.1 bypass ratio, 21.5 compressor pressure ratio, and 1.5 fan pressure ratio at sea-level static takeoff thrust. Figures 5.5 and 5.6 contain estimated performance data for the JT9D-70/-70A engine models.

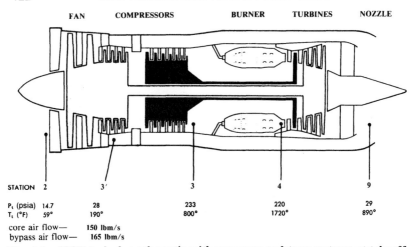

STATION	2	3′	3	4	9
P$_t$ (psia)	14.7	28	233	220	29
T$_t$ (°F)	59°	190°	800°	1720°	890°

core air flow— 150 lbm/s
bypass air flow— 165 lbm/s

Fig. 5.1 JT8D turbofan schematic with pressures and temperatures at takeoff thrust (courtesy of Pratt and Whitney).

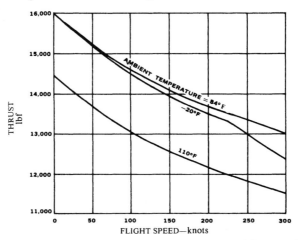

Fig. 5.2 JT8D-17 turbofan takeoff thrust (courtesy of Pratt and Whitney).

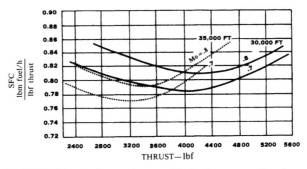

Fig. 5.3 JT8D-17 turbofan cruise specific fuel consumption (courtesy of Pratt and Whitney).

Figures 1.3 and 1.4 are other examples of Pratt and Whitney turbofan engines. As already noted in Chap. 1, the values of some of the design parameters for the separate flow PW2037 (Fig. 1.3) are 5.8 bypass ratio, 32 compressor pressure ratio, and 1.4 fan pressure ratio; for the mixed flow afterburning, F100 (Fig. 1.4), they are 0.78 bypass ratio, 25 compressor pressure ratio, and 3.0 fan pressure ratio, all at sea-level takeoff conditions.

The Garrett TFE731 two-spool medium-bypass ratio separate flow turbofan engine of Fig. 5.7 serves the business aircraft type market. In contrast to the preceding engines, this engine has a geared fan, a centrifugal high-pressure compressor, and a reverse flow combustion chamber. The TFE731-5 engine model at sea-level static takeoff thrust has a 3.33 bypass ratio, 14.4 compressor pressure ratio, and 1.55 fan pressure ratio. Performance curves for the TFE731-5 are shown in Figs. 5.8 and 5.9. The thrust specific fuel consumption of this engine model at sea-level (73.4 F day) takeoff thrust is 0.484 (lbm fuel/h)/lbf thrust and 0.802 (lbm fuel/h)/lbf thrust at 0.8 Mach number and 40,000 ft.

Figure 5.10 is a cutaway view of the Garrett ATF3 three-spool medium-bypass ratio mixed flow turbofan engine for business type aircraft. At sea-level static takeoff thrust the ATF3-6A engine model has a 2.81 bypass ratio and 21.35 compressor pressure ratio. The estimated performance of this engine model is given in Figs. 5.11 and 5.12.

Figure 5.13 shows the unique engine spool arrangement of the ATF3. Note that the high-pressure centrifugal compressor spool is mounted aft of the two concentric spools containing the fan and the low-pressure compressor. What do you suppose are the advantages that led designers to this novel engine configuration?

Figure 5.14 is an installed cross-sectional view of the engine showing the gas flow paths through the engine components. Referring to Fig. 5.14, it is seen that the engine core airflow passes through the fan, the low-pressure compressor, and eight carryover ducts leading to the rear of the engine where a 180-deg turn is made into the high-pressure centrifugal compressor. From this compressor the air enters the reverse flow combustion chamber. The gases leaving the combustion chamber proceed toward the front of the engine and in turn pass through the single-, three-, and two-stage turbines which drive the high-pressure compressor, the fan, and the low-pressure compressor, respectively. After leaving the last turbine stage, the gases are split into eight 180-deg turning vane modules that exhaust the gases into the fan airflow contained in an annular duct surrounding the engine. The fan air and the turbine gases mix and exit through a common exhaust nozzle.

An installed cutaway view and a schematic of the high-bypass ratio separate flow Rolls-Royce RB.211-524 three-spool turbofan engine are shown in Figs. 5.15 and 5.16.

Here it is worthy to note that the performance curves that are obtained by the *ideal* engine *on-design* cycle analysis of this chapter differ from the *actual* engine *off-design* performance curves presented in Figs. 5.2–5.3, 5.5–5.6, 5.8–5.9, and 5.11–5.12, as indicated by the descriptive words *ideal* vs *actual*, and *on-design* vs *off-design*. Each point of an on-design cycle analysis performance curve represents the performance of a different engine

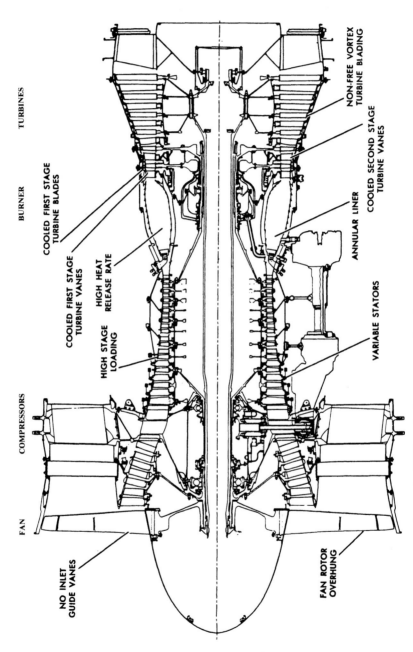

FAN   **COMPRESSORS**   **BURNER**   **TURBINES**

NO INLET
GUIDE VANES

FAN ROTOR
OVERHUNG

HIGH STAGE
LOADING

HIGH HEAT
RELEASE RATE

COOLED FIRST STAGE
TURBINE VANES

COOLED FIRST STAGE
TURBINE BLADES

VARIABLE STATORS

ANNULAR LINER

COOLED SECOND STAGE
TURBINE VANES

NON-FREE VORTEX
TURBINE BLADING

Fig. 5.4   JT9D turbofan cross section (courtesy of Pratt and Whitney).

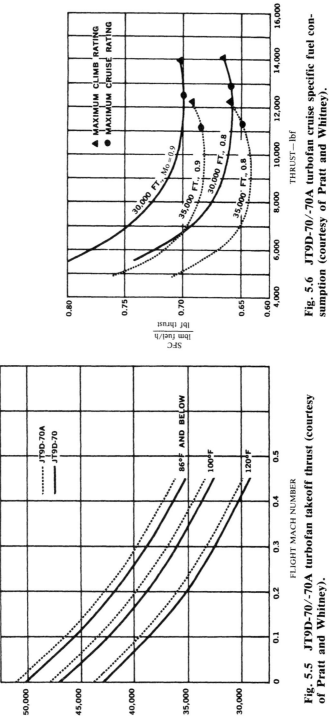

Fig. 5.6 JT9D-70/-70A turbofan cruise specific fuel consumption (courtesy of Pratt and Whitney).

Fig. 5.5 JT9D-70/-70A turbofan takeoff thrust (courtesy of Pratt and Whitney).

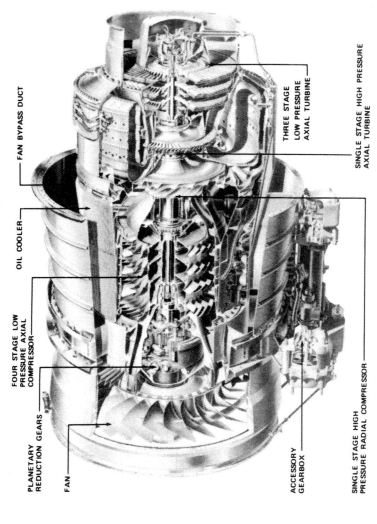

Fig. 5.7 TFE731 turbofan cutaway (courtesy of Garrett).

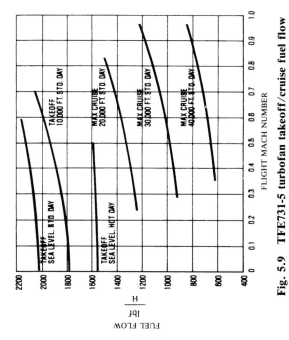

Fig. 5.9 TFE731-5 turbofan takeoff/cruise fuel flow (courtesy of Garrett).

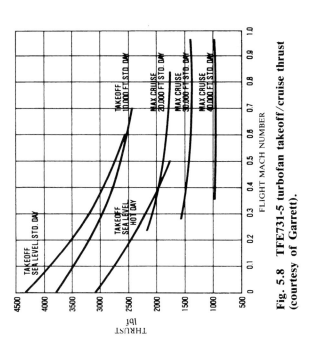

Fig. 5.8 TFE731-5 turbofan takeoff/cruise thrust (courtesy of Garrett).

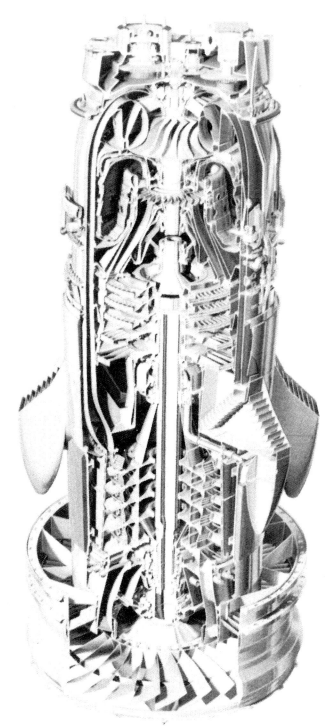

Fig. 5.10 ATF3 turbofan cutaway (courtesy of Garrett).

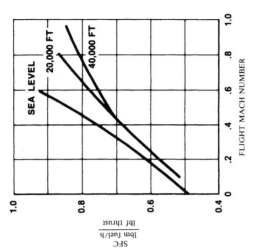

Fig. 5.12 ATF3-6A turbofan takeoff/cruise specific fuel consumption (courtesy of Garrett).

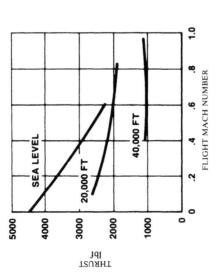

Fig. 5.11 ATF3-6A turbofan takeoff/cruise thrust (courtesy of Garrett).

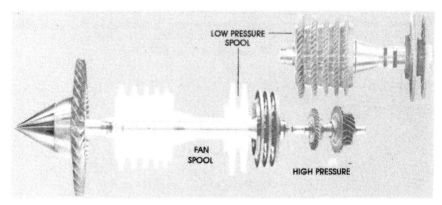

Fig. 5.13   ATF3 turbofan spool arrangement (courtesy of Garrett).

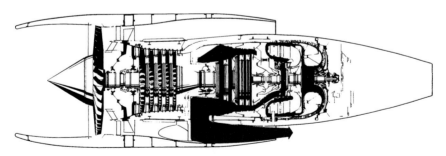

Fig. 5.14   ATF3 turbofan installed cross-sectional view (courtesy of Garrett).

Fig. 5.15 RB.211-524 turbofan installed cutaway (courtesy of Rolls-Royce).

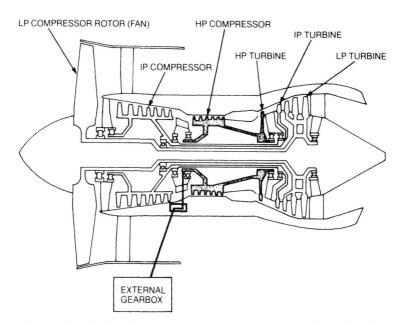

Fig. 5.16 RB.211-524 turbofan schematic (courtesy of Rolls-Royce).

from a family of engines, each of which is operating at its own design point. The performance curves in the figures referenced above, on the other hand, are for a single engine operating at off-design conditions and are the subject of Chap. 8. This chapter's engine performance is also given only in terms of specific thrust and specific fuel consumption since engine size is not specified in the analysis. Thrust and fuel mass flow rate as given in Figs. 5.8 and 5.9, for example, can be determined when engine size expressed in terms of engine air mass flow rate is known.

5.2 Notation

A systematic notation will facilitate simple manipulation of the equations to follow. Throughout this chapter the engine station numbers indicated in Fig. 5.17 will be used. The locations indicated in Fig. 5.17 are:

0	Far upstream
1	Inlet entry
2	Compressor face
3	Compressor exit
3′	Fan exit
4	Turbine entry
5	Turbine exit
6	Afterburner entry
6′	Duct afterburner entry
7	Primary nozzle entry
7′	Secondary nozzle entry
8	Primary nozzle throat
8′	Secondary nozzle throat
9	Primary nozzle exit
9′	Secondary nozzle exit

Appendix B contains the standardized gas turbine engine station identification and nomenclature system recommended by SAE, Inc. Note that the station numbers defined above conform to Sec. 2.2 of App. B, but the bypass flow stations are identified here by 3′, 6′, 7′, 8′, and 9′ in lieu of the two-digit numbering system in Sec. 2.3 of App. B.

The ratio of stagnation pressures π and ratio of stagnation temperatures τ are introduced, where

$$\pi = \frac{\text{stagnation pressure leaving component}}{\text{stagnation pressure entering component}}$$

$$\tau = \frac{\text{stagnation temperature leaving component}}{\text{stagnation temperature entering component}}$$

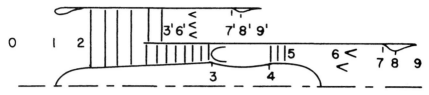

Fig. 5.17   Station numbering.

Examples

$\tau_c, \pi_c \equiv$ compressor stagnation temperature, pressure ratio

$\tau_b, \pi_b \equiv$ burner stagnation temperature, pressure ratio, etc.

Exceptions

τ_r and π_r are defined by

$$\tau_r = 1 + \frac{\gamma - 1}{2} M_0^2 = \frac{T_{t_0}}{T_0}$$

$$\pi_r = \left(1 + \frac{\gamma - 1}{2} M_0^2\right)^{\gamma/(\gamma - 1)} = \frac{p_{t_0}}{p_0} \tag{5.1}$$

Thus, freestream stagnation temperature $T_{t_0} = T_0\tau_r$; freestream stagnation pressure $p_{t_0} = p_0\pi_r$. It should be noted that τ_r and π_r represent the effects of the flight Mach number M_0.

Further Exceptions

It is often appropriate to introduce the effect of a design limitation such as the maximum allowable turbine inlet stagnation enthalpy, $C_{p_t}T_{t_4}$. The term τ_λ is thus introduced, defined by

$$\tau_\lambda \equiv C_{p_t}T_{t_4}/C_{p_c}T_0 \tag{5.2}$$

Similarly, $\tau_{\lambda_{AB}}$ and $\tau_{\lambda_{AB'}}$ will be used where the maximum stagnation enthalpy referred to is the stagnation enthalpy following the primary stream afterburner or duct afterburner, respectively.

Components

Each component will be identified by a subscript as follows:

AB = afterburner (primary stream)
AB' = afterburner (secondary stream)
b = burner

c = compressor
c' = fan
d = diffuser (or "inlet")
n = nozzle (primary stream)
n' = nozzle (secondary stream)
t = turbine

Table 5.1 gives the relationships between all defined π and τ and the corresponding temperatures, pressures, and Mach numbers.

5.3 Ideal Component Behaviors

In the analysis to follow in this chapter, ideal performance of all components will be assumed. In addition, it is assumed that the gas is calorically

Table 5.1   Temperature and Pressure Relationships for All τ and π

$$\tau_r = 1 + \frac{\gamma - 1}{2} M_0^2 \qquad\qquad \pi_r = \left(1 + \frac{\gamma - 1}{2} M_0^2\right)^{\gamma/(\gamma-1)}$$

$$\tau_\lambda = \frac{C_{p_t}}{C_{p_c}} \frac{T_{t_4}}{T_0}$$

$$\tau_{\lambda_{AB}} = \frac{C_{p_{AB}}}{C_{p_c}} \frac{T_{t_8}}{T_0}$$

$$\tau_{\lambda_{AB'}} = \frac{C_{p_{AB'}}}{C_{p_c}} \frac{T_{t_{8'}}}{T_0}$$

$$\tau_d = \frac{T_{t_2}}{T_{t_0}} \qquad \pi_d = \frac{p_{t_2}}{p_{t_0}} \qquad \tau_{AB} = \frac{T_{t_8}}{T_{t_5}} \qquad \pi_{AB} = \frac{p_{t_8}}{p_{t_5}}$$

$$\tau_c = \frac{T_{t_3}}{T_{t_2}} \qquad \pi_c = \frac{p_{t_3}}{p_{t_2}} \qquad \tau_{AB'} = \frac{T_{t_{8'}}}{T_{t_{3'}}} \qquad \pi_{AB'} = \frac{p_{t_{8'}}}{p_{t_{3'}}}$$

$$\tau_{c'} = \frac{T_{t_{3'}}}{T_{t_2}} \qquad \pi_{c'} = \frac{p_{t_{3'}}}{p_{t_2}} \qquad \tau_n = \frac{T_{t_9}}{T_{t_7}} \qquad \pi_n = \frac{p_{t_9}}{p_{t_7}}$$

$$\tau_b = \frac{T_{t_4}}{T_{t_3}} \qquad \pi_b = \frac{p_{t_4}}{p_{t_3}} \qquad \tau_{n'} = \frac{T_{t_{9'}}}{T_{t_{7'}}} \qquad \pi_{n'} = \frac{p_{t_{9'}}}{p_{t_{7'}}}$$

$$\tau_t = \frac{T_{t_5}}{T_{t_4}} \qquad \pi_t = \frac{p_{t_5}}{p_{t_4}}$$

perfect throughout, with γ and C_p constant throughout, and that the fuel mass flow is so small that the fuel-to-air ratio may be ignored in comparison to unity. Under these circumstances, the following relationships for the components are valid.

Diffuser

To a very high degree of approximation the flow through the diffuser may be considered to be adiabatic. In addition, when the flow is ideal it may also be considered to be isentropic. Thus with Eq. (2.57),

$$\tau_d = 1 \qquad \text{and} \qquad \pi_d = 1 \tag{5.3}$$

Compressor or Fan

The compressor or fan pressure ratio is usually selected as a design choice and hence may be considered prescribed. For an ideal process, the process will be isentropic, so [again utilizing Eq. (2.57)]

$$\tau_c = \pi_c^{(\gamma-1)/\gamma} \qquad \text{and} \qquad \tau_{c'} = \pi_{c'}^{(\gamma-1)/\gamma} \tag{5.4}$$

Combustor or Afterburner

For an ideal burner the stagnation pressure remains constant. It may be noted, as shown in Sec. 2.18, that this assumption implies burning at very low Mach number. Then

$$\pi_b = 1 \tag{5.5}$$

Turbine

Here, as with the compressor, the ideal process is an isentropic process so that

$$\tau_t = \pi_t^{(\gamma-1)/\gamma} \tag{5.6}$$

Nozzle

As with the diffuser, the flow through nozzles is very nearly adiabatic and is ideally isentropic. Thus,

$$\tau_n = 1 \qquad \text{and} \qquad \pi_n = 1 \tag{5.7}$$

5.4 The Ideal Thermodynamic Cycle

An ideal turbojet is considered in this section and its behavior as a heat engine investigated. (See Fig. 5.18.)

The pressure/specific volume and temperature/entropy diagrams for this ideal engine are indicated in Figs. 5.19 and 5.20. These diagrams represent

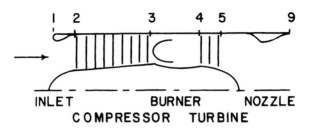

Fig. 5.18   Ideal turbojet.

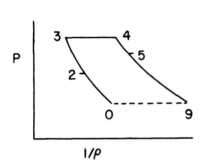

Fig. 5.19   Pressure-specific volume
diagram.

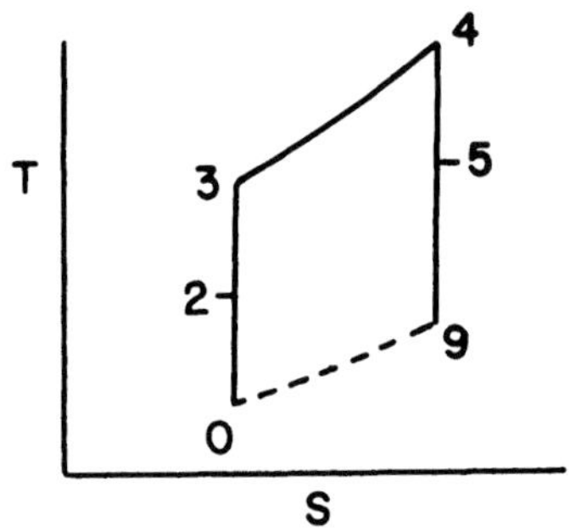

Fig. 5.20   Temperature-entropy
diagram.

"Brayton cycle," which consists of

$0 \rightarrow 3$ Isentropic compression
$3 \rightarrow 4$ Constant-pressure combustion (equivalent to a constant-pressure heat interaction)
$4 \rightarrow 9$ Isentropic expansion
$9 \rightarrow 0$ Constant-pressure "heat rejection"

When viewed as a thermal engine, the "work" of the engine, in the thermodynamic sense, appears as the change in kinetic energy between the incoming and outgoing fluid. It can be noted, for example, that if the engine were to be utilized for ground power, the kinetic energy of the jet could be extracted by a further turbine that in turn would supply a mechanical work interaction. Not all of the work of the turbojet engine appears as useful work (supplied to the aircraft), however, because the force from the engine provides work to the aircraft in an amount proportional to the flight speed.

Performance parameters of direct utility to the aircraft designer are the thrust F and specific fuel consumption S. The specific fuel consumption is measured as the milligrams of fuel flow per second divided by the thrust in Newtons [or alternatively $S = $ (lbm fuel/h)/lbf thrust]. Clearly, S is, in some sense, the inverse of the overall engine efficiency and both the thermodynamic efficiency of the engine and the efficiency of transmitting the work of the engine into work on the aircraft [or equivalently of the "propulsive efficiency" (Sec. 5.6)] are of importance.

The thermal efficiency η_{th} of this ideal engine may be obtained directly by writing

$$\eta_{th} = 1 - \frac{C_p(T_9 - T_0)}{C_p(T_{t_4} - T_{t_3})} = 1 - \frac{T_0}{T_{t_3}} \frac{[(T_9/T_0) - 1]}{[(T_{t_4}/T_{t_3}) - 1]}$$

Note, however,

$$\left(\frac{T_{t_3}}{T_0}\right)^{\gamma/(\gamma-1)} = \frac{p_{t_3}}{p_0} = \frac{p_{t_4}}{p_9} = \left(\frac{T_{t_4}}{T_9}\right)^{\gamma/(\gamma-1)}$$

thus

$$T_{t_3}/T_0 = T_{t_4}/T_9$$

and hence

$$T_9/T_0 = T_{t_4}/T_{t_3}$$

Thus, with $T_{t_3}/T_0 = \tau_r\tau_c$,

$$\eta_{th} = 1 - 1/\tau_r\tau_c \tag{5.8}$$

Thus, it is apparent that the thermal efficiency of the ideal engine increases as the flight Mach number increases (τ_r increases) and as the compressor pressure ratio increases (τ_c increases).

5.5 The Effect of Burning at Finite Mach Number

It is of interest to consider the thermodynamic behavior of a turbojet that has ideal behavior in all components except that the burning within the combustor occurs at finite Mach number. The cycle is still to be a Brayton cycle, except now the *static* pressure is to be kept constant in the burner. Note that an inevitable loss in stagnation pressure will occur when the Mach number is finite [Eq. (2.81) and Problem 2.12].

Consider the case where "internally" the flow can be considered reversible. By this it is meant that the additional entropy gains created by burning at finite Mach number are due only to the decreased static temperature brought about by the finite Mach number, and are not due to the presence of viscous stresses. In this case the entropy gain is given in terms of the thermal addition by $ds = d'q/T$, and consequently the thermal energy added or removed during a process can be represented as the area under the process line on the temperature/entropy plot.

Three methods of obtaining the thermal efficiency are now given.

(1) Classical method. Here use $d'q = d'q_R = T\,ds$ to write

$$\eta_{th} = \frac{\int_3^4 T\,ds - \int_0^9 T\,ds}{\int_3^4 T\,ds} = 1 - \frac{\int_0^9 T\,ds}{\int_3^4 T\,ds}$$

The processes from 3–4 and 0–9 are both constant-pressure processes, so

$$T\,ds = dh - (1/\rho)\,dp = dh = C_p\,dT$$

Thus

$$\eta_{th} = 1 - \frac{C_p(T_9 - T_0)}{C_p(T_4 - T_3)} = 1 - \frac{T_0}{T_3}\frac{(T_9/T_0 - 1)}{(T_4/T_3 - 1)}$$

but

$$\frac{T_3}{T_0} = \left(\frac{p_3}{p_0}\right)^{(\gamma-1)/\gamma} = \left(\frac{p_4}{p_9}\right)^{(\gamma-1)/\gamma} = \frac{T_4}{T_9}$$

and it follows that $\eta_{th} = 1 - T_0/T_3$.

(2) Method utilizing the "flow" form of the first law. Here the "heat added" per mass is given by $C_p(T_{t_4} - T_{t_3})$. Because the process in the burner is at constant pressure, the momentum per mass will not change, so that $U_4 = U_3$. Then

$$T_{t_4} - T_{t_3} = \left(T_4 + \tfrac{1}{2}U_4^2\right) - \left(T_3 + \tfrac{1}{2}U_3^2\right) = T_4 - T_3$$

The expression for the heat rejected remains unchanged, so using the same algebra as above, $\eta_{th} = 1 - T_0/T_3$.

(3) Method using industrial bookkeeping. The stagnation pressure decrease in a burner is considered to be a loss mechanism, so it is customary to represent the performance of a burner in terms of the stagnation pressure ratio across it; $\pi_b \equiv p_{t_4}/p_{t_3}$. The thermal efficiency would thus be written in the form

$$\eta_{th} = 1 - \frac{T_9 - T_0}{T_{t_4} - T_{t_3}} = 1 - \frac{T_0}{T_{t_3}}\frac{T_9/T_0 - 1}{T_{t_4}/T_{t_3} - 1}$$

But

$$\frac{T_{t_3}}{T_0} = \left(\frac{p_{t_3}}{p_0}\right)^{(\gamma-1)/\gamma} = \left(\frac{p_{t_4}}{p_9}\frac{p_{t_3}}{p_{t_4}}\right)^{(\gamma-1)/\gamma} = \frac{T_{t_4}}{T_9}\pi_b^{-[(\gamma-1)/\gamma]}$$

So

$$\frac{T_9}{T_0} = \frac{T_{t_4}}{T_{t_3}} \pi_b^{-[(\gamma-1)/\gamma]} = \tau_b \pi_b^{-[(\gamma-1)/\gamma]}$$

Hence

$$\eta_{th} = 1 - \frac{T_0}{T_{t_3}} \frac{\tau_b \pi_b^{-[(\gamma-1)/\gamma]} - 1}{\tau_b - 1}$$

This form gives the impression that the thermal efficiency of a cycle is dependent upon the ambient static temperature and the *stagnation* temperatures in the combustor. This interpretation is unfortunate, because it is, in fact, the static temperatures that determine the efficiency of the cycle. For this particular example it is relatively easy to show (see Problem 2.12) that

$$\tau_b \pi_b^{-[(\gamma-1)/\gamma]} = \tau_b + [(\gamma-1)/2] M_3^2 (\tau_b - 1)$$

which gives $\eta_{th} = 1 - T_0/T_3$, in agreement with methods 1 and 2.

It is important to realize that the decrease in stagnation pressure brought about by burning at finite Mach number is absolutely unavoidable and is simply a reflection of the fact that thermal interaction at the reduced *static* temperature identified with the flow at finite Mach number causes a larger entropy increase. Hence, it is the static temperatures that determine the cycle thermal efficiency.

5.6 The Propulsive Efficiency, η_p

The propulsive efficiency is a measure of how well the power produced by the engine is utilized in propelling the vehicle. It is defined by

$$\eta_p = \frac{\text{power delivered to vehicle}}{\text{net (mechanical) power in the exhaust}}$$

$$= \frac{\text{thrust} \times \text{vehicle flight speed}}{\text{power produced by the engine}}$$

An ideal engine has the nozzle exit pressure equal to the ambient pressure, so the thrust is simply equal to the rate of momentum production of the engine. Thus,

$$F = \dot{m}_9 u_9 - \dot{m}_0 u_0 = \dot{m}(u_9 - u_0) \tag{5.9}$$

The power produced by the engine is $\frac{1}{2}\dot{m}(u_9^2 - u_0^2)$, so for this ideal engine

$$\eta_p = 2u_0/(u_9 + u_0) \tag{5.10}$$

In order to increase the propulsive efficiency the exit velocity u_9 should be reduced. This, of course, will come with a penalty in thrust if the mass flow is not increased. An obvious way to obtain good propulsive efficiency is to utilize a large-diameter propeller in order to move a very large mass. Turboprop engines are, in fact, highly efficient engines, but have encountered two major problems in the past. The large gearbox necessary to reduce the propeller speed has often caused weight and reliability problems, and the onset of high Mach numbers at the propeller tips (as the aircraft speed increases) has led to unacceptably low propeller efficiencies.

Lately, a resurgence of interest has occurred in "very-high-bypass-ratio turbofans," however. Thus, engines with up to 10 relatively small-radius propeller blades are envisioned. These blades will be swept backward (in the relative flow) in order to forestall the onset of high Mach number effects; because of the relatively small radius of the blades, the amount of gear reduction required will be much reduced as compared to conventional turboprops.

By ducting the "propeller" or fan, the tip Mach number problems can be avoided by diffusing the flow prior to the fan. An additional benefit occurs because the blades may be highly loaded, aerodynamically, right out to the blade tips because the cowl much reduces tip flow. Such ducted fan engines, termed turbofan engines, have been very successful when utilized for high-subsonic or low-supersonic flight regimes. Present turbofans used for subsonic flight have "bypass ratios" (the ratio of air passing through the outer duct to that passing through the core engine) of about 5 or 6, whereas it is imagined that the very-high-bypass engines discussed in the preceding paragraph will have (equivalent) bypass ratios of 25–50. (A turboprop has a bypass ratio of about 100.) A cowl for an engine with such a huge bypass ratio would not only be large and heavy, but would also present a large wetted area and projected area, with a consequent large drag penalty.

The various "tradeoffs" for such engine choices are best determined by a systematic use of cycle analysis.

5.7 Systems of Units

In the following sections, the various dimensionless quantities will be calculated in the SI system of units. However, because of the greater familiarity to some readers of the British system of units (or of the British gravitational system of units), the various formulas will also be presented in these alternate systems. Table 5.2 gives a brief list of pertinent terms for use in propulsion, together with appropriate conversion factors.

5.8 The Ideal Turbojet

Methodology of Cycle Analysis

The pertinent conservation equations will now be manipulated to obtain the performance variables specific thrust and specific fuel consumption in terms of assumed design variables, ambient conditions, and design limita-

Table 5.2 Units and Conversion Factors, British System to SI

Item	Units British System	Units SI	Conversion Factor (multiply British system to get SI value)
Length	ft	m	0.30480
Mass	slug, lbm	kg	14.594, 0.45359
Rotational speed	rpm	rad/s	$2\pi/60 = 0.10472$
Power	hp	W	745.70
Fuel heating value h	Btu/lbm	J/kg	2326.0
Specific heat C_p	Btu/(lbm · °R)	J/(kg · K)	4186.8
Gas constant R	ft^2/(s^2 · °R)	m^2/(s^2 · K)	0.16723
Specific fuel consumption S	$\dfrac{\text{lbm fuel/h}}{\text{lbf thrust}} = \dfrac{\text{lbm/h}}{\text{lbf}}$	$\dfrac{\text{mg fuel/s}}{\text{N thrust}} = \dfrac{\text{mg}}{(\text{N}\cdot\text{s})}$	28.325
Power specific fuel consumption S_P	$\dfrac{\text{lbm fuel/h}}{\text{hp}}$	$\dfrac{\text{mg/s}}{\text{W}} = \dfrac{\text{kg}}{(\text{W}\cdot\text{s})}$	0.16897
Specific thrust	$\dfrac{F}{g_0\dot{m}} = \dfrac{\text{lbf}}{(\text{lbm/s})}$	$\dfrac{F}{\dot{m}} = \dfrac{\text{N}\cdot\text{s}}{\text{kg}} = \dfrac{\text{m}}{\text{s}}$	9.8067

British system: gravitational constant $g_0 = 32.174$ lbm/lbf · ft/s^2.
SI: acceleration of gravity 9.8067 m/s^2.

tions. The methodology for all cycles to be considered will be the same. Thus, it will be found (even in the cases where losses are included) that to obtain the specific thrust both the ratio of the temperature at the nozzle exit to the ambient temperature and the ratio of the Mach number at the nozzle exit to the flight Mach number are required. By then writing the ratio of stagnation to static temperature, T_{t_9}/T_9, at the exit as a function of exit Mach number, and then further writing the stagnation temperature at the exit in terms of the products of all the component temperature ratios, an expression for the ratio T_9/T_0 will be obtained in terms of M_9 and stagnation temperature ratios. A second equation for M_9 in terms of all the component pressure ratios (and imposed exit pressure ratio p_9/p_0) is similarly obtained by writing the ratio of stagnation to static pressure, p_{t_9}/p_9, at the exit as a function of exit Mach number. The component relationships of Sec. 5.3 (or their nonideal equivalents) then allow description of the engine specific thrust in terms of the component performances and design choices.

Not all component performances are independent, however, because, for example, the turbine and compressor work interaction rates must be equated. Such a power balance will lead to evaluation of the turbine temperature and pressure ratios in terms of the chosen compressor pressure ratio and other parameters.

Finally, the specific fuel consumption is evaluated by considering an enthalpy balance across the combustor. Several example cycles are evaluated in the following sections as well as in Chap. 7, and it will be seen that the methodology described in this section is systematically applied throughout.

Cycle Analysis of the Ideal Turbojet

Ideal behavior is assumed so that the component relationships are as in Sec. 5.3. Additionally, the gas is assumed to be calorically perfect throughout, the pressure at exit is assumed equal to the ambient pressure, and the fuel-to-air ratio, $\dot{m}_f/\dot{m}_0 \equiv f$, is assumed to be much less than unity. With these assumptions (referencing Fig. 5.18)

$$F = \dot{m}(u_9 - u_0)$$

or

$$\text{Specific thrust} = \frac{F}{\dot{m}} = u_0\left(\frac{u_9}{u_0} - 1\right) = a_0 M_0\left(\frac{u_9}{u_0} - 1\right) \tag{5.11}$$

Now write

$$\left(\frac{u_9}{u_0}\right)^2 = \frac{\gamma_9 R_9 T_9}{\gamma_0 R_0 T_0}\left(\frac{M_9}{M_0}\right)^2 = \frac{T_9}{T_0}\left(\frac{M_9}{M_0}\right)^2 \tag{5.12}$$

(Note $\gamma_9 = \gamma_0 = \gamma$, $R_9 = R_0 = R$.) Now note

$$T_{t_9} = T_9\left(1 + \frac{\gamma - 1}{2}M_9^2\right) = T_0\frac{T_{t_4}}{T_0}\frac{T_{t_5}}{T_{t_4}}\frac{T_{t_9}}{T_{t_5}}$$

or

$$T_{t_9} = T_0\tau_\lambda\tau_t\tau_n = T_0\tau_\lambda\tau_t = T_9\left(1 + \frac{\gamma - 1}{2}M_9^2\right) \tag{5.13}$$

Here $\tau_n = 1$, which follows from Eq. (5.7). Also note

$$\tau_\lambda = \frac{T_{t_4}}{T_0} = \frac{T_{t_0}}{T_0}\frac{T_{t_2}}{T_{t_0}}\frac{T_{t_3}}{T_{t_2}}\frac{T_{t_4}}{T_{t_3}} = \tau_r\tau_d\tau_c\tau_b$$

Hence, with Eq. (5.3)

$$\tau_\lambda = \tau_r\tau_c\tau_b \tag{5.14}$$

Thus, note that the minimum conceivable τ_λ, which corresponds to no burning in the combustor, is

$$(\tau_\lambda)_{\min} = \tau_r\tau_c \tag{5.15}$$

For the pressures, write

$$p_{t_9} = p_9\left(1 + \frac{\gamma - 1}{2}M_9^2\right)^{\gamma/(\gamma-1)}$$

$$= p_0\frac{p_{t_0}}{p_0}\frac{p_{t_2}}{p_{t_0}}\frac{p_{t_3}}{p_{t_2}}\frac{p_{t_4}}{p_{t_3}}\frac{p_{t_5}}{p_{t_4}}\frac{p_{t_9}}{p_{t_5}} = p_0\pi_r\pi_d\pi_c\pi_b\pi_t\pi_n \qquad (5.16)$$

or noting $p_9 = p_0$ and from Eqs. (5.1) and (5.3–5.7), $\pi_d = \pi_b = \pi_n = 1$ and $\pi_r^{(\gamma-1)/\gamma} = \tau_r$, $\pi_c^{(\gamma-1)/\gamma} = \tau_c$, $\pi_t^{(\gamma-1)/\gamma} = \tau_t$, Eq. (5.16) leads to

$$1 + [(\gamma - 1)/2]M_9^2 = \tau_r\tau_c\tau_t \qquad (5.17)$$

Combination of Eqs. (5.1), (5.13), and (5.17) then gives

$$T_9/T_0 = \tau_\lambda/\tau_r\tau_c \qquad (5.18)$$

and

$$\left(\frac{M_9}{M_0}\right)^2 = \frac{\tau_r\tau_c\tau_t - 1}{\tau_r - 1} = \frac{2}{\gamma - 1}\frac{1}{M_0^2}(\tau_r\tau_c\tau_t - 1) \qquad (5.19)$$

The specific thrust then follows by combining Eqs. (5.11), (5.12), (5.18), and (5.19) to give

$$\frac{F}{\dot{m}} = a_0\left\{\left[\frac{2}{\gamma - 1}\frac{\tau_\lambda}{\tau_r\tau_c}(\tau_r\tau_c\tau_t - 1)\right]^{\frac{1}{2}} - M_0\right\} \qquad (5.20)$$

The power balance. A power balance between the compressor and turbine is used to relate the turbine temperature ratio τ_t to other variables. It follows from the first law applied to a control volume (Sec. 2.15) that for steady-state conditions the mechanical work interaction per mass is equal to the negative of the change in stagnation enthalpies across the control volume. The turbine work output must be equal to the compressor work input, so applying the first law results in

$$(\dot{m} + \dot{m}_f)C_{p_t}(T_{t_4} - T_{t_5}) = \dot{m}C_{p_c}(T_{t_3} - T_{t_2})$$

or

$$(1 + f)\frac{C_{p_t}T_{t_4}}{C_{p_c}T_0}\left(1 - \frac{T_{t_5}}{T_{t_4}}\right) = \frac{T_{t_0}}{T_0}\frac{T_{t_2}}{T_{t_0}}\left(\frac{T_{t_3}}{T_{t_2}} - 1\right)$$

where $f = \dot{m}_f/\dot{m}$ or

$$(1+f)\tau_\lambda(1-\tau_t) = \tau_r\tau_d(\tau_c - 1)$$

But here $f \ll 1$ and $\tau_d = 1$ (also $C_{p_t} = C_{p_c} = C_p$), so that

$$\tau_t = 1 - (\tau_r/\tau_\lambda)(\tau_c - 1) \tag{5.21}$$

Equation (5.21) can now be substituted into Eq. (5.20) to give, after some manipulation

$$\frac{F}{\dot{m}} = a_0 \left\{ \left[\frac{2\tau_r}{\gamma - 1} \left(\frac{\tau_\lambda}{\tau_r\tau_c} - 1 \right)(\tau_c - 1) + \frac{\tau_\lambda}{\tau_r\tau_c} M_0^2 \right]^{\frac{1}{2}} - M_0 \right\} \tag{5.22}$$

Note that when a ramjet is considered (no compressor, so $\tau_c = 1$), the equation for the specific thrust reduces to the very simple form

$$\frac{F}{\dot{m}} = a_0 M_0 \left[\sqrt{\frac{\tau_\lambda}{\tau_r}} - 1 \right] \quad \text{(ideal ramjet)} \tag{5.23}$$

Note also, from Eq. (5.22), that the thrust goes to zero (of course) for no combustion in the burner ($\tau_b = \tau_\lambda/\tau_r\tau_c = 1$).

Specific fuel consumption.   From the enthalpy balance across the combustor

$$(\dot{m} + \dot{m}_f)C_p T_{t_4} - \dot{m}C_p T_{t_3} = \dot{m}_f h$$

where h is the "heating value of the fuel."
Thus, again taking $f \ll 1$, there is obtained

$$f = (C_p T_0/h)(\tau_\lambda - \tau_r\tau_c) \tag{5.24}$$

The specific fuel consumption may hence be written

$$S = \frac{\text{mg fuel/s}}{\text{N thrust}} = \frac{\dot{m}_f}{F}(10^6) = \frac{f}{F/\dot{m}}(10^6) \tag{5.25}$$

Summary of the Equations—Ideal Turbojet (or Ramjet)

The equations are summarized here in a form suitable for calculation. The pertinent equations and terms for the British system of units will be included in brackets.

Inputs: $T_0(\text{K})[^\circ\text{R}], \gamma, h(\text{J/kg})[\text{Btu/lbm}],$

$$C_p(J/kg \cdot K)[Btu/lbm \cdot °R], \tau_\lambda, \pi_c, M_0$$

Outputs:

$$\frac{F}{\dot{m}}\left(\frac{N \cdot s}{kg}\right)\left[\frac{F}{g_0\dot{m}}, \frac{lbf}{lbm/s}\right]$$

$$S\left(\frac{mg}{N \cdot s}\right)\left[\frac{lbm\ fuel/h}{lbf\ thrust}\right]$$

$$f\left(\frac{kg\ fuel/s}{kg\ air/s}\right)\left[\frac{lbm\ fuel/s}{lbm\ air/s}\right]$$

Equations:

$$R = \frac{\gamma-1}{\gamma}C_p \ m^2/s^2 \cdot K \qquad \left[R = \frac{\gamma-1}{\gamma}C_p(2.505)(10^4)\ ft^2/s^2 \cdot °R\right]$$

(5.26)

$$a_0 = \sqrt{\gamma R T_0}\ m/s\ [ft/s] \tag{5.27}$$

$$\tau_r = 1 + \frac{\gamma-1}{2}M_0^2 \tag{5.28}$$

$$\tau_c = \pi_c^{(\gamma-1)/\gamma} \tag{5.29}$$

$$\frac{F}{\dot{m}} = a_0\left\{\left[\frac{2\tau_r}{\gamma-1}\left(\frac{\tau_\lambda}{\tau_r\tau_c} - 1\right)(\tau_c - 1) + \frac{\tau_\lambda}{\tau_r\tau_c}M_0^2\right]^{\frac{1}{2}} - M_0\right\}$$

(5.30)

$$\left[\frac{F}{g_0\dot{m}} = \frac{a_0}{32.174}\left\{\left[\frac{2\tau_r}{\gamma-1}\left(\frac{\tau_\lambda}{\tau_r\tau_c} - 1\right)(\tau_c - 1) + \frac{\tau_\lambda}{\tau_r\tau_c}M_0^2\right]^{\frac{1}{2}} - M_0\right\}\right]$$

$$f = (C_p T_0/h)(\tau_\lambda - \tau_r\tau_c) \tag{5.31}$$

$$S = \frac{f}{F/\dot{m}}(10^6) \qquad \left[S = \frac{3600f}{F/(g_0\dot{m})}\right] \tag{5.32}$$

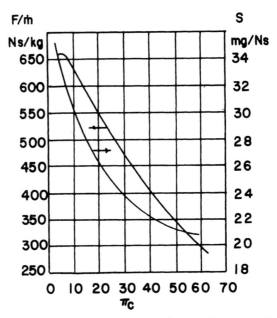

Fig. 5.21 Effect of compressor pressure ratio on performance, ideal turbojet.

Example Results—Ideal Turbojet and Ramjet

As an example of the use of the performance equations, consider the problem of selecting an appropriate compressor pressure ratio for a given flight condition. Consider a turbojet to fly at $M_0 = 2$ with the conditions $\gamma = 1.4$, $T_0 = 222.2$ K [400°R], $h = 4.4194(10^7)$ J/kg [19,000 Btu/lbm], $C_p = 1004.9$ J/kg · K [0.24 Btu/lbm · °R], $\tau_\lambda = 7$. Figure 5.21 indicates the results.

It can be seen from Fig. 5.21 that even with the assumption of ideal engine behavior some important design trends become evident. Thus, this ideal analysis indicates that the specific fuel consumption tends to decrease as the compressor pressure ratio increases. (This trend is true also when the losses are included over a large pressure ratio range, but not out to extreme pressure ratios.) The specific thrust, however, maximizes at quite low values of the compressor pressure ratio. (Note that as the compressor pressure ratio increases, the stagnation temperature at entry to the burner increases, with the consequence that the allowable fuel addition is reduced because of the restricted τ_λ.) Hence, it is evident that (ignoring afterburning effects for the moment) if a designer wanted a high-thrust lightweight engine for use in an interceptor, he would favor a low-compression-ratio engine, whereas if the engine was to be used for transport purposes where fuel consumption is of paramount importance, a heavier, higher-compression-ratio but more efficient engine would be appropriate.

Figures 5.22 and 5.23 show the expected performance of engines over a range of flight Mach numbers.

The conditions assumed for these examples are identical to those in the preceding example except that τ_λ was taken to be equal to eight for the ramjet. It is important to note that these examples illustrate the behavior of what would be a series of engines designed to fly at the illustrated Mach number at the indicated pressure ratio. In other words, the graphs represent the behavior of a family of engines, all with the same compressor pressure ratio but flying at various design Mach numbers. If an engine designed for a certain Mach number and pressure ratio is flown off-design (at a different Mach number), its compressor pressure ratio will change. These effects are analyzed in Sec. 8.2.

It is apparent in Fig. 5.22 that, in the case of the turbojets, as the flight Mach number increases the resultant increase in compressor outlet temperature restricts the allowable fuel addition, causing a reduction in specific thrust. [Note that in an actual engine the airflow would increase as M_0 increased, causing the thrust (but not specific thrust) to initially increase.] In the case of the ramjet, the thermal efficiency of the engine is so low at low Mach numbers that the specific thrust at first increases with increasing Mach numbers before eventually decreasing due to the limitation on fuel addition.

Figure 5.23 indicates that the turbojet specific fuel consumption at first increases with increase in M_0. This is because the energy required for a given velocity change from inlet to exit (and hence thrust production) increases as the flight speed increases. At very low thrusts the specific fuel

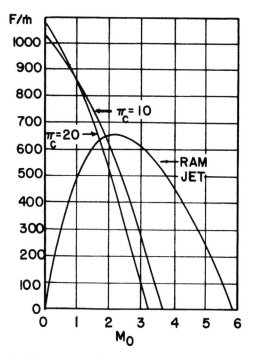

Fig. 5.22 Specific thrust vs Mach number, ideal turbojets and ramjet.

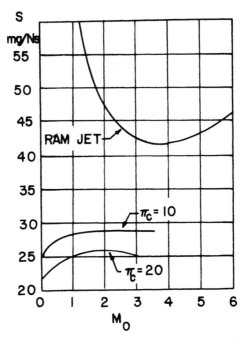

Fig. 5.23   Specific fuel consumption vs Mach number, ideal turbojets and ramjet.

consumption decreases (for this *ideal* engine) because the propulsive and thermal efficiencies increase. For the ramjet, the initial sharp drop in S reflects the increasing thermal efficiency. Later, the effect of the extra required energy for a given velocity change causes S to increase.

As a final example calculation, again consider a family of turbojets with $M_0 = 2$ and with $\pi_c = 20$. Consider the effect of varying the turbine inlet temperature or, equivalently, of varying τ_λ. Figure 5.24 indicates the results.

Figure 5.24 indicates how the specific fuel consumption decreases as the turbine inlet temperature decreases because of the increasing propulsive efficiency. However, if component losses were present, a finite fuel flow would be required as the thrust approached zero, with the result that the specific fuel consumption would increase dramatically.

5.9   Interpretation of the Behavior of the Specific Fuel Consumption

In the preceding sections the equations for S and $F/\dot{m}$ were deliberately formulated in a manner that led to easy algebraic manipulation and calculation. For interpretive purposes, write S in the alternative form

$$S = \frac{\dot{m}_f}{F}(10^6) = \frac{(h\dot{m}_f)(u_0/h)}{u_0 F}(10^6)$$

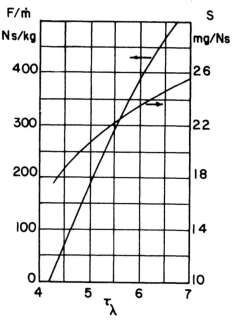

Fig. 5.24 Specific fuel consumption and specific thrust vs τ_λ, ideal turbojet.

or

$$S = \frac{u_0}{h} \frac{10^6}{\left[\dfrac{u_0 F}{(\dot{m}/2)(u_9^2 - u_0^2)}\right]\left[\dfrac{(\dot{m}/2)(u_9^2 - u_0^2)}{h\dot{m}_f}\right]} \qquad (5.33)$$

From Secs. 5.4 and 5.6 it can be seen that the expressions in brackets are just the propulsive and thermal efficiencies. That is,

$$\eta_p = \frac{u_0 F}{(\dot{m}/2)(u_9^2 - u_0^2)}$$

and

$$\eta_{th} = \frac{(\dot{m}/2)(u_9^2 - u_0^2)}{h\dot{m}_f} \qquad (5.34)$$

Hence,

$$S = \frac{a_0 M_0}{h} \frac{10^6}{\eta_p \eta_{th}} \qquad (5.35)$$

Equation (5.8) related the thermal efficiency to the thermodynamic variables by

$$\eta_{th} = 1 - \frac{1}{\tau_r \tau_c} = 1 - \frac{1}{(\pi_r \pi_c)^{(\gamma-1)/\gamma}}$$

An expression for the propulsive efficiency is obtained by combining Eqs. (5.10), (5.11), and (5.30)

$$\eta_p = \frac{2M_0}{\left[\frac{2\tau_r}{\gamma-1}\left(\frac{\tau_\lambda}{\tau_r \tau_c} - 1\right)(\tau_c - 1) + \frac{\tau_\lambda}{\tau_r \tau_c}M_0^2\right]^{\frac{1}{2}} + M_0} \qquad (5.36)$$

Equation (5.35) indicates that the behavior of the specific fuel consumption can be interpreted as a combination of three influences. Thus, as the flight Mach number increases, the required increase in energy requirement appears in the factor M_0. This effect is somewhat compensated for by the increases in η_p and η_{th} that occur for increases in M_0 for the ideal engine case. Figure 5.25 illustrates the variation in behavior of the thermal and propulsive efficiencies for the ideal turbojets and ramjets previously considered.

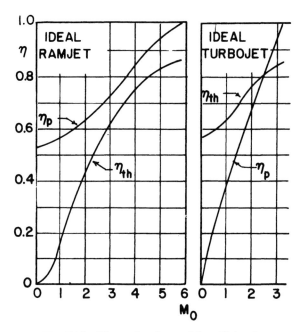

Fig. 5.25 Thermal and propulsive efficiencies.

It can be seen that the trends support the comments made in the previous section in that the low thermal efficiency dominates the ramjet performance at low M_0, whereas the rapidly increasing propulsive efficiency leads to the reduced specific fuel consumption at high Mach numbers for the turbojet.

As a final observation upon the behavior of the specific fuel consumption, it can be noted that for given $M_0(\tau_r)$ and compressor pressure ratio π_c (hence τ_c), the thermal efficiency of the ideal engine is independent of turbine inlet temperature. Thus, the observed decrease in S with τ_λ seen in Fig. 5.24 results solely from the increase in propulsive efficiency that occurs with decrease in τ_λ.

5.10 The Maximum Thrust Turbojet

In Fig. 5.21 it is evident that a maximum specific thrust occurs for a specific value of compressor pressure ratio. This specific value of compressor pressure ratio can be directly obtained from Eq. (5.22) by equating the derivative of the specific thrust with τ_c to zero. Noting that at fixed flight conditions and turbine inlet temperature, a_0, τ_r, M_0, and τ_λ are all constant, it is evident that $F/\dot{m}$ will be a maximum when

$$\frac{\partial}{\partial \tau_c}\left[\frac{2\tau_r}{\gamma-1}\left(\frac{\tau_\lambda}{\tau_r\tau_c}-1\right)(\tau_c-1)+\frac{\tau_\lambda}{\tau_r\tau_c}M_0^2\right]=0$$

or when

$$\frac{2\tau_r}{\gamma-1}\left(-1+\frac{\tau_\lambda}{\tau_r\tau_c^2}\right)-\frac{\tau_\lambda}{\tau_r\tau_c^2}M_0^2=0$$

Hence

$$\tau_c=\sqrt{\tau_\lambda}/\tau_r \qquad (5.37)$$

when $F/\dot{m}$ is a maximum. The expression for the specific thrust in this special case is hence

$$\left(\frac{F}{\dot{m}}\right)_{\text{max}}=a_0\left\{\left[\left(\frac{2}{\gamma-1}\right)\left(\sqrt{\tau_\lambda}-1\right)^2+M_0^2\right]^{\frac{1}{2}}-M_0\right\} \qquad (5.38)$$

Also, it follows from Eqs. (5.31) and (5.37) that

$$f=(C_pT_0/h)\sqrt{\tau_\lambda}\left(\sqrt{\tau_\lambda}-1\right) \qquad (5.39)$$

Temperature Relationships at Maximum Thrust

The relationship of Eq. (5.37) leads to the result that the stagnation temperature following the compressor is equal to the static temperature at nozzle exit. This can be shown as follows:

$$T_{t_3}/T_0=\tau_c\tau_r=\left(\pi_c\pi_r\right)^{(\gamma-1)/\gamma}=\left(p_{t_3}/p_0\right)^{(\gamma-1)/\gamma}$$

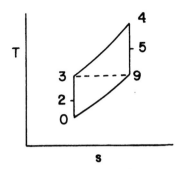

Fig. 5.26 Temperature entropy diagram for maximum thrust turbojet.

But $p_0 = p_9$ and $p_{t_3} = p_{t_4}$ so that with

$$\left(p_{t_3}/p_0 \right)^{(\gamma-1)/\gamma} = \left(p_{t_4}/p_9 \right)^{(\gamma-1)/\gamma} = T_{t_4}/T_9$$

Write

$$T_{t_3}/T_0 = \tau_c \tau_r = T_{t_4}/T_9 = \sqrt{\tau_\lambda}$$

or

$$\left(T_{t_3}/T_0 \right)^2 = T_{t_4}/T_0 = \left(T_{t_4}/T_9 \right)^2$$

Thus,

$$T_{t_3} = \sqrt{T_{t_4} T_0} = T_9 \qquad \text{Q.E.D.} \tag{5.40}$$

The temperature entropy diagram is then as indicated in Fig. 5.26. Note that the T–s diagram is very "full" for the condition $T_{t_3} = T_9$.

Example Results—Maximum Thrust Turbojet

The performance of a maximum thrust turbojet may be plotted in a similar manner to the conventional turbojet and ramjet. Figure 5.27 shows the specific thrust vs flight Mach number for a family of maximum thrust turbojets. Shown for comparison is the equivalent performance of a family of turbojets with a compressor pressure ratio of 20. Conditions are as in Figs. 5.21–5.23, namely $\gamma = 1.4$, $T_0 = 222.2$ K, $h = 4.4194(10^7)$ J/kg, $C_p = 1004.9$ J/kg · K, and $\tau_\lambda = 7$.

The related compressor pressure ratio giving maximum thrust is shown in Fig. 5.28. Note that each maximum thrust turbojet graph must terminate

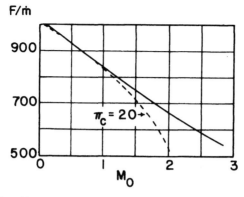

Fig. 5.27 Specific thrust vs flight Mach number, maximum thrust turbojet.

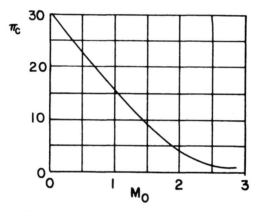

Fig. 5.28 Compressor pressure ratio for maximum thrust.

where $\pi_c = 1$. This occurs when $\tau_c = 1$, giving from Eq. (5.37) $\tau_r = \sqrt{\tau_\lambda}$. Hence

$$(M_0)_{max} = \left\{ [2/(\gamma - 1)]\left(\sqrt{\tau_\lambda} - 1\right) \right\}^{\frac{1}{2}} \qquad (5.41)$$

5.11 The Ideal Turbojet with Afterburning

A well-established and relatively simple method of increasing the thrust level of a turbojet is to "afterburn" in the duct following the turbine outlet. The additional enthalpy coupled with the nozzle pressure ratio provides a substantial thrust augmentation, although at the expense of an increase in specific fuel consumption. The mechanical arrangement and related station numbering are indicated in Fig. 5.29.

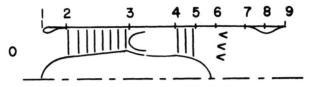

Fig. 5.29　The turbojet with afterburning.

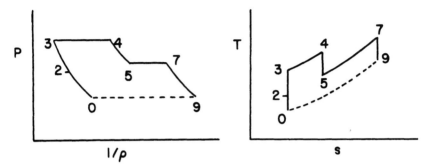

Fig. 5.30　Pressure-volume diagram, ideal afterburning turbojet.

Fig. 5.31　Temperature-entropy diagram, ideal afterburning turbojet.

When the entire cycle is considered ideal, all the component relationships of Sec. 5.3 remain true, and in addition note $\pi_{AB} = 1$. Usually the maximum temperature attainable in the afterburner is substantially higher than that attainable in the primary combustor because of the restriction placed upon the attainable temperature in the primary combustor by the presence of the turbine. Figures 5.30 and 5.31 indicate the thermodynamic cycles appropriate for the ideal afterburning turbojet.

Cycle Analysis of the Ideal Afterburning Turbojet

Again assume, as in Sec. 5.8, that all component efficiencies are perfect, that the gas is calorically perfect, that the exit static pressure is equal to the ambient pressure, and that the fuel-to-air ratio in both primary combustor and afterburner is much less than unity, with the result

$$\text{Specific thrust} = \frac{F}{\dot{m}} = a_0 M_0 \left(\frac{u_9}{u_0} - 1 \right) \qquad (5.42)$$

where as before [see Eq. (5.12)],

$$(u_9/u_0)^2 = (T_9/T_0)(M_9/M_0)^2 \qquad (5.43)$$

Now write

$$T_{t_9} = T_9\left(1 + \frac{\gamma - 1}{2} M_9^2\right) = T_0 \tau_{\lambda AB} \tag{5.44}$$

$$P_{t_9} = P_9\left(1 + \frac{\gamma - 1}{2} M_9^2\right)^{\gamma/(\gamma-1)} = p_0 \pi_r \pi_c \pi_t \tag{5.45}$$

Hence, with Eqs. (5.1), (5.4), (5.6), and (5.45)

$$1 + \frac{\gamma - 1}{2} M_9^2 = \tau_r \tau_c \tau_t \tag{5.46}$$

Equations (5.42–5.46) may now be combined to give

$$\frac{F}{\dot{m}} = a_0\left\{\left[\frac{2}{\gamma - 1}\frac{\tau_{\lambda AB}}{\tau_r \tau_c \tau_t}(\tau_r \tau_c \tau_t - 1)\right]^{\frac{1}{2}} - M_0\right\} \tag{5.47}$$

The similarity between this expression and that obtained for the turbojet without afterburning [Eq. (5.20)] is evident when it is noted that the minimum value for $\tau_{\lambda AB}$ that occurs for no afterburning is $(\tau_{\lambda AB})_{\min} = \tau_\lambda \tau_t$. With this substitution Eq. (5.47) reduces immediately to Eq. (5.20).

The power balance between the compressor and turbine remains unchanged, so that Eq. (5.21) remains valid, namely,

$$\tau_t = 1 - (\tau_r/\tau_\lambda)(\tau_c - 1)$$

The expression for the fuel-to-air ratio in the primary burner also remains as before [Eq. (5.24)]. The fuel-to-air ratio for the afterburner is obtained from an enthalpy balance across the burner to give

$$\dot{m}_{fAB} h = (\dot{m} + \dot{m}_f + \dot{m}_{fAB}) C_p T_{t_8} - (\dot{m} + \dot{m}_f) C_p T_{t_5}$$

or

$$f_{AB} = \frac{\dot{m}_{fAB}}{\dot{m}} \approx \frac{C_p T_0}{h}(\tau_{\lambda AB} - \tau_\lambda \tau_t) \tag{5.48}$$

Combining Eqs. (5.21), (5.24), and (5.48) then gives

$$f_{tot} = \frac{\dot{m}_f + \dot{m}_{fAB}}{\dot{m}} = \frac{C_p T_0}{h}(\tau_{\lambda AB} - \tau_r) \tag{5.49}$$

Summary of the Equations—Ideal Turbojet with Afterburning

Inputs: T_0(K) [°R], γ, h (J/kg) [Btu/lbm],

C_p (J/kg · K)[Btu/lbm · °R], τ_λ, $\tau_{\lambda AB}$, π_c, M_0

Outputs:
$$\frac{F}{\dot{m}}\left(\frac{\text{N}\cdot\text{s}}{\text{kg}}\right)\left[\frac{F}{g_0\dot{m}},\frac{\text{lbf}}{\text{lbm/s}}\right]$$

$$S\left(\frac{\text{mg}}{\text{N}\cdot\text{s}}\right)\left[\frac{\text{lbm fuel/h}}{\text{lbf thrust}}\right]$$

$$f_{\text{tot}}\left(\frac{\text{kg fuel/s}}{\text{kg air/s}}\right)\left[\frac{\text{lbm fuel/s}}{\text{lbm air/s}}\right]$$

Equations:

$$R=\frac{\gamma-1}{\gamma}C_p \ \text{m}^2/\text{s}^2\cdot\text{K}\qquad\left[R=\frac{\gamma-1}{\gamma}C_p(2.505)(10^4)\ \text{ft}^2/\text{s}^2\cdot\text{°R}\right]$$

(5.50)

$$a_0=\sqrt{\gamma R T_0}\ \text{m/s [ft/s]}$$

(5.51)

$$\tau_r=1+\frac{\gamma-1}{2}M_0^2$$

(5.52)

$$\tau_c=\pi_c^{(\gamma-1)/\gamma}$$

(5.53)

$$\tau_t=1-(\tau_r/\tau_\lambda)(\tau_c-1)$$

(5.54)

$$\frac{F}{\dot{m}}=a_0\left\{\left[\frac{2}{\gamma-1}\frac{\tau_{\lambda AB}}{\tau_r\tau_c\tau_t}(\tau_r\tau_c\tau_t-1)\right]^{\frac{1}{2}}-M_0\right\}$$

(5.55)

$$\left[\frac{F}{g_0\dot{m}}=\frac{a_0}{32.174}\left\{\left[\frac{2}{\gamma-1}\frac{\tau_{\lambda AB}}{\tau_r\tau_c\tau_t}(\tau_r\tau_c\tau_t-1)\right]^{\frac{1}{2}}-M_0\right\}\right]$$

Note: The minimum allowable value for $\tau_{\lambda AB}$ is $\tau_{\lambda AB}=\tau_\lambda\tau_t$.

$$f_{\text{tot}}=(C_p T_0/h)(\tau_{\lambda AB}-\tau_r)$$

(5.56)

$$S=\frac{f_{\text{tot}}}{F/\dot{m}}(10^6)\qquad\left[S=\frac{3600f_{\text{tot}}}{F/(g_0\dot{m})}\right]$$

(5.57)

Example Results—Ideal Turbojet with Afterburning

Consider a turbojet to fly at a flight Mach number of 2, with the conditions $\gamma=1.4$, $T_0=233$ K, $h=4.54\,(10^7)$ J kg^{-1}, $C_p=1005$ J kg^{-1} K^{-1}, $\tau_\lambda=7$, and $\tau_{\lambda AB}=8$. Figure 5.32 compares the performance of the turbojet over a range of compressor pressure ratios operating with and without an afterburner.

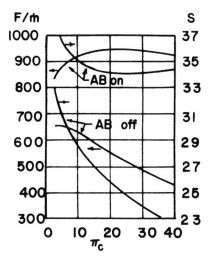

Fig. 5.32   Effect of compressor pressure ratio on performance, ideal turbojet with afterburner.

As predicted, the addition of afterburning leads to an increase in the specific thrust at the expense of an increase in the specific fuel consumption. It is of interest to note the location of the maximum in the specific thrust when afterburning is present. By formally taking the derivative of the specific thrust with τ_c [Eq. (5.55)], it is evident that the maximum occurs when the product $\tau_r\tau_c\tau_t$ reaches a maximum. Such a maximum occurs when $\pi_r\pi_c\pi_t = p_{t_5}/p_0 = p_{t_5}/p_9$ reaches a maximum and it is evident (and obvious) that the maximum thrust occurs when the nozzle pressure ratio reaches a maximum. Note also that because f_{tot} is a function of $\tau_{\lambda AB}$ and τ_r (and $C_p T_0/h$) only, the maximum in thrust corresponds to the minimum in specific fuel consumption!

The analytical expression for the compressor pressure ratio giving maximum thrust is obtained by noting with Eq. (5.54),

$$\frac{\partial \tau_c\tau_t}{\partial \tau_c} = \frac{\partial}{\partial \tau_c}\left[\tau_c - \frac{\tau_r}{\tau_\lambda}\left(\tau_c^2 - \tau_c\right)\right]$$

Hence at the maximum

$$1 - (\tau_r/\tau_\lambda)(2\tau_c - 1) = 0 \qquad \text{or} \qquad (\tau_c)_{\text{max thrust}} = (\tau_\lambda + \tau_r)/2\tau_r$$

and

$$\pi_{c\text{ max thrust}} = \left[(\tau_\lambda + \tau_r)/2\tau_r\right]^{\gamma/(\gamma-1)} \qquad (5.58)$$

In the example given above,

$$\pi_{c \text{ max thrust}} = [(7 + 1.8)/3.6]^{3.5} = 22.84$$

It would appear then that such a value for the compressor pressure ratio would be near a suitable compromise for an aircraft that was to cruise efficiently with no afterburning, but was to engage in combat with high thrust and relatively efficient afterburning fuel consumption.

5.12 The Turbofan with Separate Exhaust Streams

Cycle Analysis of the Ideal Turbofan with Separate Exhaust Streams

The methodology of cycle analysis, as described in Sec. 5.8, will again be applied. When applying the cycle analysis, the thrust of both the primary (core engine) and secondary (fan) streams must be accounted for and the turbine power output must now be equated to the power input to both the fan and compressor. It is again assumed that all component efficiencies are perfect, that the gas is calorically perfect, that the fuel-to-air ratio in the combustor is much less than unity, and that the exit static pressures of both the primary and secondary streams are equal to the ambient pressure. The nomenclature and notation are as in Fig. 5.17 and Sec. 5.2.

The total thrust may be written as the sum of the thrust contributions of the two streams to give

$$F = \dot{m}_c(u_9 - u_0) + \dot{m}_F(u_{9'} - u_0)$$

or

$$\frac{F}{\dot{m}_c + \dot{m}_F} = \frac{a_0 M_0}{1 + \alpha}\left[\left(\frac{u_9}{u_0} - 1\right) + \alpha\left(\frac{u_{9'}}{u_0} - 1\right)\right] \qquad (5.59)$$

where $\dot{m}_c$ is the mass flow rate of the primary stream, $\dot{m}_F$ the mass flow rate of the secondary stream, and $\alpha = \dot{m}_F/\dot{m}_c$ the bypass ratio.

Secondary stream.   Again,

$$(u_{9'}/u_0)^2 = (T_{9'}/T_0)(M_{9'}/M_0)^2 \qquad (5.60)$$

Thus,

$$T_{t_{9'}} = T_{9'}\left(1 + \frac{\gamma - 1}{2}M_{9'}^2\right) = T_0\tau_r\tau_{c'} \qquad (5.61)$$

$$p_{t_{9'}} = p_{9'}\left(1 + \frac{\gamma - 1}{2}M_{9'}^2\right)^{\gamma/(\gamma - 1)} = p_0\pi_r\pi_{c'} \qquad (5.62)$$

With Eqs. (5.1), (5.4), and (5.62),

$$1 + [(\gamma - 1)/2] M_{9'}^2 = \tau_r \tau_{c'} \qquad (5.63)$$

So that

$$T_{9'} = T_0 \qquad (5.64)$$

This result could have been obtained more directly by noting $p_{9'} = p_0$ and $s_{9'} = s_0$, hence $T_{9'} = T_0$. Equations (5.60), (5.63), and (5.64) then give

$$M_0 \left(\frac{u_{9'}}{u_0} - 1 \right) = \left[\frac{2}{\gamma - 1} (\tau_r \tau_{c'} - 1) \right]^{\frac{1}{2}} - M_0 \qquad (5.65)$$

Primary stream. Here, the relationships for temperatures and pressures are exactly as previously obtained for the turbojet (Sec. 5.8), so

$$M_0 \left(\frac{u_9}{u_0} - 1 \right) = \left[\frac{2}{\gamma - 1} \frac{\tau_\lambda}{\tau_r \tau_c} (\tau_r \tau_c \tau_t - 1) \right]^{\frac{1}{2}} - M_0 \qquad (5.66)$$

Power balance. For this ideal cycle, the power output from the turbine will just equal the power input to the fan and compressor. Hence,

$$\dot{m}_c C_p (T_{t_4} - T_{t_5}) = \dot{m}_c C_p (T_{t_3} - T_{t_2}) + \dot{m}_F C_p (T_{t_{3'}} - T_{t_2})$$

or

$$\tau_\lambda (1 - \tau_t) = \tau_r (\tau_c - 1) + \alpha \tau_r (\tau_{c'} - 1)$$

Thus

$$\tau_t = 1 - (\tau_r / \tau_\lambda)[(\tau_c - 1) + \alpha(\tau_{c'} - 1)] \qquad (5.67)$$

Specific fuel consumption. An enthalpy balance across the combustor gives

$$\dot{m}_f h = \dot{m}_c C_p (T_{t_4} - T_{t_3})$$

or

$$f = \dot{m}_f / \dot{m}_c = (C_p T_0 / h)(\tau_\lambda - \tau_r \tau_c) \qquad (5.68)$$

The specific fuel consumption then follows from

$$S = \frac{\dot{m}_f}{F} (10^6) = \frac{\dot{m}_f}{\dot{m}_c} \frac{1}{(\dot{m}_c + \dot{m}_F)/\dot{m}_c} \frac{1}{F/(\dot{m}_c + \dot{m}_F)} (10^6)$$

or

$$S = \frac{f(10^6)}{(1 + \alpha)[F/(\dot{m}_c + \dot{m}_F)]} \tag{5.69}$$

Combination of Eqs. (5.59) and (5.65–5.69) then gives the following summary of equations.

Summary of the Equations—Ideal Turbofan, Bypass Ratio Prescribed

Inputs: $T_0(\text{K})\ [^\circ\text{R}]$, γ, $h(\text{J/kg})\ [\text{Btu/lbm}]$

$$C_p\ (\text{J/kg} \cdot \text{K})\left[\frac{\text{Btu}}{\text{lbm} \cdot ^\circ\text{R}}\right],\ \tau_\lambda,\ \pi_c,\ \pi_{c'},\ M_0,\ \alpha$$

Outputs: $$\frac{F}{\dot{m}_c + \dot{m}_F}\left(\frac{\text{N} \cdot \text{s}}{\text{kg}}\right)\left[\frac{F}{g_0(\dot{m}_c + \dot{m}_F)},\ \frac{\text{lbf}}{\text{lbm/s}}\right]$$

$$S\left(\frac{\text{mg}}{\text{N} \cdot \text{s}}\right)\left[\frac{\text{lbm fuel/h}}{\text{lbf thrust}}\right]$$

$$f\frac{\text{kg fuel/s}}{\text{kg primary air/s}}$$

Equations:

$$R = \frac{\gamma - 1}{\gamma}C_p\ \text{m}^2/\text{s}^2 \cdot \text{K}\left[R = \frac{\gamma - 1}{\gamma}C_p(2.505)(10^4)\frac{\text{ft}^2}{\text{s}^2 \cdot ^\circ\text{R}}\right] \tag{5.70}$$

$$a_0 = \sqrt{\gamma R T_0}\ \text{m/s [ft/s]} \tag{5.71}$$

$$\tau_r = 1 + \frac{\gamma - 1}{2}M_0^2 \tag{5.72}$$

$$\tau_c = \pi_c^{(\gamma - 1)/\gamma} \tag{5.73}$$

$$\tau_{c'} = \pi_{c'}^{(\gamma - 1)/\gamma} \tag{5.74}$$

$$\tau_t = 1 - (\tau_r/\tau_\lambda)[(\tau_c - 1) + \alpha(\tau_{c'} - 1)]^{\;*} \tag{5.75}$$

$$\frac{F}{\dot{m}_c + \dot{m}_F} = \frac{a_0}{1 + \alpha}\left\{\left[\frac{2}{\gamma - 1}\frac{\tau_\lambda}{\tau_r\tau_c}(\tau_r\tau_c\tau_t - 1)\right]^{\frac{1}{2}}\right.$$

$$\left. - M_0 + \alpha\left(\left[\frac{2}{\gamma - 1}(\tau_r\tau_{c'} - 1)\right]^{\frac{1}{2}} - M_0\right)\right\} \tag{5.76}$$

(To obtain British units of lbf/lbm/s, replace a_0 in meters/second with $a_0/32.174$ where a_0 should be given in feet/second.)

$$f = (C_p T_0/h)(\tau_\lambda - \tau_r \tau_c) \tag{5.77}$$

$$S = \frac{f(10^6)}{(1+\alpha)[F/(\dot{m}_c + \dot{m}_F)]} \qquad \left[S = \frac{3600f}{(1+\alpha)[F/g_0(\dot{m}_c + \dot{m}_F)]} \right] \tag{5.78}$$

Bypass Ratio for Minimum Specific Fuel Consumption

It would be desirable to select the bypass ratio α to give the minimum specific fuel consumption possible for the given prescribed operating conditions (T_0, M_0), design limits (τ_λ), and design choices $(\pi_c, \pi_{c'})$.

It is evident from Eqs. (5.76–5.78) that the minimum value for S will occur at the maximum value of the expression

$$\frac{(1+\alpha)}{a_0} \frac{F}{\dot{m}_c + \dot{m}_F} = M_0 \frac{u_9}{u_0} - M_0 + \alpha\left(M_0 \frac{u_{9'}}{u_0} - M_0 \right)$$

Taking the derivative of this expression with α, it follows with Eqs. (5.65) and (5.66) that

$$-\frac{\partial[M_0(u_9/u_0)]^2}{\partial\alpha} = 2M_0\frac{u_9}{u_0}\left(M_0\frac{u_{9'}}{u_0} - M_0 \right) \tag{5.79}$$

But with Eqs. (5.66) and (5.67)

$$-\frac{\partial[M_0(u_9/u_0)]^2}{\partial\alpha} = \frac{2}{\gamma-1}\tau_r(\tau_{c'} - 1)$$

$$= \frac{2}{\gamma-1}(\tau_r\tau_{c'} - 1) - \frac{2}{\gamma-1}(\tau_r - 1)$$

$$= \left(M_0\frac{u_{9'}}{u_0} \right)^2 - M_0^2 \tag{5.80}$$

Thus, combining Eqs. (5.79) and (5.80),

$$M_0\frac{u_9}{u_0} = \frac{1}{2}\left(M_0\frac{u_{9'}}{u_0} + M_0 \right) \tag{5.81}$$

and

$$2\left(M_0\frac{u_9}{u_0} - M_0 \right) = \left(M_0\frac{u_{9'}}{u_0} - M_0 \right) \tag{5.82}$$

This latter form reveals that, when the bypass ratio is such as to give minimum specific fuel consumption, the thrust per mass per second of the core stream is just one-half that of the fan stream. This somewhat surprising result follows primarily because, when the efficiency of energy transmission between the two streams is very high, the propulsive efficiency is improved by expending most of the energy in providing momentum to the denser (cooler) fan stream. It should be noted that when component losses are present, the optimal thrust per mass per second of the core stream is a much higher fraction of the fan stream thrust.

The turbine temperature ratio corresponding to this optimal case τ_t^* follows from Eqs. (5.65), (5.66), and (5.81) to give

$$\tau_t^* = \frac{1}{\tau_r \tau_c} + \frac{1}{4\tau_\lambda}\left[(\tau_r \tau_{c'} - 1)^{\frac{1}{2}} + (\tau_r - 1)^{\frac{1}{2}}\right]^2 \tag{5.83}$$

The related value of the bypass ratio α^* follows from Eq. (5.75) to give

$$\alpha^* = \frac{1}{\tau_{c'} - 1}\left[\frac{\tau_\lambda}{\tau_r}(1 - \tau_t^*) - (\tau_c - 1)\right] \tag{5.84}$$

The performance variables for this optimum case may now be obtained with the summary equations (5.70–5.78), except that Eq. (5.83) would replace Eq. (5.75) and α^* would be calculated from Eq. (5.84) rather than input.

Example Results—Ideal Turbofan with Separate Exhaust Streams

An example of the use of the performance equations is the effect upon the performance parameters and the optimal bypass ratio of variations in the bypass pressure ratio. (See Figs. 5.33 and 5.34.)

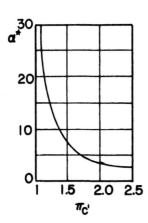

Fig. 5.33   Optimal bypass ratio vs bypass pressure ratio.

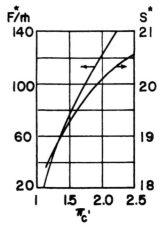

Fig. 5.34   Specific thrust and specific fuel consumption vs bypass pressure ratio.

Conditions assumed for these calculations were $\gamma = 1.4$, $T_0 = 233.3$ K, $h = 4.4194(10^7)$ J/kg, $C_p = 1004.9$ J/kg·K, $\pi_c = 20$, $\tau_\lambda = 6.5$, $M_0 = 2$. It is apparent that a tradeoff between high thrust per frontal area and low specific fuel consumption occurs. This clearly reflects the increase of propulsive efficiency with a decrease in fan exit velocity (low $\pi_{c'}$). The enormous optimal bypass ratios occurring for the very low bypass pressure ratios do not in fact occur when losses are considered (see Chap. 7). The effect of losses is to much reduce the optimal fraction of energy to be supplied to the bypass stream.

The variations of α^*, S^*, and $F^*/(\dot{m}_c + \dot{m}_F)$ with flight Mach number are shown in Figs. 5.35 and 5.36. The conditions assumed are as above with the additional value $\pi_{c'} = 2.0$.

The strong effect of flight at increasing Mach number is evident in these curves. Thus, because of the increase in the entering enthalpy of the fan and core streams, the work interaction per mass required to supply the needed fan and compressor pressure ratios increases greatly. As a result, the turbine, which has a fixed entry enthalpy, cannot supply the necessary energy to drive a large bypass ratio fan, and the bypass ratio must be decreased.

Figures 5.37 and 5.38 indicate that increasing τ_λ has a similar effect upon the optimal bypass ratio and specific thrust as does reducing the flight Mach number. (Conditions assumed are as above with $M_0 = 2$.) This is true, of course, because an increase in τ_λ gives an increase in turbine power capability. The slight increase in specific fuel consumption with τ_λ would appear to negate the thrust advantages of going to higher turbine temperatures. However, it should be noted that not only would the engine be smaller (particularly the core engine), but the specific fuel consumption will actually tend to decrease with increasing τ_λ when the component losses are included.

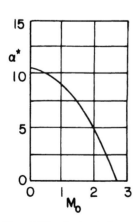

Fig. 5.35 Optimal bypass ratio vs flight Mach number.

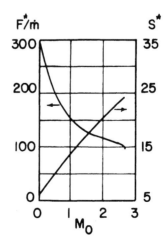

Fig. 5.36 Specific thrust and specific fuel consumption vs flight Mach number.

As a final example of the ideal turbofan with separate exhaust streams, consider the effects of varying the bypass ratio from the optimum value. Conditions are as assumed in the above examples and the results are illustrated in Fig. 5.39.

Figure 5.39 illustrates that a truly optimal choice of bypass ratio might be other than that leading to the minimum specific fuel consumption. Thus, for example, note that by selecting $\alpha = 3$ rather than $\alpha^* = 3.91$, a 21% increase in specific thrust can be obtained at a penalty of a 1.5% increase in specific fuel consumption. When the engine size, weight, and cowl diameter are all

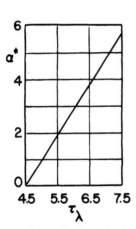

Fig. 5.37 Optimal bypass ratio vs τ_λ.

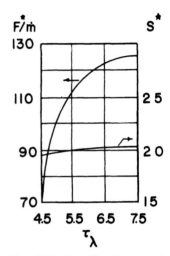

Fig. 5.38 Specific thrust and specific fuel consumption vs τ_λ.

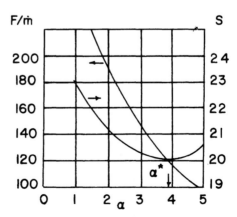

Fig. 5.39 Specific thrust and specific fuel consumption vs bypass ratio.

considered for the installation effects, it is probable that a bypass ratio lower than that corresponding to α^* would be selected.

5.13 The Ideal Turbofan with Mixed Exhaust Streams

In many installations, particularly aircraft with body-mounted engines, it is suitable and convenient to duct the primary and secondary streams through a common exit nozzle. In the event that little mixing of the streams occurs, the analysis in Sec. 5.12 would remain valid. In many applications, however, "forced mixers" are used to greatly enhance the mixing of the streams. Several benefits may accrue from mixing the streams, such as improvements in the performance parameters and reductions in the exhaust jet noise.

In order to analyze the behavior of a turbofan with ideal stream mixing, the presence of an "ideal constant-area mixer" is assumed. An ideal constant-area mixer is defined as a constant-area mixer with no sidewall friction. Analysis of such a device provides the outlet stagnation conditions as a function of the two sets of inlet conditions. When the outlet conditions are known, the performance of the engine can be determined.

Cycle Analysis of the Ideal Turbofan with Mixed Exhaust Streams

This turbofan is shown in Fig. 5.40. Again assume that all efficiencies are perfect, that the nozzle exit pressure is equal to the ambient pressure, that the gas is calorically perfect, and that the combustor fuel-to-air ratio is much less than unity. The specific thrust may hence be written

$$F/(\dot{m}_c + \dot{m}_F) = a_0 M_0 [(u_9/u_0) - 1] \tag{5.85}$$

Also

$$(u_9/u_0)^2 = (T_9/T_0)(M_9/M_0)^2 \tag{5.86}$$

and

$$T_{t_9} = T_9\left(1 + \frac{\gamma - 1}{2} M_9^2\right) \tag{5.87}$$

$$p_{t_9} = p_9\left(1 + \frac{\gamma - 1}{2} M_9^2\right)^{\gamma/(\gamma - 1)} = p_0 \pi_r \pi_{c'}\left(\frac{p_{t_9}}{p_{t_{3'}}}\right) \tag{5.88}$$

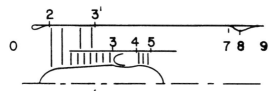

Fig. 5.40 The ideal turbofan with mixed exhaust streams.

An enthalpy balance gives

$$\dot{m}_c C_p T_{t_5} + \dot{m}_F C_p T_{t_{3'}} = (\dot{m}_c + \dot{m}_F) C_p T_{t_9}$$

or

$$T_{t_9}/T_0 = [1/(1+\alpha)](\tau_\lambda \tau_t + \alpha \tau_r \tau_{c'}) \tag{5.89}$$

Combination of Eqs. (5.85–5.89) then gives

$$\frac{F}{\dot{m}_c + \dot{m}_F} = a_0 \left\{ \left[\frac{2/(\gamma-1)}{1+\alpha} \left(1 - \frac{1}{\tau_r \tau_{c'}(p_{t_9}/p_{t_{3'}})^{(\gamma-1)/\gamma}} \right) (\tau_\lambda \tau_t + \alpha \tau_r \tau_{c'}) \right]^{\frac{1}{2}} - M_0 \right\} \tag{5.90}$$

The power balance between the compressor, fan, and turbine remains unchanged, so τ_t is still given from Eq. (5.67). The following will obtain the ratio p_{t_9}/p_{t_5}, so note here

$$\frac{p_{t_9}}{p_{t_{3'}}} = \frac{p_{t_9}}{p_{t_5}} \frac{p_{t_5}}{p_{t_{3'}}} = \frac{p_{t_9}}{p_{t_5}} \frac{\pi_c \pi_t}{\pi_{c'}} \tag{5.91}$$

The expressions for f and S remain as in Eqs. (5.68) and (5.69). These equations will be summarized following the next section.

The Ideal Constant-Area Mixer

The constant-area mixer is shown in Fig. 5.41. Consider p_{t_5}, T_{t_5}, M_5, $p_{t_{3'}}$, $T_{t_{3'}}$, and α to be prescribed. The common static pressure at the splitter p can then be obtained directly from

$$\frac{p_{t_5}}{p} = \left(1 + \frac{\gamma-1}{2} M_5^2 \right)^{\gamma/(\gamma-1)} \tag{5.92}$$

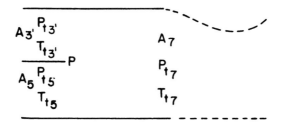

Fig. 5.41   The constant-area mixer.

from which

$$M_{3'}^2 = \frac{2}{\gamma - 1}\left[\left(\frac{p_{t_{3'}}}{p}\right)^{(\gamma-1)/\gamma} - 1\right] = \frac{2}{\gamma - 1}\left[\left(\frac{p_{t_{3'}}}{p_{t_5}}\right)^{(\gamma-1)/\gamma}\left(1 + \frac{\gamma - 1}{2}M_5^2\right) - 1\right]$$

(5.93)

Equation (2.102) for the mass flows may be utilized to give

$$\frac{A_{3'}}{A_5} = \alpha\sqrt{\frac{T_{t_{3'}}}{T_{t_5}}}\frac{M_5}{M_{3'}}\left(\frac{p_{t_{3'}}}{p_{t_5}}\right)^{-[(\gamma-1)/2\gamma]}$$

(5.94)

The momentum equation may be written as

$$A_5\left(p_5 + \rho_5 u_5^2\right) + A_{3'}\left(p_{3'} + \rho_{3'}u_{3'}^2\right) = A_7\left(p_7 + \rho_7 u_7^2\right)$$

or

$$A_5 p_5\left(1 + \gamma M_5^2\right) + A_{3'}p_{3'}\left(1 + \gamma M_{3'}^2\right) = A_7 p_7\left(1 + \gamma M_7^2\right)$$

(5.95)

The equation for the mass flow [Eq. (2.102)] may be written in the form

$$p_i A_i = \sqrt{\frac{R}{\gamma}}\sqrt{T_{t_i}}\frac{1}{M_i}\left(1 + \frac{\gamma - 1}{2}M_i^2\right)^{-\frac{1}{2}}\dot{m}_i \qquad (i = 3', 5, \text{ or } 7)$$

(5.96)

Noting from the enthalpy balance that

$$\frac{T_{t_7}}{T_{t_5}} = \frac{1 + \alpha\left(T_{t_{3'}}/T_{t_5}\right)}{1 + \alpha}$$

(5.97)

an equation for M_7 can be obtained by combining Eqs. (5.95–5.97) and may be written in the form

$$f = f(M_7) = (1 + \alpha)\left(1 + \alpha\frac{T_{t_{3'}}}{T_{t_5}}\right)\left[\frac{1}{\sqrt{f(M_5)}} + \frac{\alpha\sqrt{T_{t_{3'}}/T_{t_5}}}{\sqrt{f(M_{3'})}}\right]^{-2}$$

where

$$f(M) \equiv M^2\left(1 + \frac{\gamma - 1}{2}M^2\right)\left(1 + \gamma M^2\right)^{-2}$$

(5.98)

This is once again of the same functional form for the Mach number as occurred in the solution for the heat interaction at constant area in Sec.

2.18, so

$$M_7^2 = \frac{2f}{1 - 2\gamma f + \sqrt{1 - 2(\gamma + 1)f}} \tag{5.99}$$

Following solution for M_7, the other downstream variables follow directly. Thus, Eqs. (5.96) and (5.97) may be combined to give

$$\frac{p_{t_7}}{p_{t_5}} = \left[(1 + \alpha)\left(1 + \alpha \frac{T_{t_{3'}}}{T_{t_5}}\right)\right]^{\frac{1}{2}} \left(\frac{1 + \frac{\gamma - 1}{2}M_7^2}{1 + \frac{\gamma - 1}{2}M_5^2}\right)^{(\gamma + 1)/2(\gamma - 1)} \frac{M_5}{M_7} \frac{1}{1 + \frac{A_{3'}}{A_5}} \tag{5.100}$$

Summary of the Equations—Ideal Constant-Area Mixer

Inputs: $p_{t_{3'}}/p_{t_5}, T_{t_{3'}}/T_{t_5}, M_5, \alpha$

Outputs: $p_{t_7}/p_{t_5}, M_7$

Equations:

$$M_{3'}^2 = \frac{2}{\gamma - 1}\left[\left(\frac{p_{t_{3'}}}{p_{t_5}}\right)^{(\gamma - 1)/\gamma}\left(1 + \frac{\gamma - 1}{2}M_5^2\right) - 1\right] \tag{5.101}$$

$$\frac{A_{3'}}{A_5} = \alpha\sqrt{\frac{T_{t_{3'}}}{T_{t_5}}}\frac{M_5}{M_{3'}}\left(\frac{p_{t_{3'}}}{p_{t_5}}\right)^{-[(\gamma - 1)/2\gamma]} \tag{5.102}$$

$$f = (1 + \alpha)\left[1 + \alpha\left(\frac{T_{t_{3'}}}{T_{t_5}}\right)\right]\left[\frac{1}{\sqrt{f(M_5)}} + \frac{\alpha\sqrt{T_{t_{3'}}/T_{t_5}}}{\sqrt{f(M_{3'})}}\right]^{-2} \tag{5.103}$$

where

$$f(M) \equiv M^2\left(1 + \frac{\gamma - 1}{2}M^2\right)(1 + \gamma M^2)^{-2}$$

$$M_7 = \left[\frac{2f}{1 - 2\gamma f + \sqrt{1 - 2(\gamma + 1)f}}\right]^{\frac{1}{2}} \tag{5.104}$$

$$\frac{p_{t_7}}{p_{t_5}} = \frac{\left[(1 + \alpha)\left(1 + \alpha\frac{T_{t_{3'}}}{T_{t_5}}\right)\right]^{\frac{1}{2}}}{1 + \frac{A_{3'}}{A_5}}\frac{M_5}{M_7}\left(\frac{1 + \frac{\gamma - 1}{2}M_7^2}{1 + \frac{\gamma - 1}{2}M_5^2}\right)^{(\gamma + 1)/2(\gamma - 1)} \tag{5.105}$$

Summary of the Equations—Ideal Turbofan with Mixed Exhaust Streams

Inputs: $T_0(K) [°R], \gamma, h(J/kg) [Btu/lbm]$

$C_p(J/kg \cdot K)[Btu/lbm \cdot °R], \tau_\lambda, \pi_c, \pi_{c'}, M_0, \alpha, M_5$

Outputs: $\dfrac{F}{\dot{m}_1 + \dot{m}_2}\left(\dfrac{N \cdot s}{kg}\right) \quad \left[\dfrac{F}{g_0(\dot{m}_1 + \dot{m}_2)}, \dfrac{lbf}{lbm/s}\right]$

$S\left(\dfrac{mg}{N \cdot s}\right)\left[\dfrac{lbm\ fuel/h}{lbf\ thrust}\right], \quad f\dfrac{kg\ fuel/s}{kg\ primary\ air/s}$

Equations:

$$R = \frac{\gamma - 1}{\gamma}C_p\ m^2/s^2 \cdot K \quad \left[R = \frac{\gamma - 1}{\gamma}C_p(2.505)(10^4)\ ft^2/s^2 \cdot °R\right]$$

$$(5.106)$$

$$a_0 = \sqrt{\gamma R T_0}\ m/s\ [ft/s] \qquad (5.107)$$

$$\tau_r = 1 + \frac{\gamma - 1}{2}M_0^2 \qquad (5.108)$$

$$\tau_c = \pi_c^{(\gamma-1)/\gamma} \qquad (5.109)$$

$$\tau_{c'} = \pi_{c'}^{(\gamma-1)/\gamma} \qquad (5.110)$$

$$\tau_t = 1 - (\tau_r/\tau_\lambda)[(\tau_c - 1) + \alpha(\tau_{c'} - 1)] \qquad (5.111)$$

$$\pi_t = \tau_t^{\gamma/(\gamma-1)} \qquad (5.112)$$

$[P_{t_9}/P_{t_5}$ is then obtained from the equations for the ideal constant-area mixer. Note that $P_{t_9} = P_{t_7}$, $P_{t_3'}/P_{t_5} = \pi_{c'}/\pi_c\pi_t$, and $T_{t_3'}/T_{t_5} = \tau_r\tau_{c'}/\tau_\lambda\tau_t$].

$$\frac{P_{t_9}}{P_{t_3'}} = \frac{P_{t_9}}{P_{t_5}}\frac{\pi_c\pi_t}{\pi_{c'}} \qquad (5.113)$$

$$\frac{F}{\dot{m}_c + \dot{m}_F} = a_0\left\{\left[\frac{2/(\gamma-1)}{1+\alpha}\left[1 - \frac{1}{\tau_r\tau_{c'}\left(P_{t_9}/P_{t_3'}\right)^{(\gamma-1)/\gamma}}\right](\tau_\lambda\tau_t + \alpha\tau_r\tau_{c'})\right]^{\frac{1}{2}} - M_0\right\}$$

$$(5.114)$$

(To obtain British units of lbf/lbm/s, replace a_0 in meters/second with $a_0/32.174$ where a_0 should be given in feet/second.)

$$f = (C_p T_0/h)(\tau_\lambda - \tau_r \tau_c) \tag{5.115}$$

$$S = \frac{f(10^6)}{(1 + \alpha)[F/(\dot{m}_c + \dot{m}_F)]} \qquad \left[S = \frac{3600f}{(1 + \alpha)\{F/[g_0(\dot{m}_c + \dot{m}_F)]\}} \right] \tag{5.116}$$

Example Results—Ideal Constant-Area Mixer

In order to compare the performance of fully mixed streams and separate streams, consider the "gross thrust" capability of the two cases. Gross thrust is the thrust that would occur if all the momentum of the exit streams contributed entirely to thrust. Thus

$$F_{un} = \dot{m}_c u_{e_1} + \dot{m}_F u_{e_2} \qquad \text{for unmixed streams} \tag{5.117}$$

$$F_{mix} = (\dot{m}_c + \dot{m}_F) u_e \qquad \text{for mixed streams} \tag{5.118}$$

Noting that for all cases (Sec. 2.15)

$$u_e^2 = \frac{2\gamma}{\gamma - 1} RT_t \left[1 - \left(\frac{p_e}{p_t} \right)^{(\gamma-1)/\gamma} \right] \tag{5.119}$$

it follows immediately that

$$\frac{F_{un}}{F_{mix}} = \frac{\left[1 - \left(\frac{p_e}{p_{t_5}} \right)^{(\gamma-1)/\gamma} \right]^{\frac{1}{2}} + \alpha \sqrt{\frac{T_{t_{3'}}}{T_{t_5}}} \left[1 - \left(\frac{p_e}{p_{t_{3'}}} \right)^{(\gamma-1)/\gamma} \right]^{\frac{1}{2}}}{\left\{ (1 + \alpha) \left(1 + \alpha \frac{T_{t_{3'}}}{T_{t_5}} \right) \left[1 - \left(\frac{p_e}{p_{t_7}} \right)^{(\gamma-1)/\gamma} \right] \right\}^{\frac{1}{2}}} \tag{5.120}$$

Equations (5.101–5.105) and (5.120) allow direct calculation and comparison of the ideal constant-area mixer performance with nonmixed stream performance. As an example, consider a mixer with $\alpha = 2$, $M_5 = 0.5$, and $\gamma = 1.4$, and investigate the effects of varying the temperature ratio $T_{t_{3'}}/T_{t_5}$, stream pressure ratio $p_{t_{3'}}/p_{t_5}$, and exit pressure ratio p_e/p_{t_5}.

Several trends are evident in Figs. 5.42 and 5.43. Thus, note that a mixing benefit occurs only when there is a difference in stagnation temperatures in the streams. This is because the stagnation pressure losses identified with mixing will cause a decrease in thrust unless one of the streams has a higher stagnation enthalpy, with a consequent possible benefit of equalizing the

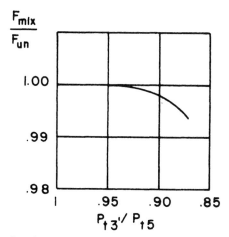

Fig. 5.42 Effect of variation in stream pressure ratio, ideal constant-area mixer.

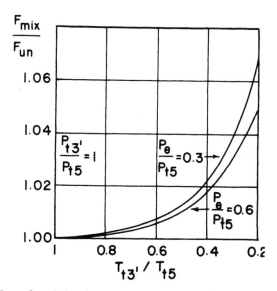

Fig. 5.43 Effect of variation in stream temperature ratio and exit pressure ratio, ideal constant-area mixer.

enthalpies. It is to be noted also that when the exit pressure is further reduced, the penalty identified with stagnation pressure loss in mixing is reduced, because the overall pressure ratio is so large.

It can be surmised, then, that real mixers (including viscous losses) may show signs of performance gains when used on nozzles with large expansion ratios and in which substantial temperature differences between the two

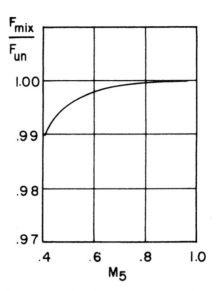

Fig. 5.44 Effect of variation in inlet Mach number, ideal constant-area mixer.

streams exist. It should be noted that "forced mixers" are being considered for transport aircraft, where the promise of performance benefits is not substantial. In such cases, the prime motivation for the use of mixers arises from the hope of substantially reducing the jet noise.

As a final example, the effect of varying the inlet Mach numbers of the mixing streams is shown in Fig. 5.44. Consider the case $\alpha = 2$, $T_{t_3}/T_{t_5} = 1$, $p_{t_3}/p_{t_5} = 0.9$, and $p_e/p_{t_5} = 0.5$. Note here that increasing the Mach number at entry to the mixer reduces the losses. This might at first appear anomalous because one might expect interactions at low Mach number to be less vigorous. It is apparent, however, that as the Mach number M_5 decreases, the pressure at the splitter plate increases, leading to a lower value of $M_{3'}$. Thus the relative velocity difference increases as M_5 decreases, leading to the greater mixing losses predicted by the ideal analysis. It should be noted that when skin-friction losses are included, there will be an opposing effect because of the Mach number squared dependence of the skin-friction losses (see Secs. 2.17 and 2.18). The design of a real mixer will involve the optimal choice of the design parameters, including not only the effects just discussed, but also consideration of the many installation effects.

Example Results—Ideal Turbofan with Mixed Exhaust Streams

To compare the performance of turbofans with and without mixing, consider a turbofan with bypass ratio α equal to two. Other conditions will be as those given for Fig. 5.39, that is: $\gamma = 1.4$, $T_0 = 420°R$, $h = 19,000$ Btu/lbm, $C_p = 0.24$ Btu/lbm · °R, $\pi_c = 20$, $\pi_{c'} = 2$, $\tau_\lambda = 6.5$, $M_0 = 2$.

The analysis utilized to obtain Fig. 5.39 yielded $\tau_r = 1.8$, $\tau_c = 2.3535$, $\tau_{c'} = 1.2190$, $\tau_t = 0.50387$, and $\pi_t = 0.090807$. These values utilized with Eqs.

(5.101–5.105), with $M_5 = 0.5$, then give $p_{t_7}/p_{t_5} = 1.0564$. Finally, Eqs. (5.106–5.116) yield for the turbofan with mixed streams

$$S = 0.741 \text{ lbm/h/lbf}$$

$$F/[g_0(\dot{m}_c + \dot{m}_F)] = 19.46 \text{ lbf/lbm/s}$$

These values are to be compared with the following values for the unmixed case

$$S = 0.754 \text{ lbm/h/lbf}$$

$$F/[g_0(\dot{m}_c + \dot{m}_F)] = 19.12 \text{ lbf/lbm/s}$$

Thus the ideal mixer, in this case, provides a thrust and specific fuel consumption improvement of almost 2%.

5.14   The Ideal Constant-Pressure Mixer

The behavior of a mixer designed with an area variation that gives constant-pressure flow will be examined in this section. It is again assumed that there is no sidewall friction. With these assumptions solution for the downstream variables becomes trivial, because the pressure is already known and the fluid momentum is conserved. As a result of the simplicity of the solution, analytical determination of the inlet conditions that lead to the maximum possible outlet stagnation pressure is straightforward.

The geometry to be considered is shown in Fig. 5.45. Because the momentum is conserved in this device, it is convenient to consider the behavior of the velocity rather than the Mach number and so a dimensionless velocity is introduced for convenience, along with other convenient groupings. Thus define,

$$V_i = u_i / \sqrt{\gamma R T_{t_5}} \qquad (i = 3', 5, 7)$$

$$\tau = T_{t_3} / T_{t_5}$$

$$S_i = (p_{t_i}/p_{t_5})^{(\gamma-1)/\gamma}$$

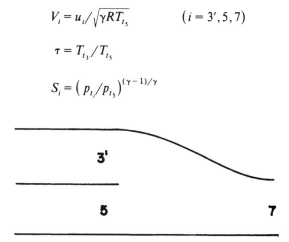

Fig. 5.45. Constant-pressure mixer.

The momentum equation gives immediately

$$V_7 = \frac{V_5 + \alpha V_{3'}}{1 + \alpha} \tag{5.121}$$

and conservation of energy gives

$$\frac{T_{t_7}}{T_{t_5}} = \frac{1 + \alpha \tau}{1 + \alpha} \tag{5.122}$$

In general

$$C_p T_{t_i} = C_p T_i + u_i^2/2$$

So that

$$\left(\frac{p}{p_{t_i}}\right)^{(\gamma-1)/\gamma} = \frac{T_i}{T_{t_i}} = 1 - \frac{\gamma - 1}{2}\frac{T_{t_5}}{T_{t_i}}V_i^2 \qquad (\text{Note } p = p_{3'} = p_5 = p_7) \tag{5.123}$$

It hence follows that

$$S_{3'} = \left(\frac{p/p_{t_5}}{p/p_{t_{3'}}}\right)^{(\gamma-1)/\gamma} = \frac{1 - [(\gamma-1)/2]V_5^2}{1 - [(\gamma-1)/2](1/\tau)V_{3'}^2} \tag{5.124}$$

Similarly, using Eqs. (5.121–5.123), there follows

$$S_7 = \frac{1 - \dfrac{\gamma - 1}{2}V_5^2}{1 - \dfrac{\gamma - 1}{2}\dfrac{(V_5 + \alpha V_{3'})^2}{(1 + \alpha)(1 + \alpha\tau)}} \tag{5.125}$$

Area Relationships

In general,

$$\dot{m}_i = \frac{p}{R}\sqrt{\gamma R T_{t_5}}\frac{1}{T_{t_i}}\frac{T_{t_i}}{T_i}V_i A_i \tag{5.126}$$

Combination of Eqs. (5.123–5.126) then gives

$$\frac{A_{3'}}{A_5} = \alpha\tau S_{3'}^{-1}\frac{V_5}{V_{3'}} \tag{5.127}$$

$$\frac{A_7}{A_5} = (1 + \alpha)(1 + \alpha\tau)S_7^{-1}\frac{V_5}{V_5 + \alpha V_{3'}} \tag{5.128}$$

Optimal Mixer Entrance Conditions

These equations allow direct solution for the outlet conditions as well as the required area variation. It is of interest to note, however, that combination of Eqs. (5.124) and (5.125) gives S_7 in terms of V_5 and the prescribed upstream stagnation conditions. This allows the analytical determination of the engine stream dimensionless velocity that gives the maximum outlet stagnation pressure.

Carrying out the required manipulations (Problem 5.15), the conditions corresponding to the maxima are found to be

$$V_{3'}/V_5 = 1 \qquad \text{or} \qquad V_{3'}/V_5 = \tau \qquad (5.129)$$

The correct root is not identified with the presence of a maximum or minimum, but rather with the possibility of the gas streams attaining the desired ratio. Thus, the maximum imaginable value for V_5 occurs as $M_5 \to \infty$. At this condition, $V_5 = \sqrt{2/(\gamma - 1)}$.

Utilizing Eqs. (5.124) and (5.129) and denoting by a subscript a the case where $V_{3'}/V_5 = 1$ and by a subscript b the case where $V_{3'}/V_5 = \tau$, it follows that

$$V_{5a}^2 = \frac{2}{\gamma - 1}\left(\frac{S_{3'} - 1}{S_{3'}/\tau - 1}\right) \qquad \text{and} \qquad V_{5b}^2 = \frac{2}{\gamma - 1}\left(\frac{S_{3'} - 1}{\tau S_{3'} - 1}\right)$$

$$(5.130)$$

It thus follows that physically allowable solutions exist for

$$\frac{V_{3'}}{V_5} = 1 \quad \text{when} \quad \frac{p_{t_{3'}}}{p_{t_5}} > 1, \tau < 1 \quad \left[\text{or when } \frac{p_{t_{3'}}}{p_{t_5}} < 1, \tau > 1\right]$$

$$\frac{V_{3'}}{V_5} = \tau \quad \text{when} \quad \frac{p_{t_{3'}}}{p_{t_5}} < 1, \tau < 1 \quad \left[\text{or when } \frac{p_{t_{3'}}}{p_{t_5}} > 1, \tau > 1\right]$$

$$(5.131)$$

The performance of the optimal mixer now follows directly by substitution of Eqs. (5.130) and (5.131) into Eq. (5.125) to give

$$S_{7a} = S_{3'}\frac{1 + \alpha\tau}{S_{3'} + \alpha\tau}, \qquad S_{7b} = S_{3'}\frac{1 + \alpha}{S_{3'} + \alpha} \qquad (5.132)$$

Further, Eqs. (5.127), (5.128), and (5.130) show that for both optimal cases

$$A_7 = A_{3'} + A_5$$

Thus the optimal constant-pressure mixer is also the optimal constant-area mixer!

These remarkably simple results for the optimal mixer output stagnation quantities allow easy determination of the optimal conceivable performance of a mixed stream turbofan. Thus, for example (Problem 5.20), combination of Eqs. (5.90), (5.91), and (5.132) leads (for case a) to the expression for the specific thrust

$$\frac{F}{\dot{m}_c + \dot{m}_F} = a_0 M_0 \left\{ \left[1 + \frac{1}{1+\alpha} \left(\frac{\tau_\lambda}{\tau_r \tau_c} - 1 \right) \frac{\tau_r \tau_c - 1}{\tau_r - 1} \right]^{\frac{1}{2}} - 1 \right\} \quad (5.133)$$

This result can be compared to the optimal results of the separate stream case (see also Problem 5.21). It is rather interesting to note that the fan pressure ratio does not appear explicitly in this expression. Thus provided only that conditions are appropriate to ensure $p_{t_3}/p_{t_5} > 1$ and $\tau < 1$, the specific thrust, in this optimal mixed case, is not a function of fan pressure ratio.

5.15 The Ideal Turbofan with Afterburning

A cycle that somewhat combines the high thrust per frontal area of a turbojet at high Mach number while providing respectable specific fuel consumption for subsonic cruise conditions is the turbofan with duct burning. The turbofan utilizes duct burning (burning in the secondary stream) for supersonic cruise, but cruises subsonically without duct burning. In many cycles, the fan and compressor require so much power extraction from the primary stream that the turbine outlet pressure is greatly reduced. The resulting low primary nozzle pressure ratio renders primary stream afterburning unattractive. In the following analysis, however, the possibility of primary stream afterburning as well as secondary stream afterburning will be included.

Cycle Analysis of the Ideal Turbofan with Afterburning

It is again assumed that all component efficiencies are perfect, that the gas is calorically perfect, that the fuel-to-air ratio in all combustors is much less than unity, and that the exit static pressures of both primary and secondary streams are equal to the ambient pressure. The nomenclature and notation are as in Fig. 5.17 and Sec. 5.2.

The expression for the total thrust remains as given by Eq. (5.59); thus,

$$\frac{F}{\dot{m}_c + \dot{m}_F} = \frac{a_0 M_0}{1+\alpha} \left[\left(\frac{u_9}{u_0} - 1 \right) + \alpha \left(\frac{u_{9'}}{u_0} - 1 \right) \right] \quad (5.134)$$

Secondary stream. Again

$$(u_{9'}/u_0)^2 = (T_{9'}/T_0)(M_{9'}/M_0)^2 \quad (5.135)$$

Now write

$$T_{t_{9'}} = T_{9'}\left(1 + \frac{\gamma - 1}{2}M_{9'}^2\right) = T_0 T_{\lambda AB'} \qquad (5.136)$$

$$P_{t_{9'}} = P_{9'}\left(1 + \frac{\gamma - 1}{2}M_{9'}^2\right)^{\gamma/(\gamma-1)} = p_0 \pi_r \pi_{c'} \qquad (5.137)$$

With Eqs. (5.1), (5.4), and (5.137)

$$1 + \frac{\gamma - 1}{2}M_{9'}^2 = \tau_r \tau_{c'} \qquad (5.138)$$

Combining Eqs. (5.136) and (5.138) gives

$$T_{9'}/T_0 = \tau_{\lambda AB'}/\tau_r \tau_{c'} \qquad (5.139)$$

so that with Eqs. (5.135), (5.138), and (5.139)

$$M_0\left(\frac{u_{9'}}{u_0} - 1\right) = \left[\frac{2}{\gamma - 1}\frac{\tau_{\lambda AB'}}{\tau_r \tau_{c'}}(\tau_r \tau_{c'} - 1)\right]^{\frac{1}{2}} - M_0 \qquad (5.140)$$

Primary stream. The relationships for temperatures and pressures here are exactly as previously obtained for the turbojet with afterburning (Sec. 5.11). So

$$M_0\left(\frac{u_9}{u_0} - 1\right) = \left[\frac{2}{\gamma - 1}\frac{\tau_{\lambda AB}}{\tau_r \tau_c \tau_t}(\tau_r \tau_c \tau_t - 1)\right]^{\frac{1}{2}} - M_0 \qquad (5.141)$$

Power balance. The power balance between the fan, compressor, and turbine remains as for the turbofan without afterburning [Eq. (5.67)], so

$$\tau_t = 1 - (\tau_r/\tau_\lambda)[(\tau_c - 1) + \alpha(\tau_{c'} - 1)] \qquad (5.142)$$

Specific fuel consumption. The specific fuel consumption could be obtained by writing an enthalpy balance for each burner separately and then summing the separate fuel consumptions. It is more direct, however, to write an enthalpy balance across the entire system and to equate the energy addition in the burners to the overall enthalpy change. Thus

$$(\dot{m}_f + \dot{m}_{fAB} + \dot{m}_{fAB'})h = \dot{m}_c C_p T_{t_9} + \dot{m}_F C_p T_{t_{9'}} - (\dot{m}_c + \dot{m}_F)C_p T_{t_2}$$

or

$$f_{tot} \equiv \frac{\dot{m}_f + \dot{m}_{fAB} + \dot{m}_{fAB'}}{\dot{m}_c} = \frac{C_p T_0}{h}\left[\tau_{\lambda AB} + \alpha\tau_{\lambda AB'} - (1 + \alpha)\tau_r\right]$$

$$(5.143)$$

The specific fuel consumption then follows from

$$S = \frac{(\dot{m}_f + \dot{m}_{fAB} + \dot{m}_{fAB'})(10^6)}{F} = \frac{f_{tot}(10^6)}{(1+\alpha)[F/(\dot{m}_c + \dot{m}_F)]} \quad (5.144)$$

These equations are summarized in the following section. It can be noted here, however, that in fact this analysis contains most of the preceding analyses of this chapter as special cases. Thus, note that if a duct burning turbofan with no primary stream burning is to be considered, $\tau_{\lambda AB}$ need only be replaced by its minimum value,

$$(\tau_{\lambda AB})_{min} = \tau_\lambda \tau_t \quad (5.145)$$

If no duct burning is present, $\tau_{\lambda AB'}$ must be replaced by its minimum value,

$$(\tau_{\lambda AB'})_{min} = \tau_r \tau_{c'} \quad (5.146)$$

If a turbojet is to be considered (with or without afterburning), the bypass ratio α must be put equal to zero; and finally if a ramjet is to be considered, α must be put equal to zero and also π_c (or τ_c) must be put equal to unity.

Summary of the Equations—Ideal Turbofan with Afterburning

Inputs: $T_0(\mathrm{K})$ [°R], γ, $h(\mathrm{J/kg})$ [Btu/lbm],

$C_p(\mathrm{J/kg \cdot K})$ [Btu/lbm · °R], τ_λ, $\tau_{\lambda AB}$, $\tau_{\lambda AB'}$, π_c, $\pi_{c'}$, M_0, α

Outputs: $\dfrac{F}{\dot{m}_c + \dot{m}_F}\left(\dfrac{\mathrm{N \cdot s}}{\mathrm{kg}}\right)$ $\left[\dfrac{F}{g_0(\dot{m}_c + \dot{m}_F)}, \dfrac{\mathrm{lbf}}{\mathrm{lbm/s}}\right]$,

$S\left(\dfrac{\mathrm{mg}}{\mathrm{N \cdot s}}\right)\left[\dfrac{\mathrm{lbm\ fuel/h}}{\mathrm{lbf\ thrust}}\right]$, $f_{tot}\dfrac{\mathrm{kg\ fuel/s}}{\mathrm{kg\ primary\ air/s}}$

Equations: The first six equations are identical to Eqs. (5.70–5.75); then:

$$\frac{F}{\dot{m}_c + \dot{m}_F} = \frac{a_0}{1+\alpha}\left\{\left[\frac{2}{\gamma-1}\frac{\tau_{\lambda AB}}{\tau_r \tau_c \tau_t}(\tau_r \tau_c \tau_t - 1)\right]^{\frac{1}{2}} - M_0\right.$$

$$\left. + \alpha\left[\left[\frac{2}{\gamma-1}\frac{\tau_{\lambda AB'}}{\tau_r \tau_{c'}}(\tau_r \tau_{c'} - 1)\right]^{\frac{1}{2}} - M_0\right]\right\} \quad (5.147)$$

(To obtain British units of lbf/(lbm/s), replace a_0 in meters per second with $a_0/32.174$ where a_0 should be given in feet/second).

$$f_{tot} = \frac{C_p T_0}{h} \left[\tau_{\lambda AB} + \alpha \tau_{\lambda AB'} - (1 + \alpha) \tau_r \right] \tag{5.148}$$

$$S = \frac{f_{tot}(10^6)}{(1 + \alpha)\left[F/(\dot{m}_c + \dot{m}_F) \right]}$$

Note: In these equations the minimum values of $\tau_{\lambda AB}$ and $\tau_{\lambda AB'}$ to be considered are

$$(\tau_{\lambda AB})_{min} = \tau_\lambda \tau_t \tag{5.149}$$

$$(\tau_{\lambda AB'})_{min} = \tau_r \tau_{c'} \tag{5.150}$$

If either of these minimum values is used, the calculation corresponds to the case where burning is not present in the respective afterburner.

Example Results—Ideal Turbofan with Afterburning

As an example of the effect of afterburning, consider the family of turbofan engines discussed in Sec. 5.12 that gave the results of Fig. 5.39. Here, again consider the addition of burning in the secondary stream such that $\tau_{\lambda AB'} = 7.5$. Conditions are hence $\gamma = 1.4$, $T_0 = 420°R$, $h = 19,000$ Btu/lbm, $C_p = 0.24$ Btu/lbm · °R, $\pi_c = 20$, $\tau_\lambda = 6.5$, $M_0 = 2$, and $\pi_{c'} = 2$. The results are indicated in Fig. 5.46.

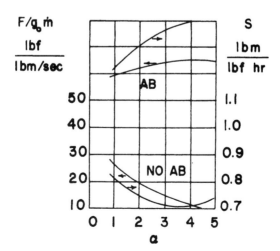

Fig. 5.46 Specific thrust and specific fuel consumption vs bypass ratio, duct burning turbofan.

Figure 5.46 illustrates the very substantial increases in specific thrust that can be obtained by the use of duct burning. The results suggest, also, that if high thrust levels are to be attained for given flight conditions, whereas efficient cruising is desired for other (lower thrust) flight conditions, the duct burning turbofan shows promise of affecting a reasonable compromise.

Problems

5.1   (a) Calculate and carefully plot the specific thrust $F/(g_0\dot{m})$, lbf/lbm/s, and specific fuel consumption S, lbm/h/lbf, vs design pressure ratio π_c for an ideal turbojet. The flight Mach number is to be 2.7, and you are given $\gamma = 1.4$, $T_0 = 400°R$, $h = 19{,}000$ Btu/lbm, $C_p = 0.24$ Btu/lbm · °R, $\tau_\lambda = 6.8$. Plot the results over the range from $\pi_c = 1$ to where the thrust goes to zero.

     (b) Obtain an analytic expression for the compressor ratio that just leads to zero thrust π_{cm} in terms of γ, M_0, and τ_λ. Check that the value calculated using the data of part (a) agrees with the graphical result of part (a).

5.2   Calculate $F/(g_0\dot{m})$ and S for a ramjet with the conditions as given for the turbojet of Problem 5.1, except that $\tau_\lambda = 8.0$.

5.3   A designer finds that because a new material has been discovered, he can increase the allowable turbine inlet temperature. He already has a turbojet design—say engine 1—and decides to redesign the engine to engine 2, so that engine 2 has the same specific thrust as engine 1 (at the same flight Mach number).

     (a) Show that the compressor stagnation temperature ratio of the new engine τ_{c_2} is related to τ_{c_1}, τ_{λ_1}, τ_r, and τ_{λ_2} by

$$\frac{\tau_{c_2}}{\tau_{c_1}} = \frac{1}{2}\left\{1 + \frac{\tau_{\lambda_1}}{\left(\tau_r\tau_{c_1}\right)^2} + \frac{\tau_{\lambda_2} - \tau_{\lambda_1}}{\tau_r\tau_{c_1}}\right.$$

$$\left. + \left[\left(1 - \frac{\tau_{\lambda_1}}{\left(\tau_r\tau_{c_1}\right)^2}\right)^2 + \frac{\tau_{\lambda_2} - \tau_{\lambda_1}}{\left(\tau_r\tau_{c_1}\right)^2}\left[\tau_{\lambda_2} - \tau_{\lambda_1} - 4 + 2\left(\tau_r\tau_{c_1} + \frac{\tau_{\lambda_1}}{\tau_r\tau_{c_1}}\right)\right]\right]^{\frac{1}{2}}\right\}$$

     (b) Given the conditions for engine 1 of $\gamma = 1.4$, $T_0 = 222$ K, $M_0 = 2$, $h = 4.42 \ (10^7)$ J/kg, $C_p = 1005$ J/kg · K, $\tau_{\lambda_1} = 7$, $\pi_{c_1} = 20$, calculate the specific thrust and specific fuel consumption of engine 1.

     (c) Given that $\tau_{\lambda_2} = 8$, calculate π_{c_2} and the specific fuel consumption of engine 2.

     (d) Compare the thermal and propulsive efficiencies of the two engines.

5.4   The designer described in Problem 5.3 decides to utilize the increased specific thrust at the same compressor pressure ratio. Calculate the resulting specific thrust and specific fuel consumption. Again compare the thermal and propulsive efficiencies of the two engines.

5.5 (a) Show that the specific thrust for an ideal afterburning turbojet operating at maximum thrust is given by

$$
\frac{F}{\dot{m}} = a_0 \left\{ \left[\frac{2\tau_{\lambda AB}}{\gamma - 1} \left(1 - \frac{4\tau_\lambda}{(\tau_\lambda + \tau_r)^2} \right) \right]^{\frac{1}{2}} - M_0 \right\}
$$

For the case $\gamma = 1.4$, $T_0 = 220$ K, $C_p = 1005$ J/kg · K, $h = 4.42 \,(10^7)$ J/kg, $M_0 = 2.5$.

(b) For $\tau_\lambda = 7$ plot the specific thrust and specific fuel consumption vs $\tau_{\lambda AB}$ over the range $7 \leq \tau_{\lambda AB} \leq 9$.

(c) For $\tau_{\lambda AB} = 8$ plot the specific thrust and specific fuel consumption vs τ_λ over the range $6.5 \leq \tau_\lambda \leq 7.5$.

5.6 (a) A family of ideal afterburning turbojets with compressor pressure ratio $\pi_c = 10$ is to be considered. Calculate and plot the specific fuel consumption and specific thrust over the Mach number range $0 \leq M_0 \leq 3.5$. Take $\gamma = 1.4$, $T_0 = 210$ K, $C_p = 1005$ J/kg · K, $h = 4.42 \,(10^7)$ J/kg, $\tau_\lambda = 7$, and $\tau_{\lambda AB} = 8$.

(b) What condition will determine the maximum Mach number at which these engines will be able to fly? Determine the numerical value of this maximum Mach number.

5.7 (a) Consider an ideal turbofan engine with the conditions $M_0 = 2$, $\pi_c = 20$, $\pi_{c'} = 2.2$, $\tau_\lambda = 7$, $\alpha = 3$, $\gamma = 1.4$. Find the Mach number at the exit of the primary stream M_9 and at the exit of the secondary stream $M_{9'}$.

(b) Find M_9 and $M_{9'}$ if $\alpha = 6$, all other parameters remaining unchanged.

5.8 The "optimal" ideal turbofan was found to have

$$
2(u_9 - u_0) = u_{9'} - u_0
$$

This optimum was obtained by choosing the bypass ratio α to give minimum specific fuel consumption for given π_c and $\pi_{c'}$. Actually, the specific fuel consumption S can be considered an analytical function of α, τ_c, and $\tau_{c'}$, all of which can be chosen independently. A true minimum in S would occur when α, τ_c, and $\tau_{c'}$ were all chosen to make the partial derivatives of S with all three variables be zero.

(a) Show that when S is minimized with respect to $\tau_{c'}$ for given τ_c and α, it is required to have $u_9 = u_{9'}$.

(b) When the joint minimum of S with α and $\tau_{c'}$ is selected, what is the relationship required between u_9, $u_{9'}$, and u_0? What is the related value of the bypass ratio α? Explain, in physical terms, why it is that this limit exists.

5.9   (a) Show that the propulsive efficiency of an ideal fan jet that has been optimized to give $\alpha = \alpha^*$ may be written in the form

$$\eta_p = \frac{2(1 + 2\alpha^*)}{1 + (u_9/u_0)(1 + 4\alpha^*)}$$

(b) Calculate and plot α^*, u_9/u_0, and η_p vs $\pi_{c'}$ over the range $1 < \pi_{c'} \leq 3.0$ for the conditions $M_0 = 0.85$, $\gamma = 1.4$, $\pi_c = 30$, and $\tau_\lambda = 7$.

5.10   The ram rocket is a device contemplated for use over large Mach number ranges. It is hoped to achieve high Mach number performance similar to that of a ramjet, but to have the advantage of static thrust.

The device operates as indicated in Fig A. Thus, the rocket plume mixes with the incoming air in an ideal constant-area mixer prior to expansion through the nozzle. The airstream itself has fuel added and burned between stations 2 and 3. For simplicity assume $p_0 = p_9$, $\dot{m}_f/\dot{m} \ll 1$, $\gamma = \gamma_R$, and $C_p = C_{pR}$. We define $\beta = \dot{m}_R/\dot{m}$, $T_{t_4}/T_{t_3} = \tau_m$, $p_{t_4}/p_{t_3} = \pi_m$, $T_{t_R}/T_0 = \tau_{\lambda R}$, and $T_{t_3}/T_0 = \tau_\lambda$.

(a) Find an expression for $F/(\dot{m}U_0)$ in terms of τ_λ, τ_m, τ_r, π_m, β, and γ.

(b) Find an expression for τ_m in terms of β, τ_λ, and $\tau_{\lambda R}$.

(c) Describe how you would solve for π_m in terms of the other variables. What additional information is necessary in order to solve for π_m?

5.11   The turborocket is another concept for airbreathing propulsion. Figure B indicates the concept, wherein a rocket chamber drives a small

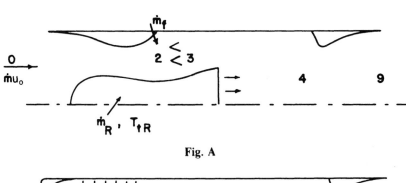

Fig. A

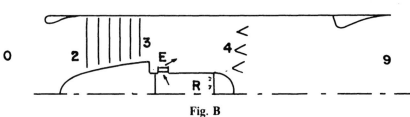

Fig. B

turbine which in turn drives the compressor. The products of combustion and the incoming air mix between stations 3 and 4 and pass through the common exit nozzle. Additional combustion may or may not occur in the afterburner indicated.

It may be assumed that (1) $\gamma = \gamma_R$ and $C_p = C_{pR}$; (2) $p_{tE} = p_{t_3}$ and no stagnation pressure drop occurs in the mixing chamber; (3) $p_0 = p_9$ and all component performances are ideal. We define $\tau_m = T_{t_4}/T_{t_3}$, $\tau_{\lambda R} = T_{t_R}/T_0$, $\tau_{\lambda b} = T_{t_9}/T_0$, $\delta = (p_0/p_{tR})^{(\gamma-1)/\gamma}$ and $\beta = \dot{m}_R/\dot{m}$.

(a) Show that in the case where no afterburning occurs

$$\frac{F}{a_0 \dot{m}} = (1 + \beta)\left[\frac{2}{\gamma - 1}\tau_m(\tau_r\tau_c - 1)\right]^{\frac{1}{2}} - M_0$$

(b) Show that

$$\tau_m = \frac{1 - \delta\tau_r}{1 - \delta\tau_r\tau_c + (\tau_r/\tau_{\lambda R})(\tau_c - 1)}$$

(c) If afterburning occurs (with no further mass addition), find an expression for $F/(a_0\dot{m})$ in terms of M_0, τ_r, γ, β, δ, τ_c, $\tau_{\lambda R}$, and $\tau_{\lambda b}$.

5.12   Figure C shows a concept for an "aft fan" engine with "aft-burning boost." The idea is that without the aft burner on, the engine will behave as a conventional fan jet, but extra thrust will be available when the aft burner is utilized. Ideal behavior of all components may be assumed with $\dot{m}_f/\dot{m} \ll 1$, etc. In addition, it is to be assumed that when the burner is on, all pressure ratios in the core engine remain unchanged, and the bypass ratio α remains unchanged. We define $\tau_{ta} = T_{t_{4u}}/T_{t_4}$, $\tau_{tb} = T_{t_5}/T_{t_{4b}}$, $\tau_{\lambda b} = T_{t_{4b}}/T_0$ and reference the zero aft-burning case with an additional 0 subscript.

(a) Show that

$$\frac{u_9/u_0}{(u_9/u_0)_0} = \left[\frac{\tau_{\lambda b}}{\tau_\lambda - \tau_r(\tau_c - 1)}\right]^{\frac{1}{2}}$$

(b) For conditions $M_0 = 2.2$, $\gamma = 1.4$, $C_p = 0.24$, Btu/lbm·°R, $h = 19{,}500$ Btu/lbm, $T_0 = 400°R$, $\tau_\lambda = 7$, $\pi_{c_0'} = 1.5$, and $\pi_c = 15$, find α^*,

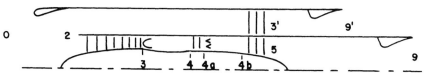

Fig. C

$F/[g_0(\dot{m}_c + \dot{m}_F)]$, and S for the aft burner off case.

(c)   Find $F/[g_0(\dot{m}_c + \dot{m}_F)]$ and S (based on total fuel flow) for the configuration of part (b), but with $\tau_{\lambda b} = 6.8$.

5.13   A ramjet has ideal performance in every component except that the combustion occurs at finite Mach number. Assuming that the combustion occurs at constant static pressure, show that the specific thrust is given by

$$F/\dot{m} = a_0 M_0 \left\{ \left[\tau_b - (\tau_b - 1)(M_3/M_0)^2 \right]^{\frac{1}{2}} - 1 \right\}$$

where $\tau_b \equiv T_{t_4}/T_{t_3}$.

Note: You might get some help from Problem 2.12.

5.14   A ramjet has ideal performance in every component except that combustion occurs at finite (fixed) Mach number M_3.

(a)   Show that the specific thrust is given by

$$\frac{F}{\dot{m}} = a_0 M_0 \left[\left\{ \tau_b \left[1 - \frac{2}{(\gamma - 1) M_0^2} \left(\tau_b^{[(\gamma-1)/2] M_3^2} - 1 \right) \right] \right\}^{\frac{1}{2}} - 1 \right]$$

(b)   Show that the thermal efficiency for this cycle is given by

$$\eta_{th} = 1 - \frac{T_0}{T_{t_3}} \frac{\tau_b^{1 + [(\gamma-1)/2] M_3^2} - 1}{\tau_b - 1}$$

5.15   In the text it was shown for an ideal constant-pressure mixer

$$S_7 = \frac{1 - \dfrac{\gamma - 1}{2} V_5^2}{1 - \dfrac{\gamma - 1}{2} \dfrac{(V_5 + \alpha V_{3'})^2}{(1 + \alpha)(1 + \alpha\tau)}}$$

and

$$S_{3'} = \frac{1 - \dfrac{\gamma - 1}{2} V_5^2}{1 - \dfrac{\gamma - 1}{2} \dfrac{1}{\tau} V_{3'}^2}$$

The combination of these two equations can be considered to give S_7 as a function of V_5 and the prescribed ratios of upstream stagnation quantities α, τ, and $S_{3'}$.

Show that the maximum in S_7 with V_5 occurs when

$$(V_{3'}/V_5)^2 - (1 + \tau)(V_{3'}/V_5) + \tau = 0$$

and hence when $V_{3'}/V_5 = 1$ or τ.

Note: You may find it helps to solve for $V_{3'}^2$ from the second equation above and to solve for $2V_{3'}(\partial V_{3'}/\partial V_5)$ prior to taking the derivative $\partial S_7/\partial V_5$.

5.16   Consider the comparison in behavior of the gross thrusts of mixed vs unmixed streams when in the mixed case an ideal optimal constant-pressure mixer is used. With notation as in the text, but defining in addition

$$S_a = \left(\frac{p_0}{p_{t_5}}\right)^{(\gamma - 1)/\gamma} \qquad (a = \text{ambient})$$

(a)   Show that the ratio of mixed to unmixed gross thrusts is

$$\frac{F_{\text{mix}}}{F_{\text{un}}} = \frac{\{(1 + \alpha)(1 + \alpha\tau)[1 - S_a/S_7]\}^{\frac{1}{2}}}{(1 - S_a)^{\frac{1}{2}} + \alpha\sqrt{\tau}(1 - S_a/S_{3'})^{\frac{1}{2}}}$$

(b)   Show that the value of τ that just causes the ratio $F_{\text{mix}}/F_{\text{un}}$ to equal unity is given by

$$\tau = \frac{1 - S_a}{1 - S_a/S_{3'}} \qquad \text{for case a} \qquad \left(\frac{p_{t_{3'}}}{p_{t_5}} > 1, \frac{T_{t_{3'}}}{T_{t_5}} < 1\right)$$

$$\tau = \frac{1 - S_a/S_{3'}}{1 - S_a} \qquad \text{for case b} \qquad \left(\frac{p_{t_{3'}}}{p_{t_5}} < 1, \frac{T_{t_{3'}}}{T_{t_5}} < 1\right)$$

(c)   Show that the result of part (b) implies, in addition, that for case a the exit velocities of the separate stream case would be equal and, in fact, equal to the entrance and exit velocities of the mixer if it was used!

(d)   Plot the curve of τ vs $p_{t_{3'}}/p_{t_5}$ for the "breakeven" cases obtained in part (b) over the range $0.5 \le p_{t_{3'}}/p_{t_5} \le 1.5$ for the two cases $p_a/p_{t_5} = 0.3$ and 0.5 (take $\gamma = 1.4$).

Indicate on the graph those regions in which incorporating a mixer would be beneficial and in which mixing would be detrimental.

5.17   Figure D shows a novel concept to be considered for development. The device is to operate by having the air in the "secondary" stream expand through the turbine (which would be the fan in a conventional cycle). The turbine is coupled directly to the compressor that compresses the

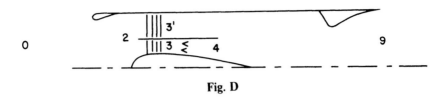

Fig. D

core stream. Combustion then occurs followed by expansion to the entrance of the optimal constant-pressure mixer and, finally, expansion through the nozzle.

(a) Is this optimal mixer case a or case b?

(b) Show that the performance of the engine can be described by the following system of equations:

Inputs: $\gamma, h, C_p, T_0, \tau_\lambda, M_0, \pi_c, \alpha$

Outputs: $\dfrac{F}{g_0(\dot{m}_c + \dot{m}_F)}, \quad S, \quad \eta_{th}$

Equations:

$$\tau_r = 1 + \frac{\gamma - 1}{2} M_0^2$$

$$\tau_c = \pi_c^{(\gamma - 1)/\gamma}$$

$$\frac{T_{t_9}}{T_{t_4}} = \tau_m = \frac{1}{(1 + \alpha)\tau_\lambda}\left[\tau_\lambda + \tau_r(1 + \alpha - \tau_c)\right]$$

$$\left(\frac{p_{t_9}}{p_{t_4}}\right)^{(\gamma - 1)/\gamma} = S_m = \frac{1 + \alpha - \tau_c}{1 + (\alpha - 1)\tau_c}$$

$$a_0 = \sqrt{(\gamma - 1)C_p T_0}$$

$$\left(M_0\frac{u_9}{u_0}\right)^2 = \frac{2}{\gamma - 1}\tau_\lambda\tau_m\left(1 - \frac{1}{\tau_r\tau_c S_m}\right)$$

$$\frac{F}{g_0(\dot{m}_c + \dot{m}_F)} = \frac{a_0}{g_0}\left(M_0\frac{u_9}{u_0} - M_0\right)$$

$$S = 3600\,\frac{C_p T_0}{h}\,\frac{\tau_\lambda - \tau_r\tau_c}{(1 + \alpha)\{F/[g_0(\dot{m}_c + \dot{m}_F)]\}}$$

$$\eta_{th} = 1 - \frac{(1 + \alpha)[(\tau_\lambda\tau_m/\tau_r\tau_c S_m) - 1]}{\tau_\lambda - \tau_r\tau_c}$$

(c) Investigate the predicted performance and determine whether the concept provides competition to conventional gas turbines in any Mach number range. (Note that τ_λ is not as restricted as it would be for a burner followed by a turbine.)

5.18   Consider a constant-pressure mixer of the form indicated in Fig. E. A supplemental flow is introduced along the sidewall of amount $S\dot{m}$. This supplemental flow is injected such that it introduces no axial momentum. Also $V = u/\sqrt{\gamma R T_{t_1}}$ and $\tau = T_{t_s}/T_{t_1}$.
(a) Find an expression for p_{tm}/p_{t_1} in terms of γ, V_1, S, and τ.
(b) Find an expression for A_2/A_1 in terms of γ, S, τ, and p_{tm}/p_{t_1}.
(c) Calculate p_{tm}/p_{t_1} and A_2/A_1 for the case $\gamma = 1.3$, $S = 0.1$, $\tau = 0.4$, $V_1 = 0.5$.

5.19   Consider an ideal turbofan with conditions $\gamma = 1.4$, $T_0 = 215$ K, $C_p = 1000$ J/kg·K, $h = 4.43 \ (10^7)$ J/kg, $\tau_\lambda = 7$, $\tau_{\lambda AB} = \tau_{\lambda AB'} = 8$ (with burners on), $\pi_c = 15$, $\pi_{c'} = 2.0$, and $M_0 = 2.5$.
(a) Find α^*, $F/(\dot{m}_c + \dot{m}_F)$, and S^*.
(b) With α fixed at α^*, find $F/(\dot{m}_c + \dot{m}_F)$ and S for the cases (1) core afterburner on, (2) fan afterburner on, and (3) core and fan afterburner on.
(c) With α raised to $(\alpha^* + 1)$, find $F/(\dot{m}_c + \dot{m}_F)$ and S for the three cases of part (b) as well as for the case with both afterburners off.

5.20   Consider an ideal turbofan engine that incorporates an ideal constant-pressure mixer to mix the fan and core streams prior to expulsion through the nozzle. The fan stream stagnation pressure is higher than that of the core stream, and the fan stream stagnation temperature is lower than that of the core stream.
Assuming that the fan and core streams are mixed "optimally," show that the specific thrust is given by

$$\frac{F}{\dot{m}_c + \dot{m}_F} = a_0 M_0 \left\{ \left[1 + \frac{1}{1 + \alpha}\left(\frac{\tau_\lambda}{\tau_r \tau_c} - 1 \right)\left(\frac{\tau_r \tau_c - 1}{\tau_r - 1} \right) \right]^{\frac{1}{2}} - 1 \right\}$$

5.21   (a) Consider an ideal turbofan engine that utilizes separate expansion of the core and fan streams to ambient pressure. The bypass ratio is chosen to give minimum specific fuel consumption for prescribed $\tau_{c'}$ and τ_c.

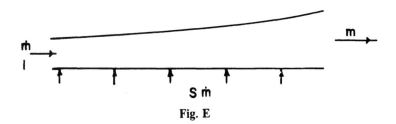

Fig. E

Show that the prescribed $\tau_{c'}$ can be eliminated in terms of the bypass ratio α (actually α^*) to give the expression for the specific thrust

$$\frac{F}{\dot{m}_c + \dot{m}_F} = a_0 M_0 \frac{(1 + 2\alpha)^2}{(1 + \alpha)(1 + 4\alpha)}$$

$$\times \left\{ \left[1 + \frac{1 + 4\alpha}{(1 + 2\alpha)^2} \left(\frac{\tau_\lambda}{\tau_r \tau_c} - 1 \right) \left(\frac{\tau_r \tau_c - 1}{\tau_r - 1} \right) \right]^{\frac{1}{2}} - 1 \right\}$$

(b) Compare the ratio of the specific thrust of this engine to the specific thrust of the engine of Problem 5.20 (and hence the ratio of their specific fuel consumptions!) for the range $2 \le \alpha \le 7$ for the case $M_0 = 0.85$, $\tau_c = 25$, and $\tau_\lambda = 6.5$.

6. COMPONENT PERFORMANCE

6.1 Introduction

In this chapter the behavior of the engine components including the nonideal effects are described. The performance of the various components are described in terms of "figures of merit," which allow cycle analyses including losses to be made efficiently. It should be noted, however, that both the accurate quantitative estimation of such figures of merit and the design of the components to reduce the losses are very demanding processes and absorb much of the industry effort. Prior to considering each individual component, the expression for the overall thrust of the engine when losses are present is considered. It is found that both internal and external losses are present and that an optimal design would be such as to minimize the combination of all losses.

6.2 The Thrust Equation

The momentum equation for a control volume (Fig. 6.1) may be written in the form:

Force on volume of fluid = rate of production of momentum

or

Pressure force + viscous force
= rate of accumulation of momentum
+ rate of convection of momentum through boundaries

Each of these terms may be represented by an appropriate area or volume integral, but for simplicity of presentation the viscous force is noted simply as F_v. The above word equation can then be written as,

$$-\iint p\,\mathbf{ds} + F_v = \frac{\partial}{\partial t}\iiint \rho\mathbf{u}\,dv + \iint \rho\mathbf{u}\mathbf{u}\cdot\mathbf{ds} \qquad (6.1)$$

Note that the vector area element ds is defined to be positive when directed outwardly, and hence the pressure force from the surface onto the fluid has a negative sign identified with it. The engine operates in steady state (or "quasi-steady state"), so the integral denoting the volumetric accumulation of momentum with time is zero.

189

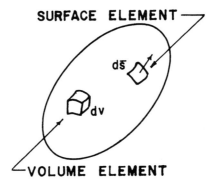

Fig. 6.1   General control volume.

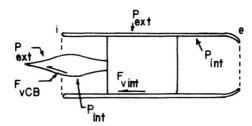

Fig. 6.2   Pressure and viscous stresses on engine.

The thrust equation may then be written

$$F_v - \iint p \, ds = \iint \rho \mathbf{u}\mathbf{u} \cdot ds \tag{6.2}$$

Usually the thrust of the engine is defined as the net force resulting from all the pressure and viscous stresses, less the external viscous drag. This means that the external viscous forces are included in the airplane drag as far as the "accounting" goes.

To find the axial component of force only, define dA as the axial projection of an area element. The thrust F_p is defined as being positive in the negative x direction, so that applying Eq. (6.2) and referencing Fig. 6.2 results in

$$F_p = \int_{i_{int}}^{e} (p_{int} - p_0) \, dA - F_{v_{int}} - \int_{ext\,CB} (p_{ext} - p_0) \, dA - F_{v\,CB}$$

$$- \int_{i_{ext}}^{e} (p_{ext} - p_0) \, dA \tag{6.3}$$

where int refers to the surfaces wetted by fluid passing through the engine, ext to the surfaces wetted by fluid passing outside the engine, and ext CB to that portion of the centerbody protruding forward of the inlet plane.

It is clear that direct evaluation of the internal viscous and pressure forces is hopeless; so the internal forces are related to the changes in fluid properties occurring from the inlet to the exit of the engine. To do this, consider the effects of the internal forces acting inward on the fluid. The forces acting on the fluid between the inlet and exit will include the pressure and viscous forces acting on the interior engine parts. Thus, again applying Eq. (6.2),

$$-F_{v_{\text{int}}} + \int_{i_{\text{int}}}^{e} (p_{\text{int}} - p_0)\,\mathrm{d}A + \int_i (p_i - p_0)\,\mathrm{d}A - \int_e (p_e - p_0)\,\mathrm{d}A$$

$$= -\int_i \rho u^2\,\mathrm{d}A + \int_e \rho u^2\,\mathrm{d}A \qquad (6.4)$$

Note here that the first two terms are identical to the same pair of terms appearing in Eq. (6.3). The effect of the minus sign that would appear because of the difference in direction of the outwardly directed area element is cancelled since F_p is defined to be positive in the negative x direction. For simplicity in writing the equations, it is assumed that the conditions are one-dimensional at the entrance and exit to obtain, by combining Eqs. (6.3) and (6.4),

$$F_p = \dot{m}_e u_e - \dot{m}_i u_i + (p_e - p_0)A_e - (p_i - p_0)A_i - \int_{\text{ext CB}} (p_{\text{ext}} - p_0)\,\mathrm{d}A$$

$$-F_{v_{\text{CB}}} - \int_{i_{\text{ext}}}^{e} (p_{\text{ext}} - p_0)\,\mathrm{d}A \qquad (6.5)$$

This form is not particularly useful, because not only is it difficult to determine u_i without quite detailed engine information, but u_i and hence the pressure integrals over the centerbody and exterior change with a change

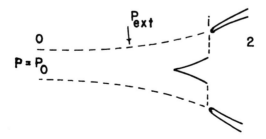

Fig. 6.3 Control volume for approaching flow.

in the design of the inlet. It is better to relate F_p to the flow conditions far upstream. The control volume scheme shown in Fig. 6.3 is one way to do this.

Again assuming one-dimensional conditions at the entrance (and far upstream), with application of Eq. (6.2) it is found

$$\dot{m}_i u_i + (p_i - p_0)A_i + \int_{\text{ext CB}}(p_{\text{ext}} - p_0)\,\mathrm{d}A + F_{v_{\text{CB}}}$$

$$= +\dot{m}_0 u_0 + \int_0^i (p_{\text{ext}} - p_0)\,\mathrm{d}A \tag{6.6}$$

and hence with Eq. (6.5)

$$F_p = \dot{m}_e u_e - \dot{m}_0 u_0 + (p_e - p_0)A_e - \int_0^i (p_{\text{ext}} - p_0)\,\mathrm{d}A - \int_{i_{\text{ext}}}^e (p_{\text{ext}} - p_0)\,\mathrm{d}A$$

$$\tag{6.7}$$

Now write

$$F_A = \dot{m}_e u_e - \dot{m}_0 u_0 + (p_e - p_0)A_e \tag{6.8}$$

$$D_{\text{add}} = \int_0^i (p_{\text{ext}} - p_0)\,\mathrm{d}A \equiv \text{additive drag} \tag{6.9}$$

$$D_{\text{ext}} = \int_{i_{\text{ext}}}^e (p_{\text{ext}} - p_0)\,\mathrm{d}A \equiv \text{external drag} \tag{6.10}$$

then

$$F_p = F_A - D_{\text{add}} - D_{\text{ext}} \tag{6.11}$$

It is usual to term F_A the uninstalled thrust and F_p the installed thrust.

Interpretation of the Terms Appearing in the Thrust Equation

Equations (6.8–6.11) are in a suitable form for design purposes, but they are at first difficult to interpret. Thus, for example, it at first seems peculiar that a pressure integral over surfaces external to the engine $(0 - i)$ could have anything to do with the forces *on* the engine. The additive drag and external drag terms are interdependent, however, and something can be learned about this interdependence by considering a "perfect engine," that is, an engine with no external viscous drag or form drag.

Consider the momentum equation for all of the fluid flowing external to the engine (Fig. 6.4). Because the flow is perfect (no shocks, no boundary layers) the fluid conditions externally are identical at 0 and ∞. The pressure

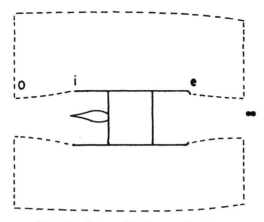

Fig. 6.4   External fluid control volume.

is p_0 around the external contour and the contour is chosen to be a streamline so that no momentum is convected through the perimeter. The momentum fluxes through the ends of the control volume are equal because the velocities are equal, so that the momentum equation reduces to the simple statement that the sum of the pressure integrals over all of the internal surfaces must be zero. Thus,

$$D_{\text{add}} + D_{\text{ext}} + \int_e^\infty (p_{\text{ext}} - p_0)\, dA = 0 \qquad \text{(perfect flow)} \quad (6.12)$$

It can be seen then, that for a correctly expanded nozzle ($p_e = p_0$, the jet parallel to the mainstream) and for perfect external flow

$$D_{\text{ext}} = -D_{\text{add}} \qquad (6.13)$$

and

$$(F_p)_{\text{perfect}} = F_A \qquad (6.14)$$

It is evident from Eq. (6.13) that when evaluating the drag terms for a real inlet the additive and external drags must be evaluated most carefully, because the net drag can be the difference between two quite large terms. (Note, of course, that in such a case the external "drag" would be negative.)

It is common practice to break the external drag into two components: D_w, the drag associated with the "front end" of the engine, and D_b, the drag associated with the back end. This is usually a reasonable approach because lip separation often dominates near the inlet and boat-tail drag near the exit. Assuming that this division is meaningful, the terms can be interpreted by considering the engine to be very long and parallel in the middle. In this

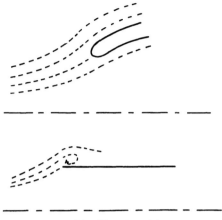

Fig. 6.5   Flow into inlets.

case, perfect flow would give p_0, etc., at the middle and the same argument as led to Eq. (6.12) would lead to

$$D_{\text{add}} + D_w = 0 \qquad \text{(perfect flow)} \qquad (6.15)$$

and

$$D_b + \int_e^\infty (p_{\text{ext}} - p_0)\, \mathrm{d}A = 0 \qquad \text{(perfect flow)} \qquad (6.16)$$

The negative forebody drag required to cancel the additive drag arises from the suction near the leading edge of the inlet. This is why supersonic inlets, with their sharp leading edges, have a large "additive drag penalty" when operated at mass flows other than their design mass flows—their sharp leading edge prevents the suction from occurring (Fig. 6.5). Note that the additive drag penalty for the supersonic inlet is much more severe than that for the subsonic inlet, even though the additive drags (in Fig. 6.5) are identical.

Similarly, separation from the trailing body of the engine prevents the diffusion that would lead to large "forward thrusts," giving boat-tail drag.

One-Dimensional Calculation of the Additive Drag

For simplicity, consider an inlet without external centerbody and obtain from Eqs. (6.6) and (6.9),

$$D_{\text{add}} = \dot{m}_i u_i - \dot{m}_0 u_0 + (p_i - p_0) A_i \qquad (6.17)$$

Noting that $\dot{m}_i = \dot{m}_0 = \rho_i u_i A_i$ it follows that

$$D_{add} = A_i p_i \left[\frac{\rho_i u_i^2}{p_i} \left(1 - \frac{u_0}{u_i} \right) + \left(1 - \frac{p_0}{p_i} \right) \right]$$

or

$$\frac{D_{add}}{A_i p_0} = \frac{p_i}{p_0} \left[\gamma M_i^2 \left(1 - \frac{u_0}{u_i} \right) + \left(1 - \frac{p_0}{p_i} \right) \right]$$

The flow from far upstream to the inlet may be considered to be isentropic so that

$$\frac{T_i}{T_0} = \left(\frac{p_i}{p_0} \right)^{(\gamma-1)/\gamma} = \frac{1 + \frac{\gamma-1}{2} M_0^2}{1 + \frac{\gamma-1}{2} M_i^2} \qquad (6.18)$$

also

$$\frac{u_0}{u_i} = \left(\frac{T_0}{T_i} \frac{M_0^2}{M_i^2} \right)^{\frac{1}{2}}$$

so that

$$\frac{D_{add}}{A_i p_0} = \left(\frac{T_i}{T_0} \right)^{\gamma/(\gamma-1)} \left[\gamma M_i^2 \left(1 - \frac{M_0}{M_i} \sqrt{\frac{T_0}{T_i}} \right) + 1 - \left(\frac{T_0}{T_i} \right)^{\gamma/(\gamma-1)} \right] \qquad (6.19)$$

Thus, when the inlet "design Mach number" (that Mach number existing at the inlet plane for the given engine setting) is known, the dimensionless drag can be obtained as a function of the flight Mach number. It is to be noted that, when the flight Mach number is equal to the design Mach number, the inlet swallows its projected image. As a result, no curvature exists in the entering stream tube and the additive drag is zero. Figure 6.6 shows the behavior of the dimensionless drag with flight Mach number for an inlet with design Mach number M_i equal to 0.6 ($\gamma = 1.4$).

A physical feel for the significance of the additive drag follows by considering it on a large engine, say at sea level (14.7 psi) and start of takeoff ($M_0 = 0$). If the inlet diameter is 7 ft, the additive drag is

$$D_{add} = \frac{\pi}{4} (49)(144)(14.7)(0.179) = 14,580 \text{ lbf}$$

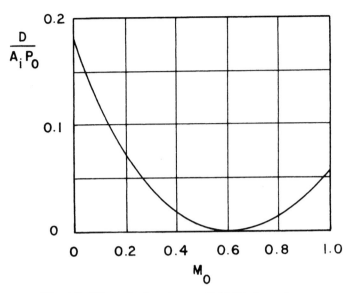

Fig. 6.6   Dimensionless drag vs flight Mach number.

A good inlet would, of course, recover most of this large force through lip suction, but the magnitude of the force tends to serve as a warning that great care must be taken to ensure that the inlet is designed to recover the additive drag without serious penalty.

Some Realities of the Determination of Inlet Drag

It has been emphasized in the preceding section that inlet drag must, in some cases, be determined as the difference between two large quantities, the additive drag and the external pressure drag. If various cowl shapes are to be compared for their drag characteristics, in some way the (net) inlet drag must be obtained accurately. One method of determining inlet drag is to heavily instrument a given cowl with static pressure taps and then to integrate the axial force implied by such measurements.

The internal flow must also be simulated accurately in such a technique in order to give the proper additive drag and correct boat-tail/jet interactions. Even assuming correct internal flow simulation, the location of the contact point of the stagnation streamline must be very accurately determined and the upstream shape of the streamline and pressure at each location accurately estimated through the use of a compressible flow calculation.[1]

It is evident that the one-dimensional approximation for the additive drag will have serious shortcomings in accuracy for several reasons. First, as is evident from Fig. 6.7, the projected area of the stagnation stream tube increases abruptly in the immediate neighborhood of the cowl. It is in this regime that the one-dimensional estimate of the static pressure is poorest, because the local static pressure is approaching the stagnation pressure, whereas the one-dimensional estimate of the static pressure will be that

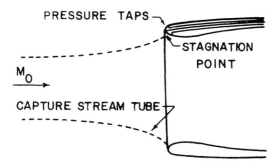

Fig. 6.7 Instrumented cowl and stagnation streamline.

corresponding to the Mach number for one-dimensional flow through the area A_i. The location of the area A_i itself is not precisely determined, because in the one-dimensional approximation it is consistent to assume A_i is the minimum area of the inlet, whereas in a two-dimensional calculation A_i would be that area within the locus of stagnation points and would change with the operating condition.

If the inlet is fully instrumented on the internal surface to give accurate static pressures and a station is available (station 2 in Fig. 6.3) where the internal flow may be assumed to be parallel and hence at constant static pressure, then a momentum balance may be applied to give the following expression for the additive drag D_{add}:

$$D_{\mathrm{add}} = \int_0^i (p - p_a)\, \mathrm{d}A = \int_2 (p_2 - p_a)\, \mathrm{d}A - \int_i^2 (p - p_a)\, \mathrm{d}A + \dot{m}(u_2 - u_0)$$

$$+ \text{``corrections''}$$

The term "corrections" here refers to those contributions arising from the viscous stresses and possibly from the "tunnel corrections" required if the ambient pressure must be adjusted to account for blockage effects. Even given that the corrections can be accurately estimated, it is evident that the additive drag is determined from the contributions of several large quantities. The cowl drag is the difference of the lip suction and the additive drag, and it is hence imperative to accurately measure each separate quantity.

6.3 Averages

It is usual, particularly when conducting performance calculations, to refer to average quantities at a particular location. This is especially the case with stagnation properties. In fact, it can be extremely important to understand the implications of the particular averaging process being utilized, and to this end, three different averaging techniques will be applied to a "step profile" in order to reveal the effects that averaging methods can have (see Fig. 6.8). The three averaging methods (for the stagnation pressure) to be considered are: the mass, stream thrust, and continuity averages.

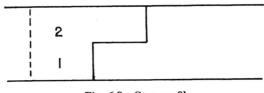

Fig. 6.8 Step profile.

The Mass Average

The mass average stagnation pressure is defined by

$$p_{t_{\text{m.a.}}} = \int p_t \, d\dot{m} \Big/ \int d\dot{m} \qquad (6.20)$$

For the case of the step profile this reduces to

$$p_{t_{\text{m.a.}}} = \frac{\dot{m}_1 p_{t_1} + \dot{m}_2 p_{t_2}}{\dot{m}_1 + \dot{m}_2} = p_{t_1} \frac{1 + \alpha\left(p_{t_2}/p_{t_1}\right)}{1 + \alpha} \qquad (6.21)$$

Stream Thrust Average

The stream thrust stagnation pressure $p_{t_{\text{s.t.}}}$ is defined as that stagnation pressure that would exist if the flow was allowed to completely mix (hence the sometimes used expression "mixed-out average") in a frictionless constant-area mixer. It is assumed here (and in virtually all averaging methods) that the appropriate stagnation temperature average is the mass average. (If such was not assumed we would find ourselves breaking the first law of thermodynamics, let alone the second law!)

The appropriate equations for this case have already been worked out in Sec. 5.13, leading to Eq. (5.105). It is necessary to prescribe the Mach number of one of the streams (say M_1) in order to fix the static pressure at the splitter plate. Utilizing the relationship for $p_{t_{\text{m.a.}}}$ given above, and introducing $\pi \equiv p_{t_2}/p_{t_1}$ and $\tau \equiv T_{t_2}/T_{t_1}$, the expression for $p_{t_{\text{s.t.}}}$ for the two-stream case can be arranged as a hierarchy of equations. Thus,

Inputs: $\qquad\qquad\qquad \pi, \tau, M_1, \alpha$

Equations:

$$M_2^2 = \frac{2}{\gamma - 1}\left[\pi^{(\gamma-1)/\gamma}\left(1 + \frac{\gamma-1}{2}M_1^2\right) - 1\right] \qquad (6.22)$$

$$\frac{A_2}{A_1} = \alpha\sqrt{\tau}\,\frac{M_1}{M_2}\pi^{-[(\gamma-1)/2\gamma]} \qquad (6.23)$$

$$\phi_{\text{s.t.}} = \frac{(1+\alpha)(1+\alpha\sqrt{\tau})}{\left[1/\sqrt{\phi_1} + \alpha\sqrt{\tau}/\sqrt{\phi_2}\right]^2} \qquad (6.24)$$

where:

$$\phi_i \equiv \frac{M_i^2 \left(1 + \frac{\gamma - 1}{2} M_i^2 \right)}{\left(1 + \gamma M_i^2 \right)^2} \tag{6.25}$$

$$\Delta = \sqrt{1 - 2(\gamma + 1)\phi_{\text{s.t.}}} \tag{6.26}$$

$$M_{\text{s.t.}} = \left(\frac{2\phi_{\text{s.t.}}}{1 - 2\gamma\phi_{\text{s.t.}} + \Delta} \right)^{\frac{1}{2}} \tag{6.27}$$

$$\frac{p_{t_{\text{s.t.}}}}{p_{t_{\text{m.a.}}}} = \frac{1 + \alpha}{1 + \alpha\pi} \left[(1 + \alpha)(1 + \alpha\tau) \right]^{\frac{1}{2}} \left[\frac{1 + \frac{\gamma - 1}{2} M_{\text{s.t.}}^2}{1 + \frac{\gamma - 1}{2} M_1^2} \right]^{(\gamma + 1)/2(\gamma - 1)} \frac{M_1}{M_{\text{s.t.}}} \frac{1}{1 + \frac{A_2}{A_1}} \tag{6.28}$$

Continuity Average

The continuity average stagnation pressure $p_{t_{\text{c.a.}}}$ is defined as that stagnation pressure calculated with the assumption that true one-dimensional flow exists at the station and given (measured) values of p, $\dot{m}$, A, and T_t. Here, of course, T_t is the mass-averaged stagnation temperature.

From Eq. (2.103) note

$$\dot{m} = \sqrt{\frac{2\gamma}{R(\gamma - 1)}} \frac{A}{\sqrt{T_t}} p^{1/\gamma} \left(y^2 - p^{[(\gamma - 1)/\gamma]} y \right)^{\frac{1}{2}} \tag{6.29}$$

where $y \equiv p_t^{(\gamma - 1)/\gamma}$.

This is easily inverted to give

$$y \equiv p_{t_{\text{c.a.}}}^{(\gamma - 1)/\gamma} = \frac{p^{(\gamma - 1)/\gamma}}{2} \left[1 + \sqrt{1 + \frac{2(\gamma - 1)}{\gamma} \frac{RT_t}{A^2} \frac{\dot{m}^2}{p^2}} \right] \tag{6.30}$$

In practice this just gives a formula for obtaining $p_{t_{\text{c.a.}}}$ in terms of the measured quantities. However, because the actual conditions existing in the stream are "known," $p_{t_{\text{c.a.}}}$ can be calculated in terms of p_{t_1}, p_{t_2}, etc. Thus, noting for the ith stream that

$$A_i = \sqrt{\frac{R(\gamma - 1)}{2\gamma}} \left\{ \dot{m} \frac{\sqrt{T_t}}{p} \left(\frac{p_t}{p} \right)^{-[(\gamma - 1)/2\gamma]} \left[\left(\frac{p_t}{p} \right)^{(\gamma - 1)/\gamma} - 1 \right]^{-\frac{1}{2}} \right\}_i$$

and

$$A = A_1 + A_2$$

and

$$T_t = \frac{1 + \alpha\tau}{1 + \alpha} T_{t_1}$$

it follows that

$$p_{t_{c.a.}}^{(\gamma-1)/\gamma} = \frac{p^{(\gamma-1)/\gamma}}{2}$$

$$\times \left\{ 1 + \left[1 + \frac{4(1+\alpha)(1+\alpha\tau)\left(\dfrac{p_{t_1}}{p}\right)^{(\gamma-1)/\gamma}}{\left\{ \left[\left(\dfrac{p_{t_1}}{p}\right)^{(\gamma-1)/\gamma} - 1 \right]^{-\frac{1}{2}} + \dfrac{\alpha\sqrt{\tau}}{\pi^{(\gamma-1)/2\gamma}} \left[\left(\dfrac{p_{t_2}}{p}\right)^{(\gamma-1)/\gamma} - 1 \right]^{-\frac{1}{2}} \right\}^2} \right]^{\frac{1}{2}} \right\}$$

$$(6.31)$$

then

$$\left(\frac{p_{t_{c.a.}}}{p_{t_{m.a.}}}\right)^{(\gamma-1)/\gamma} = \left(\frac{1+\alpha}{1+\alpha\pi}\right)^{(\gamma-1)/\gamma} \frac{1}{2} \left(\frac{p_{t_1}}{p}\right)^{-[(\gamma-1)/\gamma]}$$

$$\times \left\{ 1 + \left[1 + \frac{4(1+\alpha)(1+\alpha\tau)\left(\dfrac{p_{t_1}}{p}\right)^{(\gamma-1)/\gamma}\left[\left(\dfrac{p_{t_1}}{p}\right)^{(\gamma-1)/\gamma} - 1\right]}{\left\{ 1 + \dfrac{\alpha\sqrt{\tau}}{\pi^{(\gamma-1)/2\gamma}} \left[\dfrac{\left(\dfrac{p_{t_1}}{p}\right)^{(\gamma-1)/\gamma} - 1}{\pi^{(\gamma-1)/\gamma}\left(\dfrac{p_{t_1}}{p}\right)^{(\gamma-1)/\gamma} - 1} \right]^{\frac{1}{2}} \right\}^2} \right]^{\frac{1}{2}} \right\}$$

$$(6.32)$$

This form is appropriate for calculation because, given M_1, p_{t_1}/p follows, as does p_{t_2}/p with π, etc. The formula can be obtained explicitly in terms of π, τ, M_1, and α (for comparison with the stream thrust average, etc.), however, and after some manipulation there is obtained

$$\left(\frac{p_{t_{c.a.}}}{p_{t_1}}\right)^{(\gamma-1)/\gamma} = 1 + \frac{1+(\gamma-1)M_1^2}{2+(\gamma-1)M_1^2}$$

$$\times\left\{-1+\left[1+\frac{(\gamma-1)M_1^2\left[2+(\gamma-1)M_1^2\right]}{\left[1+(\gamma-1)M_1^2\right]^2}\right.\right.$$

$$\left.\left.\times\left(\frac{(1+\alpha)(1+\alpha\tau)}{\left\{1+\dfrac{\alpha\sqrt{\tau}}{\pi^{(\gamma-1)/2\gamma}}\left[\dfrac{\dfrac{\gamma-1}{2}M_1^2}{\pi^{(\gamma-1)/2\gamma}\left(1+\dfrac{\gamma-1}{2}M_1^2\right)-1}\right]^{\frac{1}{2}}\right\}^2}-1\right)\right]^{\frac{1}{2}}\right\} \quad (6.33)$$

This has been written in this way to make it clear that when π and τ go to one, the ratio is unity.

Comments

It is important to comprehend the implications of these various averaging techniques, because many serious problems have arisen in industry through improper interpretation. Thus note the following.

Mass average. This is the thermodynamically appropriate average in the sense that p_t so defined would follow the dictates of the second law. Thus, for example, if two streams of the same stagnation temperature continued to flow adiabatically, the stagnation pressure so defined would decrease. (This is because the entropy is increasing.) Note that this would be true even if the sidewall friction was negligible, because the mixing process generates entropy.

In practice, however, it is the very tendency of the stagnation pressure, mass averaged, to decrease that makes its use unpopular in some circumstances. Thus, for example, if a wake traverse is taken close behind a cascade of aerofoils, the stagnation pressure so measured would be higher than that measured some distance further downstream. Many companies feel that it is more appropriate to utilize the stream thrust average, which, of course, does not change (in a constant-area channel). Note the stream thrust average hence represents the "mixed-out" limit of the mass-averaged stagnation pressure.

Stream thrust average. As discussed above, this average is often utilized because it tends to represent a somewhat conservative value of the stagnation pressure for use in performance analysis. In utilizing this definition, note that all other properties should also be defined as those that would exist at the exit from a constant-area ideal mixer. For example, the static pressure is *not* that actually measured in the channel, but rather is that calculated from the constant-area mixer equations.

Continuity average. This average is popular in some cases where, because of physical restraints or adverse fluid properties (extreme temperatures etc.), direct stagnation probe traverses cannot be taken. It is easy to verify that for flows with uniform stagnation temperatures the continuity-averaged stagnation pressure is the lowest of the three. What this means is that, for flow in a constant-area duct with no sidewall friction (and uniform stagnation temperature), the mass-averaged stagnation pressure goes down, the stream-thrust-averaged stagnation pressure remains the same, and the continuity-averaged stagnation pressure goes up!

There was a case in a highly respected research laboratory where, because of the difficulty in utilizing upstream stagnation pressure probes, the upstream stagnation pressure was determined utilizing the continuity average. A careful traverse of the outlet flow (utilizing the mass average) then revealed that the cascade being tested had the remarkable property that it increased the stagnation pressure!

Example 6.1

A flow exists where unknown to the investigators a separation exists. This flow can be approximated as a flow with "freestream" stagnation pressure p_t through the area A_B (area with blockage) and with no flow in the blocked

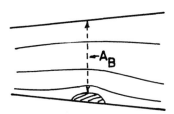

Fig. 6.9   Channel with blockage.

region. See Fig. 6.9. Obviously, the mass-averaged stagnation pressure is just p_t.

Denoting by the subscript 0 conditions that would exist if no blockage were present:

(1) Show that the ratio $p_{t_{c.a.}}/p_{t_{m.a.}}$ can be calculated from the following hierarchy of equations

Inputs: $\qquad \gamma, M_0, A_B/A_0$

Equations:

$$f(p_0) = \frac{\dfrac{\gamma-1}{2}M_0^2}{\left(1 + \dfrac{\gamma-1}{2}M_0^2\right)^{(\gamma+1)/(\gamma-1)}} \qquad (6.34)$$

$$f(p_B) = f(p_0)\left(\frac{A_0}{A_B}\right)^2 \qquad (6.35)$$

$$\frac{p}{p_t} = \left[1 - \frac{f(p_B)}{(p/p_t)^{2/\gamma}}\right]^{\gamma/(\gamma-1)} \qquad (6.36)$$

$$\frac{p_{t_{c.a.}}}{p_{t_{m.a.}}} = \frac{p}{p_t}\left\{\frac{1 + \left[1 + 4\dfrac{f(p_0)}{(p/p_t)^2}\right]^{\frac{1}{2}}}{2}\right\}^{\gamma/(\gamma-1)} \qquad (6.37)$$

(2) The limitingly small area at blockage will occur when $M_B \to 1$, hence show for this case

$$\left(\frac{A_{BL}}{A_0}\right)^2 = \frac{M_0^2}{\left[\dfrac{2}{\gamma+1}\left(1 + \dfrac{\gamma-1}{2}M_0^2\right)\right]^{(\gamma+1)/(\gamma-1)}} \qquad (6.38)$$

and show that for this case

$$\frac{p}{p_t} = \left(\frac{2}{\gamma+1}\right)^{\gamma/(\gamma-1)}$$

6.4 The Inlet

The remainder of this chapter will cover the internal behavior of the engine; the inlet will be considered first.

Inlet losses arise because of the presence of wall friction, shock waves, and regimes of separated flow. See Fig. 6.10. All of these loss mechanisms

Fig. 6.10 Internal losses in an inlet.

result in a reduction in stagnation pressure so that

$$\pi_d < 1 \qquad (6.39)$$

Virtually all inlets are adiabatic to a very high degree of approximation, so

$$\tau_d = 1 \qquad (6.40)$$

The design of subsonic inlets is dominated by the requirements to retard separation at extreme angle of attack and high air demand (as would occur in a two-engine aircraft with engine failure at takeoff) and to retard the onset of both internal and external shock waves in transonic flight. These two requirements tend to be in conflict, because a somewhat "fat" lower inlet lip best suits the high-angle-of-attack requirement, whereas a thin inlet lip best suits the high Mach number requirement. Modern development for the best compromise design is greatly aided by the advent of high-speed electronic computation, which allows analytical estimation of the complex flowfields and related losses.

Supersonic Inlets

Estimation of the losses within supersonic inlets is an easier task than for subsonic inlets for the simple reason that the major losses occur across shock waves, and hence may be estimated using the relatively simple shock wave formulas. More exacting estimates require estimation of the boundary-layer and separation losses.

In order to gain an understanding of some of the physical processes involved, a deceptively simple apparatus—the fixed geometry inlet—is considered (see Fig. 6.11). When the performance penalties of such an inlet are understood, it becomes evident why more complicated geometries are considered in the higher Mach number ranges. Thus, first postulate the existence of an inlet designed to isentropically retard ("diffuse") the flow from a flight Mach number $M_0 > 1$, through Mach 1 at the throat, and subsequently to $M < 1$ at station 2.

Unfortunately, the inlet suggested (and illustrated in Fig. 6.12) will not behave as postulated because (1) such an inlet cannot be "started" by conventional flight practice, and (2) the flow in such an inlet is unstable.

In order to comprehend the "starting problem," consider the behavior of the flow in the inlet at a flight Mach number M_0, less than the design flight

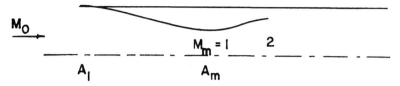

Fig. 6.11   Fixed geometry inlet.

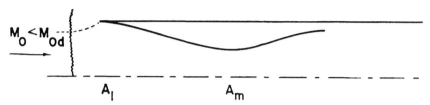

Fig. 6.12   Fixed geometry inlet in unstarted condition.

Mach number, M_{0d}. It is apparent that the inlet-to-throat area ratio A_1/A_m must have a unique value in order to bring the flow from M_{0d} at A_1 down to $M = 1$ at A_m. When the flight Mach number is less than M_{0d}, this fixed area ratio A_1/A_m is too large, and the inlet cannot swallow the mass flow approaching the area A_1. As a result, a normal shock wave appears in front of the inlet and the subsonic flow behind the normal shock wave is partially spilled around the inlet.

It is to be noted that even when the inlet is accelerated to give $M_0 = M_{0d}$, the normal shock wave present in front of the inlet will decrease the stagnation pressure so that the throat will still be unable to pass the desired mass flow and the shock wave will remain in front of the inlet. It is possible to consider diving the aircraft in order to increase the flight Mach number sufficiently to have the shock wave pass into the inlet. Once this state is reached, a slight movement of the shock wave further into the inlet will cause a decrease in the local Mach number entering the shock wave. As a consequence, the stagnation pressure behind the shock wave will increase slightly, leading to a larger mass flow capability for the throat that in turn leads to a reduction in the pressure behind the shock wave and the shock wave "snapping" through the throat. Following this rather demanding maneuver, the pilot could then carefully decelerate to $M_0 = M_{0d}$, at which time the Mach number at the throat would be unity and the inlet would be shock free.

Unfortunately, even if the difficult maneuvering described in the preceding paragraph was carried out, the inlet would not be stable to small flow disturbances. Thus, if an upstream gust occurred to momentarily reduce the flight Mach number to less than M_{0d}, A_1/A_m would again be too large to swallow the flow supplied and a shock wave would form within the inlet and snap to the outside, thereby "unstarting" the inlet. Similarly, if a downstream disturbance such as a momentary decrease in engine air demand occurred, the throat would be back pressured, again causing a shock to form and the inlet to unstart.

The Kantrowitz-Donaldson Inlet

In an early paper[2] it was suggested that in view of the operational and stability problems inherent in the fixed geometry "ideal" inlet described in the preceding paragraphs, it would be appropriate to size a fixed geometry inlet so that the shock wave in front of the engine would be swallowed just as the inlet reached flight Mach number M_{0_d}. This, as will be shown quantitatively in the following, would require a larger throat area and would lead to a Mach number at the throat M_m larger than unity. It was proposed that the inlet be operated with the engine air demand such as to cause the normal shock wave to stabilize slightly downstream of the throat. (That is to say, the inlet would be operated in a slightly "supercritical" condition.) By operating in this condition, the inlet would be made stable to both upstream and downstream (small) disturbances. It is to be noted that the presence of the shock wave will, of course, introduce stagnation pressure losses and that, by operating with the shock wave slightly downstream of the throat to enhance stability, an additional penalty is paid because the local Mach number approaching the shock wave will be larger than that at the throat and hence the shock wave will be stronger.

Analysis

Note that the very best performance for a fixed throat inlet will occur when the throat is sized so that, with a normal shock at the inlet face, the inlet will *just* swallow the air coming from the shock. As a result the shock wave will "just" enter the inlet and hence be swallowed as described earlier. Then it is postulated that the shock wave be stabilized "at" the throat, even though that configuration would be unstable to downstream disturbances.

Figure 6.13 shows conditions at the initiation of the shock swallowing process. The stagnation pressure ratio across the normal shock wave is given by[3]

$$\frac{p_{t_1}}{p_{t_0}} = \left(\frac{\dfrac{\gamma+1}{2}M_0^2}{1+\dfrac{\gamma-1}{2}M_0^2} \right)^{\gamma/(\gamma-1)} \left[1 + \frac{2\gamma}{\gamma+1}\left(M_0^2 - 1 \right) \right]^{-[1/(\gamma-1)]} \qquad (6.41)$$

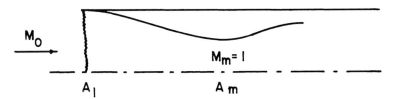

Fig. 6.13   Inlet at instant of starting.

Because the Mach number at the throat, at this instant of starting, is $M_m = 1$, Eq. (2.105) gives

$$\frac{A_m}{A_1} = M_0 \left[\frac{2}{\gamma + 1} \left(1 + \frac{\gamma - 1}{2} M_0^2 \right) \right]^{-[(\gamma+1)/2(\gamma-1)]} \frac{p_{t_0}}{p_{t_1}} \qquad (6.42)$$

and hence

$$\frac{A_m}{A_1} = M_0^{-[(\gamma+1)/(\gamma-1)]} \left[\frac{2}{\gamma + 1} \left(1 + \frac{\gamma - 1}{2} M_0^2 \right) \right]^{\frac{1}{2}} \left[1 + \frac{2\gamma}{\gamma + 1} \left(M_0^2 - 1 \right) \right]^{1/(\gamma-1)}$$

$$(6.43)$$

This area ratio, which is the minimum area ratio that will allow the inlet to "self-start" at the Mach number M_0, is referred to as the Kantrowitz-Donaldson contraction ratio.

When the shock wave just enters the inlet, it will continue toward the throat, as described earlier. The limitingly good performance of the inlet will occur when the shock is stabilized right at the throat. In order to calculate the corresponding stagnation pressure ratio, it is necessary to obtain the Mach number at the throat M_{m_1} immediately upstream of the shock wave. See Fig. 6.14.

The flow prior to the shock wave is isentropic, so Eq. (2.105) may be utilized to obtain M_{m_1}. Thus write

$$F(M_{m_1}) = \frac{M_0}{M_{m_1}} \left(\frac{1 + \frac{\gamma - 1}{2} M_{m_1}^2}{1 + \frac{\gamma - 1}{2} M_0^2} \right)^{(\gamma+1)/2(\gamma-1)} - \frac{A_m}{A_1} = 0 \qquad (6.44)$$

This equation is solved by Newtonian iteration. Thus, noting

$$F' = \frac{M_0}{M_{m_1}^2} \left(\frac{M_{m_1}^2 - 1}{1 + \frac{\gamma - 1}{2} M_{m_1}^2} \right) \left(\frac{1 + \frac{\gamma - 1}{2} M_{m_1}^2}{1 + \frac{\gamma - 1}{2} M_0^2} \right)^{(\gamma+1)/2(\gamma-1)}$$

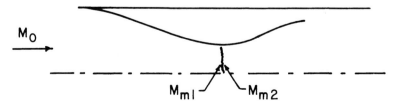

Fig. 6.14   Ideal fixed geometry inlet in started condition.

Table 6.1   Fixed Geometry Normal Shock Inlet Performance ($\gamma = 1.4$)

M_0	A_m/A_1	M_m	π_d
1.2	0.977	1.10	0.999
1.6	0.893	1.40	0.957
2.0	0.822	1.75	0.834
2.4	0.770	2.11	0.670
2.8	0.733	2.47	0.511

leads to

$$\left(M_{m_1}\right)_{j+1} = \left(M_{m_1}\right)_j - \left(F/F'\right)_j$$

The iteration is to be initiated with a suitable supersonic first guess for M_{m_1}. Following solution for M_{m_1}, the stagnation pressure ratio

$$\frac{p_{t_{m_2}}}{p_{t_{m_1}}} = \frac{p_{t_2}}{p_{t_0}} = \pi_d$$

follows directly from Eq. (6.41) with M_0 replaced by M_{m_1}. If the Mach number immediately following the shock M_{m_2} is desired, it follows from

$$M_{m_2}^2 = \frac{1 + \dfrac{\gamma - 1}{2} M_{m_1}^2}{\gamma M_{m_1}^2 - \dfrac{\gamma - 1}{2}} \tag{6.45}$$

The on-design performance of a family of fixed throat inlets, calculated using the above equations, is tabulated in Table 6.1 for $\gamma = 1.4$.

Off-Design Performance— Fixed Geometry

It is evident from Table 6.1 that the design performance of a fixed geometry normal shock inlet degrades considerably for flight Mach numbers in excess of about $M_0 = 1.6$. It is the off-design performance that most restricts the acceptable performance of this class of inlet, however. First, consider the performance of a given fixed throat inlet as the flight Mach number is varied from $M_0 = 1$ up to $M_0 > M_{0d}$ and then returned to $M_0 = 1$. Assume that the engine demand is adjusted to keep the normal shock at the throat, if possible.

The resulting performance is depicted in Fig. 6.15. Thus, as M_0 increases, π_d corresponds to that for a normal shock at the flight Mach number. At $M_0 = M_{0d}$ the shock snaps into the inlet, giving the design value of π_d; then as M_0 continues to increase, the shock wave at the throat strengthens, leading to increased shock losses.

The performance of the inlet upon reduction of M_0 repeats that with increasing M_0, except when M_0 passes through M_{0d}. With good engine matching and no incoming disturbances, the shock wave can continue to be stabilized at the throat. This situation continues (but becomes progressively more unstable) until the throat Mach number reaches unity and the shock snaps into the freestream.

The variation of inlet pressure ratio with mass flow rate at fixed flight Mach number is shown in Fig. 6.16. If engine mass flow demand is decreased from the design value, the inlet plenum pressure will rise, forcing the shock wave forward and hence unstarting the inlet. When engine mass flow demand increases, the inlet mass flow remains at the value passing through the throat, but the normal shock progresses further into the inlet. As a consequence the stagnation pressure decreases (leading to an increase in "corrected mass flow").

The off-design characteristics discussed here are quite unsuitable and for this reason fixed geometry, normal shock inlets are used only for aircraft with low supersonic Mach number capability. Inlet concepts offering increased performance over a wider operating range were illustrated in Fig. 1.6.

The external compression inlet offers the advantages of relatively simple construction, short axial length, and good off-design performance. Note that the final normal shock wave need not unstart in such an inlet, but rather need only move sufficiently far forward to allow the required spillage. The inlet ramp angles must be adjustable if the "shock-on-lip" design is to be

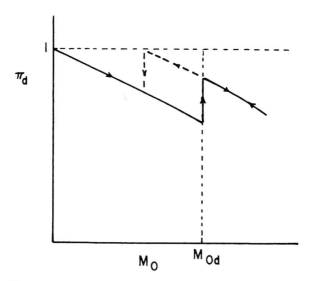

Fig. 6.15 Fixed geometry inlet off-design behavior with variation in flight Mach number.

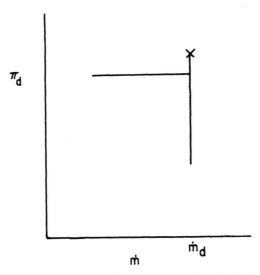

Fig. 6.16   Fixed geometry inlet off-design behavior with variation in mass flow rate.

maintained at other than a single design flight Mach number. A disadvantage of the external compression inlet arises when used at high flight Mach numbers because the required flow turning is so great that the external cowl angle becomes excessive and generates strong external shocks with consequent high drag.

The internal compression inlet does not suffer from the onset of excessive external drags at high design flight Mach numbers, but does have its own disadvantages. Thus, the geometry is such that excessive inlet lengths must be utilized, and the off-design characteristics can be unacceptable if sophisticated variable geometry is not employed. In order to "start" the inlet, the shock system must first be swallowed and the geometry then varied to locate the normal shock near the throat. Quick-acting throat and subsonic diffuser bleed systems must also be provided to prevent the sudden disgorging of the shock system (inlet "unstart") with a change in engine air demand.

The mixed-flow inlet design offers a useful compromise to these two designs for use at high flight Mach numbers. Although it suffers from some of the disadvantages of both, it offers the possibility of decreased inlet length and geometrical complication compared to the internal compression inlet.

Estimation of Losses in Supersonic Inlets

An upper limit to supersonic inlet performance can be estimated by assuming that all the losses occur across the shock waves.

Figure 6.17 indicates the geometry and nomenclature for shock wave interaction, and includes an example external compression inlet with three

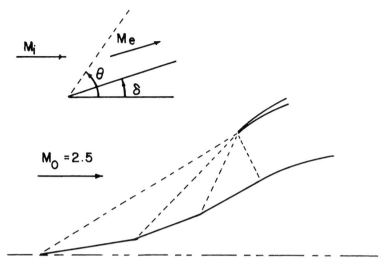

Fig. 6.17   Shock wave nomenclature and example of external compression inlet.

oblique shock waves followed by a normal shock wave. With the availability of desk-top computers or calculators with branching and looping capability, it is now more convenient to calculate desired quantities directly, rather than to refer to tables. To this end, the following summary of equations is suggested for estimation of inlet shock losses.

Summary— Shock Losses

Input for oblique shocks: M_i, δ, γ

Equations:

$$y = \frac{1}{M_i^2} + \left(\frac{1}{M_i^2} + \frac{\gamma + 1}{2} - y\right)\left(\frac{y}{1 - y}\right)^{\frac{1}{2}} \tan\delta \qquad (6.46)$$

To solve this equation, assume $y = 1/M_i^2$ on the right side and calculate a new value of y. Use this new value of y as an updated value on the right side. Continue until appropriate convergence. (Note y is actually $\sin^2\theta$.)

$$\frac{p_{t_e}}{p_{t_i}} = \left[\frac{\gamma + 1}{2\gamma M_i^2 y - (\gamma - 1)}\right]^{1/(\gamma - 1)}\left[\frac{(\gamma + 1) M_i^2 y}{2 + (\gamma - 1) M_i^2 y}\right]^{\gamma/(\gamma - 1)} \qquad (6.47)$$

$$M_e = \left\{\frac{4 + 4(\gamma - 1) M_i^2 y + (\gamma + 1)^2 M_i^4 y - 4\gamma M_i^4 y^2}{[2\gamma M_i^2 y - \gamma + 1][2 + (\gamma - 1) M_i^2 y]}\right\}^{\frac{1}{2}} \qquad (6.48)$$

The equivalent expressions for the normal shock wave have already been presented as Eqs. (6.41) and (6.45).

It is to be noted that two limiting values of the ramp angle occur, δ^* which is that ramp angle just leading to $M_e = 1$ and δ_{max}, the maximum ramp angle for which an oblique shock solution exists. It can be shown[3] that

$$(\sin^2\theta)_{\delta_{max}} = \frac{1}{4\gamma M_i^2}((\gamma + 1)M_i^2 - 4$$

$$+ \left\{(\gamma + 1)[(\gamma + 1)M_i^4 + 8(\gamma - 1)M_i^2 + 16]\right\}^{\frac{1}{2}}) \qquad (6.49)$$

$$\sin^2\theta^* = \frac{1}{4\gamma M_i^2}((\gamma + 1)M_i^2 - (3 - \gamma)$$

$$+ \left\{(\gamma + 1)[(\gamma + 1)M_i^4 - 2(3 - \gamma)M_i^2 + \gamma + 9]\right\}^{\frac{1}{2}}) \qquad (6.50)$$

For both cases the related value of δ follows from

$$\tan\delta = \frac{2\cot\theta(M_i^2\sin^2\theta - 1)}{2 + M_i^2(\gamma + 1 - 2\sin^2\theta)} \qquad (6.51)$$

The related value of stagnation pressure follows from Eq. (6.47) (with $y \equiv \sin^2\theta$).

Effect of Distortion in Inlets

It is important to note that a major problem with inlets centers about the lack of one-dimensionality of the flow. The shock system and wall friction lead to areas of reduced stagnation pressure at the exit from the diffuser. The static pressure is very nearly constant across the diffuser exit, so a reduced stagnation pressure corresponds to a reduced axial velocity. Such a reduced velocity will cause the rotor blade to encounter a sudden increase in the angle of attack with the consequent possibility of a blade stall. In addition, the shock system is unstable and produces time-varying fluctuations in the stagnation pressure.

As a result of these distortion effects, engine compressors must be designed with sufficient tolerance for distortion to operate without stalling when distortion is present. Unfortunately, as is to be made evident in Sec. 8.1, increased tolerance can be provided only at the expense of decreased design performance.

6.5 The Compressor

Compressors are, to a high degree of approximation, adiabatic, so that the work interaction across the compressor per mass is just equal to the change in the stagnation enthalpy per mass. Assuming that the gas is calorically perfect across the compressor makes it easy to relate the temperature change to the desired pressure change. There are, in fact, three related definitions of efficiency of use in describing compressor behavior, each of which is described in the following sections.

The Compressor Efficiency, η_c

The compressor efficiency is defined by

$$\eta_c = \frac{\text{ideal work interaction for a given pressure ratio}}{\text{actual work interaction for a given pressure ratio}}$$

$$= \frac{C_p\left(T_{t_{3i}} - T_{t_2}\right)}{C_p\left(T_{t_3} - T_{t_2}\right)} = \frac{\left[\left(T_{t_{3i}}/T_{t_2}\right) - 1\right]}{\left[\left(T_{t_3}/T_{t_2}\right) - 1\right]} \equiv \frac{\tau_{ci} - 1}{\tau_c - 1}$$

The ideal process is an isentropic process so that from Eq. (2.57)

$$\frac{T_{t_{3i}}}{T_{t_2}} = \left(\frac{p_{t_3}}{p_{t_2}}\right)^{(\gamma_c - 1)/\gamma_c} \equiv \pi_c^{(\gamma_c - 1)/\gamma_c}$$

hence

$$\eta_c = \frac{\pi_c^{(\gamma_c - 1)/\gamma_c} - 1}{\tau_c - 1} \tag{6.52}$$

Thus, for example, if the desired pressure ratio is given and the compressor efficiency estimated, the stagnation temperature ratio and hence required work interaction may be obtained.

The Compressor Polytropic Efficiency, e_c

This efficiency, which is related to the compressor efficiency, is defined by

$$e_c = \frac{\text{ideal work interaction for a given differential pressure change}}{\text{actual work interaction for a given differential pressure change}}$$

The concept embodied in the use of the polytropic efficiency is that if it may be assumed that the stage efficiencies are constant throughout a given compressor (it will be shown shortly that the stage efficiency is very nearly the polytropic efficiency), then by assuming that e_c is a constant, the effect of increasing the compressor pressure ratio (by adding stages) upon the compressor efficiency may be estimated. Thus, when conducting a cycle analysis the most appropriate compressor pressure ratio may be selected. So,

$$e_c = \frac{dh_{ti}}{dh_t} \tag{6.53}$$

The Gibbs equation (2.12) gives

$$T_t \, ds_t = dh_t - (1/\rho_t) \, dp_t$$

but for the ideal process $ds_{ti} = 0$, so that

$$dh_{ti} = (1/\rho_t) \, dp_t \tag{6.54}$$

Thus, writing $dh_t = C_p \, dT_t$, Eqs. (6.53) and (6.54) together with the equation of state give

$$e_c = \frac{dp_t}{\rho_t C_p \, dT_t} = \frac{R}{C_p} \frac{dp_t/p_t}{dT_t/T_t}$$

hence

$$\frac{dT_t}{T_t} = \frac{\gamma_c - 1}{\gamma_c e_c} \frac{dp_t}{p_t} \tag{6.55}$$

Assuming that e_c is constant over the range of interest (this is similar to assuming that each stage efficiency is the same), Eq. (6.55) may be integrated immediately to give

$$\frac{T_{t_3}}{T_{t_2}} = \left(\frac{p_{t_3}}{p_{t_2}} \right)^{(\gamma_c-1)/\gamma_c e_c} \quad \text{or} \quad \tau_c = \pi_c^{(\gamma_c-1)/\gamma_c e_c} \tag{6.56}$$

Equations (6.52) and (6.56) give an equation relating η_c, e_c, and the compressor pressure ratio π_c,

$$\eta_c = \frac{\pi_c^{(\gamma_c-1)/\gamma_c} - 1}{\pi_c^{(\gamma_c-1)/\gamma_c e_c} - 1} \tag{6.57}$$

This relationship is plotted in Fig. 6.18.

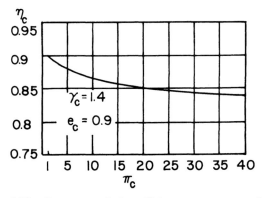

Fig. 6.18 Compressor design efficiency vs pressure ratio.

It can be seen that the compressor efficiency goes down with the increase in the compressor pressure ratio for a given fixed value of e_c. It must be emphasized here that this behavior of compressor efficiency with pressure ratio reflects the estimated behavior of a family of compressors designed to different pressure ratios (and hence incorporating different numbers of stages). Such a curve in no way reflects the expected behavior for a given compressor operating off-design.

The Compressor Stage Efficiency

The compressor stage efficiency is defined in a completely analogous manner to the compressor efficiency, except that the reference pressure ratio is that of the stage itself. Thus,

$$\eta_{c_j} = \frac{\left(\pi_{c_j}\right)^{(\gamma_c-1)/\gamma_c} - 1}{\tau_{c_j} - 1} \tag{6.58}$$

where the subscript j refers to the jth stage of N total stages.

To relate the stage efficiencies to the compressor efficiency, note

$$\frac{T_{t_j}}{T_{t_{j-1}}} = \tau_{c_j} = 1 + \frac{1}{\eta_{c_j}}\left[\left(\pi_{c_j}\right)^{(\gamma_c-1)/\gamma_c} - 1\right]$$

and hence

$$\tau_c = \frac{T_{tN}}{(T_t)_0} = \prod_{j=1}^{N} \tau_{c_j} = \prod_{j=1}^{N} \left\{1 + \frac{1}{\eta_{c_j}}\left[\left(\pi_{c_j}\right)^{(\gamma_c-1)/\gamma_c} - 1\right]\right\} \tag{6.59}$$

Thus

$$\eta_c = \frac{\pi_c^{(\gamma_c-1)/\gamma_c} - 1}{\prod_{j=1}^{N} \left\{1 + \left(1/\eta_{c_j}\right)\left[\left(\pi_{c_j}\right)^{(\gamma_c-1)/\gamma_c} - 1\right]\right\} - 1} \tag{6.60}$$

Equation (6.60) gives a method of predicting the efficiency of a compressor from the (possibly measured) efficiencies of the individual stages. Note, of course, that

$$\pi_c = \prod_{j=1}^{N} \pi_{c_j} \tag{6.61}$$

In the special case where all stage efficiencies (η_s) and stage pressure ratios are equal, Eq. (6.60) reduces to the special form

$$\eta_c = \frac{\pi_c^{(\gamma_c-1)/\gamma_c} - 1}{\left[1 + (1/\eta_s)\left(\pi_c^{(\gamma_c-1)/\gamma_c N} - 1\right)\right]^N - 1} \tag{6.62}$$

Relationship between the Compressor, Stage, and Polytropic Efficiencies

Equations (6.57), (6.60), and (6.62) give analytical relationships between the various definitions of efficiency, but it should be of interest to see if η_s formally approaches e_c as the number of stages, for a given pressure ratio, gets very large. (That is, the pressure ratio per stage approaches a differential pressure ratio per stage.)

For convenience write

$$y \equiv \pi_c^{(\gamma_c-1)/\gamma_c\eta_s} \tag{6.63}$$

and

$$1 + (1/\eta_s)(y^{\eta_s/N} - 1) = x^{1/N} \tag{6.64}$$

Noting that $y^{\eta_s/N} = \exp(\eta_s/N\ell ny) = 1 + (\eta_s/N)\ell ny + O(1/N^2)$, the left side of Eq. (6.64) becomes

$$\left[1 + \frac{1}{N}\ell ny\right] \tag{6.65}$$

Then writing

$$x^{1/N} = \exp(1/N\ell nx) = 1 + (1/N)\ell nx + O(1/N^2) \tag{6.66}$$

Comparison of Eqs. (6.62–6.66) shows that

$$\eta_c \to \frac{\pi_c^{(\gamma_c-1)/\gamma_c} - 1}{\pi_c^{(\gamma_c-1)/\gamma_c\eta_s} - 1} \qquad \text{as} \qquad N \to \infty$$

and hence $\eta_s \to e_c$ as N becomes large.

As an example calculation, say a 16-stage compressor of $\pi_c = 25$ is to be constructed. The compressor efficiency is to be estimated from the measured stage efficiency η_s. Note that $\pi_s = 25^{1/16} = 1.223$ and say η_s is measured at 0.93. Then from

$$\eta_s = \frac{\pi_s^{(\gamma_c-1)/\gamma_c} - 1}{\pi_s^{(\gamma_c-1)/\gamma_c e_c} - 1}$$

it follows that

$$e_c = \frac{[(\gamma_c-1)/\gamma_c]\ell n\,\pi_s}{\ell n\left[1 + (1/\eta_s)(\pi_s^{(\gamma_c-1)/\gamma_c} - 1)\right]} = 0.932$$

Two estimates for η_c are obtained from Eqs. (6.57) and (6.62), to give $\eta_c = 0.897$ or 0.896, respectively. It can be noted that if e_c was assumed to be equal to the measured η_s, Eq. (6.57) would have given $\eta_c = 0.895$.

The point of these manipulations is that use of the polytropic efficiency allows rapid estimation of the compressor efficiency. In addition, for very rapid preliminary estimates it is sufficient to assume the polytropic efficiency is equal to the stage efficiency and then to utilize Eq. (6.57) to estimate η_c.

6.6 The Burner

The burner is usually approximated as having adiabatic combustion because no heat transfer is assumed at the boundaries. There are two measures of the burner performance, incomplete combustion and stagnation pressure loss. The burner efficiency η_b is defined as

$$\eta_b = \frac{1}{\dot{m}_f h} \left[(\dot{m} + \dot{m}_f) h_{t_4} - \dot{m} h_{t_3} \right]$$

$$= \frac{1}{\dot{m}_f h} \left[(\dot{m} + \dot{m}_f) C_{p_t} T_{t_4} - \dot{m} C_{p_c} T_{t_3} \right] \qquad (6.67)$$

where h is the "heating value" of the fuel, h_t the stagnation enthalpy $(C_p T_t)$, and $\dot{m}_f$ the fuel mass flow rate.

The stagnation pressure loss arises from two effects, the viscous losses in the combustion chamber and the stagnation pressure losses due to enthalpy addition at finite Mach number. These effects are combined for the purposes of performance analysis in the burner stagnation pressure ratio π_b, where

$$\pi_b = p_{t_4}/p_{t_3} < 1 \qquad (6.68)$$

As with the inlet, there are many important limitations brought about by the lack of one-dimensionality of the flow (hot spots). These effects appear in the preliminary cycle analysis only indirectly through the required reduction in average combustor outlet temperature T_{t_4}.

The Behavior of the Thermodynamic Properties across the Combustor

In the cycle analysis to follow in Chap. 7, it will be assumed that the gas prior to the combustor is calorically perfect with properties C_{p_c}, γ_c, etc. Similarly, the gas following the combustor will be assumed calorically perfect with properties C_{p_t}, γ_t, etc. In very-high-temperature engines this latter assumption is not highly accurate, and it should be understood that, if highly accurate results are required, the real gas tables should be used. The tendencies, and even magnitudes (provided γ_t is selected in the appropriate range), of the relatively simple calculations to follow are quite suitable for preliminary cycle analysis.

When considering changes in the thermodynamic properties C_p and γ, it should be remembered that these two properties are related. Thus note

$$C_p = \frac{\gamma}{\gamma - 1} R = \frac{\gamma}{\gamma - 1} \frac{R_u}{M} \tag{6.69}$$

where R_u is the universal gas constant and M the molecular "weight."

Thus, if the chemical reaction in the combustion chamber causes the vibrational modes to be excited but does not cause appreciable dissociation, and also if the rather small percentage of fuel addition does not significantly change the molecular weight, then M would be approximately constant. In this case, a reduction in γ_t is directly related to an increase in C_{p_t} by the formula

$$\frac{C_{p_t}}{C_{p_c}} = \frac{\gamma_t}{\gamma_c} \frac{\gamma_c - 1}{\gamma_t - 1} \tag{6.70}$$

This approximation will be used throughout Chaps. 7 and 8.

One-Dimensional Estimation of the Burner Stagnation Pressure Ratio

Consider the effect of flameholder drag and enthalpy increase at finite Mach number on the stagnation properties in a combustion chamber (Fig. 6.19). For simplicity, consider a constant-area duct and assume that the drag may be estimated by relating the drag loss to the incoming dynamic pressure. (Such an analysis tends to be most suitable to the description of flow in a constant-area afterburner.)

To simply analyze the combustion chamber behavior, assume the gases to be calorically perfect at the entrance and exit to the chamber, and in addition assume that the mass addition of fuel is extremely small compared to the air mass flow. The momentum equation may then be written

$$p_3 + \rho_3 u_3^2 = p_4 + \rho_4 u_4^2 + \left(\tfrac{1}{2}\rho_3 u_3^2\right) C_D$$

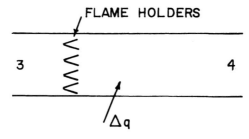

Fig. 6.19   Constant-area combustion chamber.

from which

$$p_3\left[1 + \frac{\rho_3 u_3^2}{p_3}\left(1 - \frac{C_D}{2}\right)\right] = p_4\left(1 + \frac{\rho_4 u_4^2}{p_4}\right)$$

or

$$\frac{p_4}{p_3} = \frac{1 + \gamma_c M_3^2\left[1 - (C_D/2)\right]}{1 + \gamma_t M_4^2} \qquad (6.71)$$

The state equation and continuity equation ($\rho_3 u_3 = \rho_4 u_4$) may be combined to give

$$\left(\frac{T_4}{T_3}\right)^2 = \left(\frac{\rho_3}{\rho_4}\frac{p_4}{p_3}\right)^2 = \left(\frac{u_4}{u_3}\right)^2\left(\frac{p_4}{p_3}\right)^2$$

hence

$$\frac{T_4}{T_3} = \frac{u_4^2}{\gamma_t R T_4}\frac{1}{u_3^2/(\gamma_c R T_3)}\left(\frac{p_4}{p_3}\right)^2\frac{\gamma_t}{\gamma_c} = \frac{\gamma_t}{\gamma_c}\frac{M_4^2}{M_3^2}\left(\frac{p_4}{p_3}\right)^2 \qquad (6.72)$$

Also

$$\frac{T_{t_4}}{T_{t_3}} = \frac{T_4}{T_3}\frac{1 + \dfrac{\gamma_t - 1}{2}M_4^2}{1 + \dfrac{\gamma_c - 1}{2}M_3^2} \qquad (6.73)$$

Combining Eqs. (6.71–6.73) then gives

$$\frac{M_4^2\left(1 + \dfrac{\gamma_t - 1}{2}M_4^2\right)}{\left(1 + \gamma_t M_4^2\right)^2} = \frac{\gamma_c}{\gamma_t}\frac{M_3^2\left(1 + \dfrac{\gamma_c - 1}{2}M_3^2\right)}{\left[1 + \gamma_c M_3^2\left(1 - \dfrac{C_D}{2}\right)\right]^2}\frac{T_{t_4}}{T_{t_3}} \equiv \chi \qquad (6.74)$$

This is an equation for M_4 in terms of the upstream variables and prescribed stagnation temperature ratio T_{t_4}/T_{t_3}. This is once again of the same functional form for the Mach number as occurred in the solution for the heat interaction at constant area (Sec. 2.18), so as before

$$M_4^2 = \frac{2\chi}{1 - 2\gamma_t\chi + \left[1 - 2(\gamma_t + 1)\chi\right]^{\frac{1}{2}}} \qquad (6.75)$$

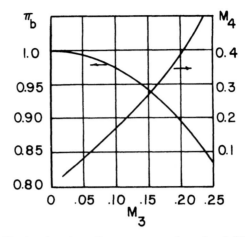

Fig. 6.20   Combustion stagnation pressure ratio and exit Mach number.

Following solution for M_4, the desired stagnation pressure ratio π_b is given by

$$\frac{p_{t_4}}{p_{t_3}} \equiv \pi_b = \frac{\left(1 + \dfrac{\gamma_t - 1}{2} M_4^2\right)^{\gamma_t/(\gamma_t - 1)}}{\left(1 + \dfrac{\gamma_c - 1}{2} M_3^2\right)^{\gamma_c/(\gamma_c - 1)}} \frac{p_4}{p_3} \tag{6.76}$$

where p_4/p_3 is given by Eq. (6.71).

As an example calculation consider an aircraft flying at Mach number 0.8 with compressor pressure ratio $\pi_c = 15$. Given $T_{t_4}/T_0 = 7$, $\gamma_c = 1.4$, $\gamma_t = 1.3$, $e_c = 0.905$, and $C_D = 1.5$, the result is

$$\frac{T_{t_4}}{T_{t_3}} = \frac{T_{t_4}/T_0}{\left(T_{t_3}/T_{t_2}\right)\left(T_{t_2}/T_0\right)} = \frac{T_{t_4}/T_0}{\tau_c \tau_r}$$

Here $\tau_r = 1.128$, and $\tau_c = (15)^{1/(3.5)(0.905)} = 2.354$, so $T_{t_4}/T_{t_3} = 2.637$. Direct calculation then yields the results of Fig. 6.20.

It is evident from Fig. 6.20 that for these conditions the inlet Mach number must be restricted to a value of 0.15 or less if the combustor pressure loss is not to become excessive. This restriction to low required Mach numbers, particularly in afterburners, can lead to design limitations upon the required burner cross-sectional area.

6.7   The Turbine

Unlike compressors, modern turbines are almost always cooled, at least in the first several high-pressure stages. Cooling is accomplished by passing air directly from the compressor to the turbine blades where any one of several cooling methods may be employed. The "accounting" of cooling losses is best carried out by considering the cooling air and mainstream air to be a combined multiple-stream adiabatic flow. In this section, the so-called adiabatic efficiencies will be described and estimation of the effect of cooling will be considered to be separately determined.

The Turbine Efficiency, η_t

The turbine efficiency is defined in a manner analogous to that of the compressor efficiency to give

$$\eta_t = \frac{\text{actual work interaction for a given pressure ratio}}{\text{ideal work interaction for a given pressure ratio}}$$

$$= \frac{C_p\left(T_{t_4} - T_{t_5}\right)}{C_p\left(T_{t_4} - T_{t_{5_i}}\right)} = \frac{1 - \tau_t}{1 - \tau_{t_i}}$$

The ideal process is isentropic so

$$\eta_t = \frac{1 - \tau_t}{1 - \pi_t^{(\gamma_t - 1)/\gamma_t}} \tag{6.77}$$

The Turbine Polytropic Efficiency, e_t

Again, analogous to the compressor polytropic efficiency, define

$$e_t = \frac{\text{actual work interaction for a given differential pressure change}}{\text{ideal work interaction for a given differential pressure change}}$$

Thus

$$e_t = \frac{\mathrm{d}h_t}{\mathrm{d}h_{t_i}} = \frac{C_p\,\mathrm{d}T_t}{(1/\rho_t)\,\mathrm{d}p_t}$$

or

$$e_t = \frac{\gamma_t}{\gamma_t - 1} \frac{\mathrm{d}T_t/T_t}{\mathrm{d}p_t/p_t}$$

Hence, if e_t may be assumed approximately constant, integration gives

$$\frac{T_{t_5}}{T_{t_4}} = \left(\frac{p_{t_5}}{p_{t_4}}\right)^{e_t(\gamma_t - 1)/\gamma_t} \quad \text{or} \quad \tau_t = \pi_t^{e_t(\gamma_t - 1)/\gamma_t} \tag{6.78}$$

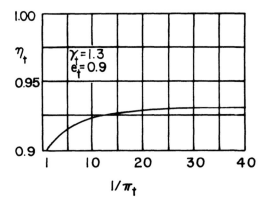

Fig. 6.21   Turbine design efficiency vs pressure ratio.

Equations (6.77) and (6.78) then give

$$\eta_t = \frac{1 - \pi_t^{e_t(\gamma_t - 1)/\gamma_t}}{1 - \pi_t^{(\gamma_t - 1)/\gamma_t}} = \frac{1 - \tau_t}{1 - \tau_t^{1/e_t}} \tag{6.79}$$

This relationship is plotted in Fig. 6.21.

It can be seen that the turbine efficiency increases as the turbine design expansion ratio $(1/\pi_t)$ increases. This occurs because energy that is not extracted in a given stage due to inefficiencies remains partly available to the succeeding stages.

The Turbine Stage Efficiency

The turbine stage efficiency is defined in a completely analogous manner to the turbine efficiency, to give

$$\eta_{t_j} = \frac{1 - \tau_{t_j}}{1 - \left(\pi_{t_j}\right)^{(\gamma_t - 1)/\gamma_t}} \tag{6.80}$$

Noting

$$\tau_{t_j} = 1 - \eta_{t_j}\left[1 - \left(\pi_{t_j}\right)^{(\gamma_t - 1)/\gamma_t}\right] \quad \text{and} \quad \tau_t = \prod_{j=1}^{N} \tau_{t_j}$$

it follows that

$$\eta_t = \frac{1 - \prod_{j=1}^{N}\left\{1 - \eta_{t_j}\left[1 - \left(\pi_{t_j}\right)^{(\gamma_t - 1)/\gamma_t}\right]\right\}}{1 - \pi_t^{(\gamma_t - 1)/\gamma_t}} \tag{6.81}$$

Table 6.2 Component Figures of Merit

Ideal Behavior	Actual Behavior	"Figures of Merit"
	Inlet	
Isentropic, hence	Adiabatic, not isentropic, hence	π_d
$\tau_d = 1$, $\pi_d = 1$	$\tau_d = 1$, $\pi_d \neq 1$	
	Compressor	
Isentropic, hence	Adiabatic, not isentropic	
$-\Delta w_m = C_{p_c} T_{t_2}(\tau_c - 1)$	$-\Delta w_m = C_{p_c} T_{t_2}(\tau_c - 1)$	$\eta_c = \dfrac{\pi_c^{(\gamma_c - 1)/\gamma_c} - 1}{\tau_c - 1}$
$\pi_c = \tau_c^{\gamma_c/(\gamma_c - 1)}$	$\pi_c = [1 + \eta_c(\tau_c - 1)]^{\gamma_c/(\gamma_c - 1)}$	
	$\tau_c = 1 + \dfrac{1}{\eta_c}\left[\pi_c^{(\gamma_c - 1)/\gamma_c} - 1\right]$	$\eta_c = \dfrac{\pi_c^{(\gamma_c - 1)/\gamma_c} - 1}{\pi_c^{(\gamma_c - 1)/\gamma_c e_c} - 1}$
	Burner	
No stagnation pressure loss, complete combustion	Stagnation pressure loss, incomplete combustion	π_b
$\pi_b = 1$	$\pi_b \neq 1$	$\eta_b =$
		$\dfrac{(\dot{m} + \dot{m}_f) C_{p_t} T_{t_4} - \dot{m} C_{p_c} T_{t_3}}{\dot{m}_f h}$
$(\dot{m} + \dot{m}_f) C_{p_4} T_{t_4}$ $- \dot{m} C_{p_3} T_{t_3} = \dot{m}_f h$	$(\dot{m} + \dot{m}_f) C_{p_t} T_{t_4} =$ $\dot{m} C_{p_c} T_{t_3} + \eta_b \dot{m}_f h$	
	Turbine	
Isentropic	Adiabatic, not isentropic	
$\Delta w_m = C_{p_t} T_{t_4}(1 - \tau_t)$	$\Delta w_m = C_{p_t} T_{t_4}(1 - \tau_t)$	$\eta_t = \dfrac{1 - \tau_t}{1 - \pi_t^{(\gamma_t - 1)/\gamma_t}}$
$\pi_t = \tau_t^{\gamma_t/(\gamma_t - 1)}$	$\pi_t = \left[1 - \dfrac{1}{\eta_t}(1 - \tau_t)\right]^{\gamma_t/(\gamma_t - 1)}$	
	$\tau_t = \left[1 - \eta_t\left(1 - \pi_t^{(\gamma_t - 1)/\gamma_t}\right)\right]$	$\eta_t = \dfrac{1 - \tau_t}{1 - \tau_t^{1/e_t}}$
	Note: Cooling to be considered separately	
	Nozzle	
Isentropic	Adiabatic	π_n
$\tau_n = 1$, $\pi_n = 1$	$\tau_n = 1$, $\pi_n \neq 1$	

In the special case when all stage efficiencies η_s and stage pressure ratios are equal, Eq. (6.81) reduces to

$$\eta_t = \frac{1 - \left\{ 1 - \eta_s \left[1 - (\pi_t)^{(\gamma_t - 1)/\gamma_t N} \right] \right\}^N}{1 - \pi_t^{(\gamma_t - 1)/\gamma_t}} \tag{6.82}$$

6.8 The Nozzle

The major loss mechanisms in a nozzle are usually identified with the pressure imbalance at the exit caused by over- or underexpansion. The degree of over- or underexpansion is often selected for the best balance of internal (exit pressure) and external (boat-tail drag) losses. In any case, the boat-tail losses are separately accounted for when the cowl drag is determined and the effect of exit pressure imbalance is included in the expression for F_A [Eq. (6.8)].

For convenience all losses occurring from the turbine exit to nozzle exit are included. Thus,

$$\pi_n = p_{t_9}/p_{t_5} < 1 \tag{6.83}$$

For engines without afterburners, π_n is usually very nearly unity (0.99 or larger), but when the flameholder ducts are present π_n can be much lower (~ 0.97). When afterburning is present π_n can be estimated by the analysis of Sec. 6.6.

Table 6.3   Typical Ranges of Parameters

T_0 °R	$380 \rightarrow 580$ (high altitude $\rightarrow$ very hot day, sea level)
T_{t_4} °R	$2400 \rightarrow 2900$
$T_{t_8}, T_{t_8'}$ °R	$2700 \rightarrow 3400$ (upper limit temperatures, not necessarily appropriate for best performance)
$\gamma_t, \gamma_{AB}, \gamma_{AB'}$	$1.35 \rightarrow 1.25$ (γ goes down as T goes up)
h Btu/lbm	$18,500 \rightarrow 19,500$
$C_{p_c} \approx 0.24$	$\dfrac{\text{Btu}}{\text{lbm °R}}$
π_d Subsonic	$0.98 \rightarrow 0.998$
π_d Supersonic	May approximate design value with formula such as $\pi_d = 1.006 - 0.016 M_0^2$
π_b	$0.93 \rightarrow 0.98$
η_b	$0.96 \rightarrow 0.998$
$\pi_{AB}, \pi_{AB'}$	$0.93 \rightarrow 0.99$ (high values if burner ring in place but no burning present)
$\eta_{AB}, \eta_{AB'}$	$0.92 \rightarrow 0.98$
$\pi_n, \pi_{n'}$	$0.99 \rightarrow 0.998$ (if afterburner present, as $\pi_{AB}, \pi_{AB'}$)
η_m	$1 \rightarrow 0.9$ (0.9 implies substantial power takeoff)
e_c	$0.86 \rightarrow 0.94$
$e_{c'}$	$0.85 \rightarrow 0.92$
e_t	$0.85 \rightarrow 0.92$

Usually nozzles are very nearly adiabatic, so that

$$\tau_n = 1 \tag{6.84}$$

In some cases of afterburning, the nozzle may be cooled with compressor air, in which case it would be necessary to consider a two-stream analysis. Such an analysis is straightforward conceptually, although somewhat tedious algebraically, and will not be included in this book.

6.9 Summary of Component Figures of Merit

The loss mechanisms and their measures considered in the preceding sections may be summarized as in Table 6.2. Table 6.3 provides typical ranges of parameters found to be appropriate for present day technology. The table is given in British units, see Table 5.2 for conversion factors for SI units.

References

[1]Bober, J. L., "Use of Potential Flow Theory to Evaluate Subsonic Inlet Data from a Simulator-Powered Nacelle at Cruise Conditions," NASA TN D-7850. Dec. 1974.

[2]Kantrowitz, A. and Donaldson, C. du P., "Preliminary Investigation of Supersonic Diffusers," NACA ACR L5020, 1945.

[3]"Equations, Tables and Charts for Compressible Flow," NACA Rept. 1135.

Problems

6.1   Calculate and plot the dimensionless additive drag $D_{add}/A_i p_0$ over a range of inlet design Mach numbers, $0.2 \le M_i \le 1$, for the two flight Mach numbers $M_0 = 0$ and 0.85. Use a one-dimensional estimate for the drag.

6.2   Verify that Eqs. (6.22–6.28) are correct.

6.3   Obtain Eq. (6.33).

6.4   Obtain Eqs. (6.34–6.38).

6.5   A flow within a duct closely approximates a "step profile." Denoting conditions in the stream by subscripts 1 or 2 and given $\alpha = \dot{m}_2/\dot{m}_1 = 1.2$, $p_{t_2}/p_{t_1} = 1.3$, $T_{t_2}/T_{t_1} = 0.6$, and $M_1 = 0.5$, evaluate the ratios $p_{t_{c.a.}}/p_{t_{m.a.}}$ and $p_{t_{s.t.}}/p_{t_{m.a.}}$.

6.6   An internal compression inlet with variable geometry is designed to stabilize a normal shock "at" the throat at a value of M_{m_1} for all flight Mach numbers above $M_0 = M_1 = 1.5$. (See Fig. A.) The flow elsewhere in the inlet may be considered to be isentropic.

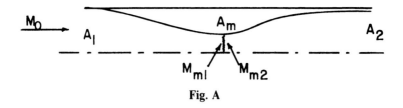

Fig. A

(a) Show that the dimensionless thrust on the inlet $F/A_1 p_1$ may be calculated from the hierarchy of equations

$$M_{m_2}^2 = \frac{1 + \frac{\gamma - 1}{2} M_{m_1}^2}{\gamma M_{m_1}^2 - \frac{\gamma - 1}{2}}$$

$$M_2 = \frac{A_1}{A_2} M_1 \frac{M_{m_2}}{M_{m_1}} \left(\frac{1 + \frac{\gamma - 1}{2} M_{m_1}^2 \; 1 + \frac{\gamma - 1}{2} M_2^2}{1 + \frac{\gamma - 1}{2} M_{m_2}^2 \; 1 + \frac{\gamma - 1}{2} M_1^2} \right)^{(\gamma + 1)/2(\gamma - 1)}$$

$$\frac{p_{t_2}}{p_{t_1}} = \left[\frac{\frac{\gamma + 1}{2} M_{m_1}^2}{1 + \frac{\gamma - 1}{2} M_{m_1}^2} \right]^{\gamma/(\gamma - 1)} \left[1 + \frac{2\gamma}{\gamma + 1} \left(M_{m_1}^2 - 1 \right) \right]^{-1/(\gamma - 1)}$$

$$\frac{p_{t_i}}{p_i} = \left(1 + \frac{\gamma - 1}{2} M_i^2 \right)^{\gamma/(\gamma - 1)} \qquad (i = 1 \text{ or } 2)$$

$$\frac{p_2}{p_1} = \frac{p_{t_2}}{p_{t_1}} \frac{p_{t_1}/p_1}{p_{t_2}/p_2}$$

$$\frac{F}{A_1 p_1} = \frac{p_2 A_2}{p_1 A_1} \left(1 + \gamma M_2^2 \right) - \left(1 + \gamma M_1^2 \right)$$

(b) Evaluate $F/A_1 p_1$ over the range $1.5 \le M_1 \le 3.0$ for the case $M_{m_1} = 1.3$, $\gamma = 1.4$, and $A_2/A_1 = 1$.

(c) Evaluate F for the case $M_1 = 3.0$ for conditions as in part (b), and for $T_0 = 400°R$, $p_0 = p_1 = 138$ lbf/ft², and an inlet diameter of 5 ft.

(d) If the engine specific thrust is 40 lbf/lbm/s, what is the ratio of this "inlet thrust" to the engine thrust?

6.7   (a) You visit a facility and are shown an aircraft inlet design. You are told that when in operation the Mach number just downstream of the shock located "at" A_m is $M_{m_2} = 0.83$, that $A_1/A_m = 1.20$, that $\gamma = 1.39$, and that the flight Mach number is "secret." Obtain an estimate of M_1 to three significant figures.

(b) Can this inlet operate if its geometry is fixed?

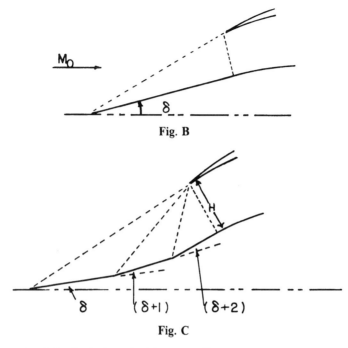

Fig. B

Fig. C

6.8 An inlet is designed with a two-dimensional ramp that creates an oblique shock. The oblique shock is followed by a normal shock just preceding the cowl. (See Fig. B.)

(a) With the assumption that the only losses occur across the shock waves, calculate π_d for the cases: $\gamma = 1.4$, $M_0 = 1.9$, and $\delta = 8$, 13, and 18 deg.

(b) Compare the performance of the 8 deg ramp to the 13 deg ramp if the aircraft is flown at $M_0 = 1.4$.

6.9 A supersonic inlet is constructed with three two-dimensional ramps of progressively increasing slope change as indicated in Fig. C (angles measured in degrees). Prior to entry into the inlet a normal shock occurs.

(a) Obtain π_d for the cases $M_0 = 2.5$ and $\delta = 9$, 10, and 11 deg.

(b) Select the inlet from part (a) that gives the best performance and *accurately* draw the required ramp geometries to cause all of the shocks to intersect at the cowl lip. The inlet height H is to be 4 cm on the drawing.

6.10 A supersonic inlet is constructed with two two-dimensional ramps. Following the second oblique shock, a fixed throat inlet is used for internal compression (Fig. D). The inlet is designed to self-start when the flight Mach number M_0 is 2.2. For the case where $\delta_1 = 7$ deg and $\delta_2 = 8$ deg ($\gamma = 1.4$):

(a) Obtain π_d assuming the inlet starts. Also, obtain the required A_m/A_1.

(b) Obtain π_d assuming the inlet does not start.

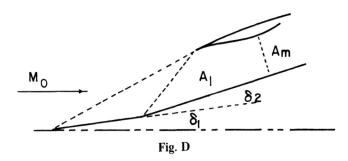

Fig. D

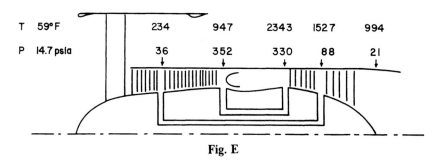

| T | 59°F | | 234 | 947 | 2343 | 1527 | 994 |
| P | 14.7 psia | | 36 | 352 | 330 | 88 | 21 |

Fig. E

6.11   In the text, the efficiency of an inlet is defined as its stagnation pressure ratio π_d. Thermodynamically, an inlet may be considered to be a compressor that adiabatically compresses the air from p_0 to p_{t_2}. Thus, defining the "classical inlet efficiency" η_I the same way as η_c was defined, show that:

(a)
$$\eta_I = \frac{\tau_r \pi_d^{(\gamma-1)/\gamma} - 1}{\tau_r - 1}$$

(b)
$$e_I = 1 + \ell n \pi_d^{(\gamma-1)/\gamma}/\ell n \tau_r$$

(c) Evaluate η_I and e_I for an inlet that obtains $\pi_d = 0.88$ when $M_0 = 2.8$.

6.12   The conditions existing throughout the core stream of a commercial turbofan engine (for takeoff setting) are indicated in Fig. E.

(a)   Given $\gamma_c = 1.4$, $\gamma_t = 1.3$, estimate η_{CL}, η_{CH}, e_{CL}, e_{CH}, η_{tL}, η_{tH}, e_{tL}, and e_{tH}. Note that $\pi_{CL} = 36/14.7$, $\tau_{CL} = (234 + 459.7)/(59 + 459.7)$, etc.

(b)   Some of the resulting efficiencies might seem inappropriate. Explain what might be incorrect in your assumptions inherent in estimating the efficiencies as in part (a).

6.13   A compressor of pressure ratio 32 has a measured efficiency of 89%.

(a)   What is the compressor polytropic efficiency?

(b)   If the compressor is a "two-spool" compressor with the high-pressure spool contributing a pressure ratio of 6, estimate the low-pressure compressor efficiency η_L and the high-pressure compressor efficiency η_h.

6.14   A compressor is composed of 15 stages, each with a pressure ratio of 1.24 and a stage efficiency of 0.92.

(a)   What is the compressor pressure ratio?

(b)   What is the compressor efficiency?

(c)   What is the polytropic efficiency?

(d)   Evaluate the compressor efficiency assuming that the polytropic efficiency is equal to the stage efficiency.

6.15   A constant-area burner has an entering Mach number of M_3 = 0.1. Obtain and plot π_b and M_4 vs T_{t_4}/T_{t_3} in the range $1.5 < T_{t_4}/T_{t_3} < 3.5$. Take $\gamma_c = 1.4$, $\gamma_t = 1.28$, and $C_D = 1.5$.

6.16   A constant cross-sectional area afterburner has $C_D = 1.7$, $\gamma = 1.25$ ($\gamma \equiv \gamma_c \equiv \gamma_t$), and an entering Mach number of 0.3.

(a)   Determine the burner stagnation temperature ratio $(T_{t_7}/T_{t_5})_{\text{choke}}$ that will just cause choking in the (constant-area) channel.

(b)   Plot π_{AB} vs T_{t_7}/T_{t_5} over the range $1 \le T_{t_7}/T_{t_5} \le (T_{t_7}/T_{t_5})_{\text{choke}}$.

6.17   A compressor of pressure ratio $\pi_c = 30$ and efficiency $\eta_c = 0.89$ is driven by a turbine with $e_t = 0.90$. Assume no leakage between the compressor and turbine and that $f = 0.02$, $\gamma_c = 1.4$, $\gamma_t = 1.3$, $T_{t_2} = 460°R$, and $T_{t_4} = 2600°R$. Determine π_t and η_t.

6.18   A nozzle has expansion ratio $p_{t_7}/p_9 = P$, γ_t, and π_n. Find an expression for T_9 in terms of γ_t, π_n, P, and T_{t_7}.

7. NONIDEAL CYCLE ANALYSIS

7.1 Introduction

In this chapter cycle analysis is again applied to several example engine types. The methodology will remain as in Sec. 5.8, the only difference between the results of this chapter and those of Chap. 5 arising because of the nonideal component processes assumed and because of the use of different thermodynamic properties following the primary burner and afterburners. The notation is that already introduced in Sec. 5.2.

7.2 The Turbojet

The performance equations for the turbojet will now be developed. It will be assumed that the gas is calorically perfect up to the compressor outlet with properties γ_c, C_{p_c}, etc. The gas will also be assumed calorically perfect following the burner with properties γ_t, C_{p_t}, etc. If an afterburner is in operation, the gas following the afterburner will again be assumed calorically perfect with properties γ_{AB}, $C_{p_{AB}}$, etc. It will be assumed that the gas constant remains unchanged throughout so that Eq. (6.70) (or the equivalent for $C_{p_{AB}}$ and γ_{AB}) remains valid. All components will be considered to be adiabatic (no turbine cooling) and the efficiencies of compressor and turbine will be described through the use of constant polytropic efficiencies. Finally, the effects of gas leakage and the use of drawn-off air for auxiliary power will not be included. The reference stations are indicated in Fig. 7.1.

The analysis to follow will be developed in a form suitable for use whether or not an afterburner is present or in operation.

Cycle Analysis of the Turbojet with Losses

The expression for the uninstalled thrust is as given by Eq. (6.8). That is, with the station numbering of Fig. 7.1,

$$F_A = \dot{m}_9 u_9 - \dot{m}_0 u_0 + (p_9 - p_0)A_9 \tag{7.1}$$

Because gas leakage and the use of auxiliary air is ignored,

$$\dot{m}_9 = \dot{m}_0 + \dot{m}_f + \dot{m}_{f_{AB}} = \dot{m}_0(1 + f + f_{AB}) \tag{7.2}$$

Also note

$$\frac{1}{\dot{m}_0}(p_9 - p_0)A_9 = (1 + f + f_{AB})\frac{A_9 p_9}{\rho_9 u_9 A_9}\left(1 - \frac{p_0}{p_9}\right)$$

$$= \frac{(1 + f + f_{AB})}{u_9/u_0}\frac{u_0}{u_0^2/(\gamma_c R_c T_0)}\frac{R_9 T_9}{\gamma_c R_c T_0}\left(1 - \frac{p_0}{p_9}\right)$$

or

$$\frac{1}{\dot{m}_0}(p_9 - p_0)A_9 = (1 + f + f_{AB})\frac{u_0}{\gamma_c M_0^2}\frac{1}{u_9/u_0}\left(1 - \frac{p_0}{p_9}\right)\frac{T_9}{T_0} \qquad (7.3)$$

In Eq. (7.3) $R_9 = R_c$ has been utilized. Equations (7.1–7.3) then give

$$\frac{F_A}{\dot{m}_0} = a_0\left[(1 + f + f_{AB})\left(M_0\frac{u_9}{u_0}\right) - M_0\right.$$

$$+ (1 + f + f_{AB})\frac{1}{\gamma_c[M_0(u_9/u_0)]}\frac{T_9}{T_0}\left(1 - \frac{p_0}{p_9}\right)\Bigg] \qquad (7.4)$$

$$M_0^2\left(\frac{u_9}{u_0}\right)^2 = \frac{\gamma_{AB} R_{AB}}{\gamma_c R_c}\frac{T_9}{T_0} M_9^2 \qquad (7.5)$$

Then noting

$$T_{t_9} = T_9\left(1 + \frac{\gamma_{AB} - 1}{2} M_9^2\right) = T_0\frac{C_{p_c}}{C_{p_{AB}}}\tau_{\lambda_{AB}} \qquad (7.6)$$

and

$$p_{t_9} = p_9\left(1 + \frac{\gamma_{AB} - 1}{2} M_9^2\right)^{\gamma_{AB}/(\gamma_{AB}-1)} = p_0\pi_r\pi_d\pi_c\pi_b\pi_t\pi_n \qquad (7.7)$$

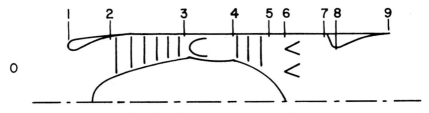

Fig. 7.1   Turbojet reference stations.

write

$$M_9^2 = \frac{2}{\gamma_{AB} - 1} \left[\left(\frac{p_{t_9}}{p_9} \right)^{(\gamma_{AB} - 1)/\gamma_{AB}} - 1 \right] \tag{7.8}$$

and

$$M_0^2 \left(\frac{u_9}{u_0} \right)^2 = \frac{\gamma_{AB} R_{AB}}{\gamma_c R_c} \frac{C_{p_c}}{C_{p_{AB}}} \frac{2}{\gamma_{AB} - 1} \tau_{\lambda_{AB}} \left[1 - \left(\frac{p_{t_9}}{p_9} \right)^{-[(\gamma_{AB} - 1)/\gamma_{AB}]} \right]$$

or

$$M_0^2 \left(\frac{u_9}{u_0} \right)^2 = \frac{2}{\gamma_c - 1} \tau_{\lambda_{AB}} \left[1 - \left(\frac{p_{t_9}}{p_9} \right)^{-[(\gamma_{AB} - 1)/\gamma_{AB}]} \right] \tag{7.9}$$

Also note

$$\frac{T_9}{T_0} = \frac{\left(C_{p_c}/C_{p_{AB}} \right) \tau_{\lambda_{AB}}}{\left(p_{t_9}/p_9 \right)^{(\gamma_{AB} - 1)/\gamma_{AB}}} \tag{7.10}$$

Equations (7.9) and (7.10) allow determination of the principal terms in the expression for the specific thrust, except that the turbine pressure ratio π_t must be obtained from the power balance between the turbine and compressor. Thus,

$$\dot{m}_0 C_{p_c} \left(T_{t_3} - T_{t_2} \right) = \left(\dot{m}_0 + \dot{m}_f \right) C_{p_t} \eta_m \left(T_{t_4} - T_{t_5} \right)$$

Dividing by $\dot{m}_0 C_{p_c} T_0$ and rearranging

$$\tau_t = 1 - \frac{1}{\eta_m (1 + f)} \frac{\tau_r}{\tau_\lambda} (\tau_c - 1) \tag{7.11}$$

and then

$$\pi_t = \tau_t^{\gamma_t/(\gamma_t - 1)e_t} \tag{7.12}$$

It should be noted here that the mechanical efficiency η_m will normally be very high (approximately unity) for most gas turbine engines. However, by retaining η_m in the equation, a convenient method of accounting for the auxiliary power takeoff is provided.

Finally, expressions are obtained for the fuel-to-air ratios from the enthalpy balances across the appropriate burners.

Primary burner. An enthalpy balance gives

$$\dot{m}_0 C_{p_c} T_{t_3} + \eta_b \dot{m}_f h = \left(\dot{m}_0 + \dot{m}_f \right) C_{p_t} T_{t_4}$$

Dividing by $\dot{m}_0 C_{p_c} T_0$ and rearranging, there is obtained

$$f = \frac{\tau_\lambda - \tau_r \tau_c}{\left(h\eta_b/C_{p_c}T_0\right) - \tau_\lambda} \tag{7.13}$$

Afterburner. An enthalpy balance gives

$$(\dot{m}_0 + \dot{m}_f)C_{p_t}T_{t_5} + \eta_{AB}\dot{m}_{f_{AB}}h = (\dot{m}_0 + \dot{m}_f + \dot{m}_{f_{AB}})C_{p_{AB}}T_{t_8}$$

Dividing by $\dot{m}_0 C_{p_c} T_0$ and rearranging,

$$f_{AB} = (1+f)\frac{\tau_{\lambda_{AB}} - \tau_\lambda \tau_t}{\left(h\eta_{AB}/C_{p_c}T_0\right) - \tau_{\lambda_{AB}}} \tag{7.14}$$

The specific fuel consumption may then be written as

$$S = \frac{(\dot{m}_f + \dot{m}_{f_{AB}})}{F_A}(10^6) = \frac{f + f_{AB}}{F_A/\dot{m}_0}(10^6) \tag{7.15}$$

Equations (7.1–7.15) may now be arranged in an order to allow direct calculation of the desired performance variables in terms of the imposed flight conditions, design limits, design choices, and component efficiencies. The equations are summarized in the next section.

Summary of the Equations— Turbojet with Losses

Inputs: $T_0(K)[°R], \gamma_c, \gamma_t, \gamma_{AB}, C_{p_c}, C_{p_t},$

$C_{p_{AB}}(J/kg \cdot K)[Btu/lbm \cdot °R], h \ (J/kg)[Btu/lbm],$

$\pi_d, \pi_b, \pi_n, \eta_b, \eta_{AB}, \eta_m, e_c, e_t,$

$p_9/p_0, \tau_\lambda, \tau_{\lambda_{AB}}, \pi_c, M_0$

Outputs: $\dfrac{F_A}{\dot{m}_0}\left(\dfrac{N \cdot s}{kg}\right)\left[\dfrac{F_A}{g_0\dot{m}_0}, \dfrac{lbf}{lbm/s}\right]$

$S\left(\dfrac{mg}{N \cdot s}\right)\left[\dfrac{lbm \ fuel/h}{lbf \ thrust}\right], f, f_{AB}, \text{etc.}$

Equations:

$$R_c = \frac{\gamma_c - 1}{\gamma_c}C_{p_c} \ m^2/s^2 \cdot K \quad \left[R_c = \frac{\gamma_c - 1}{\gamma_c}C_{p_c}(2.505)(10^4) \ ft^2/s \cdot °R\right] \tag{7.16}$$

$$a_0 = \sqrt{\gamma_c R_c T_0} \ \text{m/s [ft/s]} \tag{7.17}$$

$$\tau_r = 1 + \frac{\gamma_c - 1}{2} M_0^2 \tag{7.18}$$

$$\pi_r = \tau_r^{\gamma_c/(\gamma_c - 1)} \tag{7.19}$$

$$\tau_c = \pi_c^{(\gamma_c - 1)/\gamma_c e_c} \tag{7.20}$$

$$f = \frac{\tau_\lambda - \tau_r \tau_c}{(h\eta_b/C_p T_0) - \tau_\lambda} \tag{7.21}$$

$$\tau_t = 1 - \frac{1}{\eta_m(1 + f)} \frac{\tau_r}{\tau_\lambda}(\tau_c - 1) \tag{7.22}$$

$$\pi_t = \tau_t^{\gamma_t/(\gamma_t - 1)e_t} \tag{7.23}$$

$$\frac{P_{t_9}}{P_9} = \frac{P_0}{P_9} \pi_r \pi_d \pi_c \pi_b \pi_t \pi_n \tag{7.24}$$

$$\frac{T_9}{T_0} = \frac{(C_{p_t}/C_{p_{AB}})\tau_{\lambda_{AB}}}{(P_{t_9}/P_9)^{(\gamma_{AB} - 1)/\gamma_{AB}}} \tag{7.25}$$

Note: If no afterburning is present, $\tau_{\lambda_{AB}}$ should be put equal to $\tau_\lambda \tau_t$. Also, $C_{p_{AB}}$ and γ_{AB} should be put equal to C_{p_t} and γ_t.

$$f_{AB} = (1 + f)\frac{\tau_{\lambda_{AB}} - \tau_\lambda \tau_t}{(h\eta_{AB}/C_p T_0) - \tau_{\lambda_{AB}}} \tag{7.26}$$

$$M_0 \frac{u_9}{u_0} = \left\{ \frac{2}{\gamma_c - 1} \tau_{\lambda_{AB}}\left[1 - \left(\frac{P_{t_9}}{P_9}\right)^{-[(\gamma_{AB} - 1)/\gamma_{AB}]} \right] \right\}^{\frac{1}{2}} \tag{7.27}$$

$$\frac{F_A}{\dot{m}_0} = a_0 \left\{ (1 + f + f_{AB})\left(M_0 \frac{u_9}{u_0} \right) - M_0 + \frac{(1 + f + f_{AB})(T_9/T_0)}{\gamma_c[M_0(u_9/u_0)]}\left(1 - \frac{P_0}{P_9} \right) \right\} \tag{7.28}$$

(To obtain British units of lbf/lbm/s, replace a_0 in meters/second with $a_0/32.174$ where a_0 should be given in feet/second,

$$S = \frac{(f + f_{AB})}{F_A/\dot{m}_0}(10^6) \qquad \left[S = \frac{3600(f + f_{AB})}{F_A/g_0\dot{m}_0} \right] \tag{7.29}$$

Example Results— Turbojet with Losses

As a first example consider a turbojet to be designed for flight at Mach 2 without an afterburner. A range of possible pressure ratios is considered and the following characteristics assumed:

$T_0 = 233.3$ K $\gamma_c = 1.4$ $e_c = 0.92, 0.89$
$\quad$ [420°R] $\gamma_t = 1.35$ $e_t = 0.91$
$C_{p_c} = 996.5$ J/kg · K $\pi_d = 0.9425$ $p_9/p_0 = 1$
$\quad$ [0.238 Btu/lbm · °R] $\pi_b = 0.98$ $h = 4.5357(10^7)$ J/kg
$C_{p_t} = 1098.2$ J/kg · K $\pi_n = 0.99$ $\quad$ [19,500 Btu/lbm]
$\quad$ [0.262 Btu/lbm · °R] $\eta_b = 0.97$ $\tau_\lambda = 7.7$
$\quad$ $\eta_m = 0.99$

Figure 7.2 indicates that the general trend in the performance of the turbojet with losses is quite similar to that of the ideal turbojet depicted in Fig. 5.21. The most notable difference in the trend is that when losses are present a minimum exists in the specific fuel consumption, whereas in the ideal case the specific fuel consumption continues to decrease with increasing compressor pressure ratio.

If the engine were to be used without afterburning, the designer would again be faced with the choice of selecting an engine with low thrust, large pressure ratio, and low specific fuel consumption as compared to one with high thrust, low pressure ratio, and high specific fuel consumption. Another aspect of the designer's dilemma becomes apparent when comparing the curves obtained for the two different compressor efficiencies. Thus, for example, if a designer chose a compression ratio of 30 for use in a supersonic transport because his compressor design group had promised a

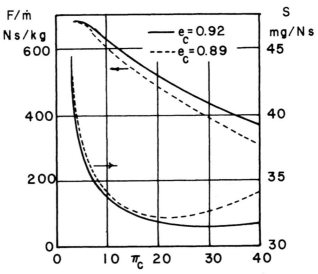

Fig. 7.2   Effect of compressor pressure ratio.

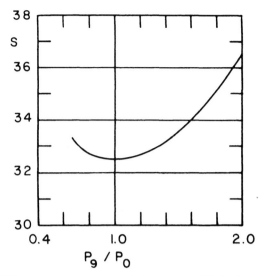

Fig. 7.3   Effect of exit pressure variations on specific fuel consumption.

compressor with $e_c = 0.92$ and then the group delivered a compressor with $e_c = 0.89$, it can be seen that the choice of $\pi_c = 30$ would be quite inappropriate. That is, such a compressor would be of a higher pressure ratio than that leading to minimum fuel consumption. Thus, the designer would have a compressor that was heavier (and more expensive) than that leading to a minimum specific fuel consumption, and he would also have lower thrust per frontal area.

The effects of nozzle off-design can be investigated by considering the engine to have the same parameters as those indicated above, but to have various values of p_9/p_0. Figure 7.3 shows the effect upon specific fuel consumption of varying the exit pressure for an engine with $\pi_c = 16$, $e_c = 0.92$, and other parameters as given above.

It is apparent from Fig. 7.3 that variations in the pressure ratio p_9/p_0 in the neighborhood of $p_9/p_0 = 1$ do not strongly affect the specific fuel consumption. Hence, it is of great importance to consider this mild sensitivity of S to p_9/p_0 when designing a nozzle so as to select that nozzle giving the best combination of internal and external performances.

In order to assess the effects of afterburning, an engine was considered with the same characteristics as those described above, but with the addition that $e_c = 0.92$, $\eta_{AB} = 0.96$, $\pi_n = 0.96$ burner on or $= 0.98$ burner off, $\tau_{\lambda_{AB}} = 8.8$, and $\gamma_{AB} = \gamma_t$. Figure 7.4 shows the results.

7.3   The Turbofan

The performance equations for the turbofan will be developed with the same, or equivalent, assumptions as utilized for the analysis of the turbojet in Sec. 7.2. The equations will be developed to allow for both primary

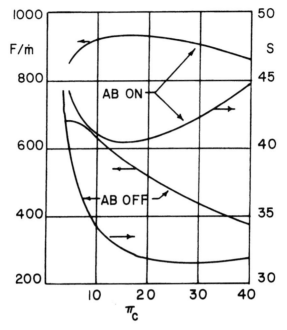

Fig. 7.4   Performance of turbojet with afterburning.

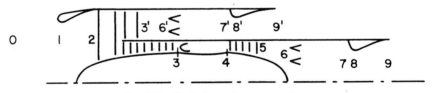

Fig. 7.5   Turbofan reference stations.

stream afterburning and/or secondary stream afterburning. Only the case of separate exhaust streams will be considered here, as it is hoped that the techniques of cycle analysis as developed herein together with the results of Sec. 5.13 will enable the interested reader to rapidly generate an analysis for the mixed stream case if so desired. The reference stations are indicated in Fig. 7.5.

Cycle Analysis of the Turbofan with Losses

The expression for the uninstalled thrust can be written as the sum of the thrust contributions of the primary and secondary streams to give

$$F_A = (\dot{m}_c + \dot{m}_f + \dot{m}_{f_{AB}})u_9 - \dot{m}_c u_0 + (p_9 - p_0)A_9 + (\dot{m}_F + \dot{m}_{f_{AB'}})u_{9'}$$

$$-\dot{m}_F u_0 + (p_{9'} - p_0)A_{9'} \tag{7.30}$$

This expression can then be manipulated in a manner completely analogous to that of Sec. 7.2 to give

$$\frac{F_A}{\dot{m}_c + \dot{m}_F} = \frac{a_0}{1+\alpha} \left\{ (1 + f + f_{AB}) \left(M_0 \frac{u_9}{u_0} \right) - M_0 \right.$$

$$+ (1 + f + f_{AB}) \frac{1}{\gamma_c [M_0(u_9/u_0)]} \frac{T_9}{T_0} \left(1 - \frac{p_0}{p_9} \right)$$

$$+ \alpha \left[(1 + f_{AB'}) \left(M_0 \frac{u_{9'}}{u_0} \right) - M_0 + (1 + f_{AB'}) \frac{1}{\gamma_c [M_0(u_{9'}/u_0)]} \frac{T_{9'}}{T_0} \left(1 - \frac{p_0}{p_{9'}} \right) \right] \right\}$$

$$(7.31)$$

Secondary stream. The temperature and pressure relationships for the secondary stream may be written

$$T_{t_{9'}} = T_{9'} \left(1 + \frac{\gamma_{AB'} - 1}{2} M_{9'}^2 \right) = T_0 \frac{C_{p_c}}{C_{p_{AB'}}} \tau_{\lambda_{AB'}} \tag{7.32}$$

$$p_{t_{9'}} = p_{9'} \left(1 + \frac{\gamma_{AB'} - 1}{2} M_{9'}^2 \right)^{\gamma_{AB'}/(\gamma_{AB'} - 1)} \tag{7.33}$$

where

$$\frac{p_{t_{9'}}}{p_{9'}} = \frac{p_0}{p_{9'}} \pi_r \pi_d \pi_{c'} \pi_{n'} \tag{7.34}$$

Hence noting that

$$M_0^2 \left(\frac{u_{9'}}{u_0} \right)^2 = \frac{\gamma_{AB'} R_{AB'} T_{9'}}{\gamma_c R_c T_0} M_{9'}^2$$

the following expression is obtained

$$M_0^2 \left(\frac{u_{9'}}{u_0} \right)^2 = \frac{2}{\gamma_c - 1} \tau_{\lambda_{AB'}} \left[1 - \left(\frac{p_{t_{9'}}}{p_{9'}} \right)^{-[(\gamma_{AB'} - 1)/\gamma_{AB'}]} \right] \tag{7.35}$$

Primary stream. The relationships of Sec. 7.2 leading to Eqs. (7.9) and (7.10) remain identical for use in the description of the turbofan primary stream. The expression for the turbine temperature ratio must be modified, however, to include the effect of power extraction by the fan. Thus, for the power balance of fan compressor and turbine

$$\dot{m}_c C_{p_c} (T_{t_3} - T_{t_2}) + \dot{m}_F C_{p_c} (T_{t_{3'}} - T_{t_2}) = \eta_m (\dot{m}_c + \dot{m}_f) C_{p_t} (T_{t_4} - T_{t_5})$$

Dividing this expression by $\dot{m}_c C_{p_c} T_0$ and rearranging,

$$\tau_t = 1 - \frac{1}{\eta_m(1+f)} \frac{\tau_r}{\tau_\lambda}\left[(\tau_c - 1) + \alpha(\tau_{c'} - 1)\right] \tag{7.36}$$

Also,

$$\pi_t = \tau_t^{\gamma_t/(\gamma_t - 1)e_t} \tag{7.37}$$

Expressions for the fuel-to-air ratios are obtained by considering enthalpy balances across each of the burners. The expressions for the primary burner and afterburner remain exactly as given by Eqs. (7.13) and (7.14). For the secondary afterburner

$$\dot{m}_F C_{p_c} T_{t_{3'}} + \dot{m}_{f_{AB'}}\eta_{AB'}h = \left(\dot{m}_F + \dot{m}_{f_{AB'}}\right)C_{p_{AB'}} T_{t_{8'}}$$

Hence, dividing by $\dot{m}_F C_{p_c} T_0$ and rearranging,

$$f_{AB'} = \frac{\tau_{\lambda_{AB'}} - \tau_r \tau_{c'}}{\left(\eta_{AB'}h/C_{p_c}T_0\right) - \tau_{\lambda_{AB'}}} \tag{7.38}$$

Finally, the specific fuel consumption is obtained from

$$S = \frac{\left(\dot{m}_f + \dot{m}_{f_{AB}} + \dot{m}_{f_{AB'}}\right)}{F_A}(10^6) = \frac{(f + f_{AB} + \alpha f_{AB'})}{(1 + \alpha)\left[F_A/(\dot{m}_c + \dot{m}_F)\right]}(10^6) \tag{7.39}$$

Equations (7.9), (7.10), (7.13), (7.14), and (7.30–7.39) completely describe the desired performance behavior of the turbofan. The equations are summarized in a form suitable for calculation in the following.

Summary of the Equations— Turbofan with Losses and Bypass Ratio Prescribed

Inputs: $T_0(\text{K})[^\circ\text{R}], \gamma_c, \gamma_t, \gamma_{AB}, \gamma_{AB'}, C_{p_c}, C_{p_t}, C_{p_{AB}},$

$C_{p_{AB'}}(\text{J/kg} \cdot \text{K})[\text{Btu/lbm} \cdot {^\circ}\text{R}], h\ (\text{J/kg})[\text{Btu/lbm}],$

$\pi_d, \pi_b, \pi_n, \pi_{n'}, \eta_b, \eta_{AB}, \eta_{AB'}, \eta_m,$

$e_c, e_{c'}, e_t, p_9/p_0, p_{9'}/p_0, \tau_\lambda, \tau_{\lambda_{AB}}, \tau_{\lambda_{AB'}},$

$\pi_c, \pi_{c'}, M_0, \alpha$

Outputs:
$$\frac{F_A}{\dot{m}_c + m_F} \left(\frac{N \cdot s}{kg} \right) \left[\frac{F_A}{g_0(\dot{m}_c + \dot{m}_F)} \frac{lbf}{lbm/s} \right]$$

$$S \left(\frac{mg}{N \cdot s} \right) \left[\frac{lbm\ fuel/h}{lbf\ thrust} \right], f, f_{AB}, f_{AB'}, \text{etc.}$$

Equations:

$$R_c = \frac{\gamma_c - 1}{\gamma_c} C_{p_c}\ m^2/s^2 \cdot K$$

$$\left[R_c = \frac{\gamma_c - 1}{\gamma_c} C_{p_c}(2.505)(10^4)ft^2/s^2 \cdot {}^\circ R \right] \tag{7.40}$$

$$a_0 = \sqrt{\gamma_c R_c T_0}\ m/s\ [ft/s] \tag{7.41}$$

$$\tau_r = 1 + \frac{\gamma_c - 1}{2} M_0^2 \tag{7.42}$$

$$\pi_r = \tau_r^{\gamma_c/(\gamma_c - 1)} \tag{7.43}$$

$$\tau_c = \pi_c^{(\gamma_c - 1)/\gamma_c e_c} \tag{7.44}$$

$$\tau_{c'} = \left(\pi_{c'} \right)^{(\gamma_c - 1)/\gamma_c e_{c'}} \tag{7.45}$$

$$f = \frac{\tau_\lambda - \tau_r \tau_c}{\left(h\eta_b/C_{p_c} T_0 \right) - \tau_\lambda} \tag{7.46}$$

$$\tau_t = 1 - \frac{1}{\eta_m(1+f)} \frac{\tau_r}{\tau_\lambda} [(\tau_c - 1) + \alpha(\tau_{c'} - 1)] \tag{7.47}$$

$$\pi_t = \tau_t^{\gamma_t/(\gamma_t - 1)e_t} \tag{7.48}$$

$$\frac{P_{t9}}{P_9} = \frac{P_0}{P_9} \pi_r \pi_d \pi_c \pi_b \pi_t \pi_n \tag{7.49}$$

$$\frac{T_9}{T_0} = \frac{\left(C_{p_c}/C_{P_{AB}} \right) \tau_{\lambda_{AB}}}{\left(p_{t9}/p_9 \right)^{(\gamma_{AB} - 1)/\gamma_{AB}}} \tag{7.50}$$

Note: If no primary stream afterburning is present, then $\tau_{\lambda_{AB}}$ should be put equal to $\tau_\lambda \tau_t$. Also, $C_{p_{AB}}$ and γ_{AB} would be put equal to C_{p_t} and γ_t.

$$M_0 \frac{u_9}{u_0} = \left\{ \frac{2}{\gamma_c - 1} \tau_{\lambda_{AB}} \left[1 - \left(\frac{p_{t_9}}{p_9} \right)^{-[(\gamma_{AB} - 1)/\gamma_{AB}]} \right] \right\}^{\frac{1}{2}} \qquad (7.51)$$

$$\frac{p_{t_{9'}}}{p_{9'}} = \frac{p_0}{p_{9'}} \pi_r \pi_d \pi_{c'} \pi_{n'} \qquad (7.52)$$

$$\frac{T_{9'}}{T_0} = \frac{\left(C_{p_c}/C_{p_{AB'}} \right) \tau_{\lambda_{AB'}}}{\left(p_{t_{9'}}/p_{9'} \right)^{(\gamma_{AB'} - 1)/\gamma_{AB'}}} \qquad (7.53)$$

Note: If no secondary stream afterburning is present, then $\tau_{\lambda_{AB'}}$ should be put equal to $\tau_r \tau_{c'}$. Also, $C_{p_{AB'}}$ and $\gamma_{AB'}$ would be put equal to C_{p_c} and γ_c.

$$M_0 \frac{u_{9'}}{u_0} = \left\{ \frac{2}{\gamma_c - 1} \tau_{\lambda_{AB'}} \left[1 - \left(\frac{p_{t_{9'}}}{p_{9'}} \right)^{-[(\gamma_{AB'} - 1)/\gamma_{AB'}]} \right] \right\}^{\frac{1}{2}} \qquad (7.54)$$

$$f_{AB} = (1 + f) \frac{\tau_{\lambda_{AB}} - \tau_\lambda \tau_t}{\left(h\eta_{AB}/C_{p_c} T_0 \right) - \tau_{\lambda_{AB}}} \qquad (7.55)$$

$$f_{AB'} = \frac{\tau_{\lambda_{AB'}} - \tau_r \tau_{c'}}{\left(h\eta_{AB'}/C_{p_c} T_0 \right) - \tau_{\lambda_{AB'}}} \qquad (7.56)$$

$$\frac{F_A}{\dot{m}_c + \dot{m}_F} = \frac{a_0}{1 + \alpha} \left\{ (1 + f + f_{AB}) \left(M_0 \frac{u_9}{u_0} \right) - M_0 + (1 + f + f_{AB}) \right.$$

$$\times \frac{1}{\gamma_c [M_0(u_9/u_0)]} \frac{T_9}{T_0} \left(1 - \frac{p_0}{p_9} \right)$$

$$+ \alpha \left[(1 + f_{AB'}) \left(M_0 \frac{u_{9'}}{u_0} \right) - M_0 \right.$$

$$\left. + (1 + f_{AB'}) \frac{1}{\gamma_c [M_0(u_{9'}/u_0)]} \frac{T_{9'}}{T_0} \left(1 - \frac{p_0}{p_{9'}} \right) \right] \right\} \qquad (7.57)$$

(To obtain British units of lbf/lbm/s, replace a_0 in meters/second with

$a_0/32.174$ where a_0 should be given in feet/second.)

$$S = \frac{(f + f_{AB} + \alpha f_{AB'})}{(1 + \alpha)[F_A/(\dot{m}_c + \dot{m}_F)]}(10^6) \qquad \left[S = \frac{3600(f + f_{AB} + \alpha f_{AB'})}{(1 + \alpha)[F_A/g_0(\dot{m}_c + \dot{m}_F)]} \right]$$

$$(7.58)$$

Operation with Convergent Exit Nozzles

In the usual subsonic transport application of separate stream turbofan engines, no afterburning is utilized and the pressure ratio across both the primary and secondary nozzles is not very large. Often, therefore, only convergent nozzles are employed. In such cases, if the nozzles are choked,

$$\frac{p_{t_9'}}{p_{9'}} = \left(\frac{\gamma_c + 1}{2} \right)^{\gamma_c/(\gamma_c - 1)} \qquad \text{and} \qquad \frac{p_{t_9}}{p_9} = \left(\frac{\gamma_t + 1}{2} \right)^{\gamma_t/(\gamma_t - 1)} \qquad (7.59)$$

Thus

$$\frac{p_0}{p_{9'}} = \frac{p_{t_9'}/p_{9'}}{p_{t_9'}/p_0} = \frac{[(\gamma_c + 1)/2]^{\gamma_c/(\gamma_c - 1)}}{\pi_r \pi_d \pi_{c'} \pi_{n'}} \qquad (7.60)$$

and

$$\frac{p_0}{p_9} = \frac{p_{t_9}/p_9}{p_{t_9}/p_0} = \frac{[(\gamma_t + 1)/2]^{\gamma_t/(\gamma_t - 1)}}{\pi_r \pi_d \pi_c \pi_b \pi_t \pi_n} \qquad (7.61)$$

Equations (7.60) and (7.61) would then be utilized in Eq. (7.57) to give the specific thrust. Note that the expressions are valid for a convergent nozzle only when p_9 and/or $p_{9'}$ are larger than p_0. If Eqs. (7.60) and (7.61) predict that p_0 is larger than p_9 or $p_{9'}$, the given nozzle will not be choked and in such a case the exit pressure should be taken equal to p_0.

Selection of Parameters Leading to Minimum Specific Fuel Consumption

The hierarchy of the equations contained in the preceding summary can be viewed as a functional expression for the specific fuel consumption given in terms of the design parameters π_c, $\pi_{c'}$, and α; thus,

$$S = S(\pi_c, \pi_{c'}, \alpha)$$

Formally at least, the minimum value of S could be found by obtaining the three partial derivatives of S with respect to π_c, $\pi_{c'}$, and α and then equating the expressions to zero to obtain three equations for the values of the parameters leading to the absolute minimum in S. In practice, it is relatively straightforward to obtain the minima with $\pi_{c'}$ and α, but algebraically complex to obtain the minimum with π_c. However, it is a simple matter to plot the resulting value of S vs π_c at the joint minimum with $\pi_{c'}$ and α to locate the absolute minimum. Fuel efficiency is a paramount consideration for transport aircraft, so attention will be directed to non-afterburning engines with the exit pressure matched in both streams. For such a case, it is evident that the minimum in S occurs (with $\pi_{c'}$ and α) when the denominator of Eq. (7.58) reaches a maximum, which itself occurs at the maximum of the function G given by

$$G \equiv (1+f)\sqrt{\left(M_0\frac{u_{9'}}{u_0}\right)^2 - M_0} + \alpha\left[\sqrt{\left(M_0\frac{u_{9'}}{u_0}\right)^2} - M_0\right] \quad (7.62)$$

Note the subsidiary relationships and definitions

$$\left(M_0\frac{u_{9'}}{u_0}\right)^2 = \frac{2}{\gamma_c - 1}\left[\tau_r\tau_{c'} - \frac{\tau_c^{(1-e_{c'})}}{(\pi_d\pi_{n'})^{(\gamma_c-1)/\gamma_c}}\right] \quad (7.63)$$

$$\left(M_0\frac{u_9}{u_0}\right)^2 = \frac{2\tau_\lambda}{\gamma_c - 1}\left(\tau_t - \frac{1}{\Pi}\tau_t^{-[(1-e_t)/e_t]}\right) \quad (7.64)$$

$$\tau_t = 1 - \frac{1}{\eta_m(1+f)}\frac{\tau_r}{\tau_\lambda}\left[\tau_c - 1 + \alpha(\tau_{c'} - 1)\right] \quad (7.65)$$

$$\pi_{c'} = \tau_{c'}^{\gamma_c e_{c'}/(\gamma_c-1)}, \qquad \pi_c = \tau_c^{\gamma_c e_c/(\gamma_c-1)}, \qquad \pi_t = \tau_t^{\gamma_t/(\gamma_t-1)e_t}$$

$$\tau_A \equiv \tau_r\tau_{c'}^{e_c}(\pi_d\pi_{n'})^{(\gamma_c-1)/\gamma_c}, \qquad \Pi \equiv (\pi_r\pi_d\pi_c\pi_b\pi_n)^{(\gamma_t-1)/\gamma_t} \quad (7.66)$$

It follows that

$$\frac{\partial[M_0(u_9/u_0)]^2}{\partial\tau_{c'}} = \frac{-2\alpha\tau_r}{(\gamma_c-1)\eta_m(1+f)}\left[1 + \frac{1}{\Pi}\left(\frac{1-e_t}{e_t}\right)\tau_t^{-1/e_t}\right] \quad (7.67)$$

$$\frac{\partial[M_0(u_{9'}/u_0)]^2}{\partial\tau_{c'}} = \frac{2\tau_r}{\gamma_c - 1}\left(1 - \frac{1-e_{c'}}{\tau_A}\right) \quad (7.68)$$

$$\frac{\partial[M_0(u_9/u_0)]^2}{\partial\alpha} = \frac{-2\tau_r(\tau_{c'}-1)}{(\gamma_c-1)\eta_m(1+f)}\left[1 + \frac{1}{\Pi}\left(\frac{1-e_t}{e_t}\right)\tau_t^{-1/e_t}\right] \quad (7.69)$$

Optimum Bypass Ratio with $\pi_{c'}$ and π_c Prescribed

The bypass ratio leading to minimum specific fuel consumption for prescribed $\pi_{c'}$ (and π_c) is obtained by equating $\partial G/\partial\alpha$ to zero. It is to be noted, particularly for subsonic aircraft, that this optimum is by itself a particularly useful condition. Thus in the case of subsonic aircraft, because the bypass ratios leading to minimum specific fuel consumption are very large, most of the engine thrust is contributed by the fan stream. Hence, prescription of $\pi_{c'}$ very nearly determines the specific thrust and, although this optimal solution creates the correct balance between the core and fan streams to provide the desired fan pressure ratio at minimum specific fuel consumption, the resultant configuration is quite close to that providing the minimum specific fuel consumption for the prescribed specific thrust.

It follows directly from Eq. (7.62) that

$$\frac{\partial G}{\partial\alpha} = \frac{1+f}{2}\frac{1}{M_0(u_9/u_0)}\frac{\partial[M_0(u_9/u_0)]^2}{\partial\alpha} + \sqrt{[M_0(u_{9'}/u_0)]^2} - M_0 = 0$$

hence

$$\left[M_0\left(\frac{u_9}{u_0}\right)\right]^2 = \left\{\frac{1+f}{2}\frac{1}{M_0(u_{9'}/u_0) - M_0}\frac{\partial[M_0(u_9/u_0)]^2}{\partial\alpha}\right\}^2 \quad (7.70)$$

Combination of Eqs. (7.64), (7.69), and (7.70) gives

$$\tau_t = \tau_t^* = \frac{1}{\Pi}\tau_t^{*-[(1-e_t)/e_t]}$$

$$+ \frac{1}{2(\gamma_c - 1)\tau_\lambda}\left\{\frac{\tau_r(\tau_{c'} - 1)}{\eta_m[M_0(u_{9'}/u_0) - M_0]}\left[1 + \left(\frac{1-e_t}{e_t}\right)\frac{1}{\Pi}\tau_t^{*-1/e_t}\right]\right\}^2$$

$$(7.71)$$

The numerical value of τ_t^* may be obtained from this expression by functional iteration. A suitable starting value for τ_t^*, $\tau_{t_0}^*$, is obtained from Eq. (7.71) with $e_t = 1$. Thus,

$$\tau_{t_0}^* = \frac{1}{\Pi} + \frac{1}{2(\gamma_c - 1)\tau_\lambda}\left\{\frac{\tau_r(\tau_{c'} - 1)}{\eta_m[M_0(u_9/u_0) - M_0]}\right\}^2 \quad (7.72)$$

The related value of the bypass ratio α^* can then be obtained from the power balance between the turbine, fan, and compressor. The equations necessary for performance calculation are summarized in the following section.

Summary of the Equations— Turbofan with Losses,
No Afterburning, Exit Pressures Matched, and Bypass Ratio
Optimized for Fan Pressure Ratio and
Compressor Pressure Ratio Prescribed

Inputs: T_0 (K)[°R], $\gamma_c, \gamma_t, C_{p_c}, C_{p_t}$ (J/kg·K)[Btu/lbm·°R],

h (J/kg)[Btu/lbm],

$\pi_d, \pi_b, \pi_n, \pi_{n'}, \eta_b, \eta_m, e_c, e_{c'}, e_t, \pi_c, \pi_{c'}, M_0, \tau_\lambda$

Outputs: $\dfrac{F_A}{\dot{m}_c + \dot{m}_F}\left(\dfrac{N \cdot s}{kg}\right)\left[\dfrac{F_A}{g_0(\dot{m}_c + \dot{m}_F)}\dfrac{lbf}{lbm/s}\right]$,

$S\left(\dfrac{mg}{N \cdot s}\right)\left[\dfrac{lbm\ fuel/h}{lbf\ thrust}\right]$, α^*

Equations: The first seven equations are identical to Eqs. (7.40–7.46), then

$$\Pi = \left(\pi_r \pi_d \pi_c \pi_b \pi_n\right)^{(\gamma_t - 1)/\gamma_t} \tag{7.73}$$

$$\frac{P_{t_{9'}}}{P_{9'}} = \pi_r \pi_d \pi_{c'} \pi_{n'} \tag{7.74}$$

$$M_0 \frac{u_{9'}}{u_0} = \left\{\frac{2\tau_r \tau_{c'}}{\gamma_c - 1}\left[1 - \left(\frac{P_{t_{9'}}}{P_{9'}}\right)^{-[(\gamma_c - 1)/\gamma_c]}\right]\right\}^{\frac{1}{2}} \tag{7.75}$$

$$\tau_{t_0}^* = \frac{1}{\Pi} + \frac{1}{2(\gamma_c - 1)\tau_\lambda}\left[\frac{\tau_r(\tau_{c'} - 1)}{\eta_m\left(M_0 \dfrac{u_{9'}}{u_0} - M_0\right)}\right]^2 \tag{7.76}$$

$$\tau_t^* = \frac{1}{\Pi}\tau_t^{*-[(1 - e_t)/e_t]}$$

$$+ \frac{1}{2(\gamma_c - 1)\tau_\lambda}\left\{\frac{\tau_r(\tau_{c'} - 1)}{\eta_m\left(M_0 \dfrac{u_{9'}}{u_0} - M_0\right)}\left[1 + \left(\frac{1 - e_t}{e_t}\right)\frac{1}{\Pi}\tau_{t^*}^{-1/e_t}\right]\right\}^2 \tag{7.77}$$

$$\alpha^* = \frac{\eta_m(1 + f)(1 - \tau_{t^*})\tau_\lambda}{(\tau_{c'} - 1)\tau_r} - \frac{\tau_c - 1}{\tau_{c'} - 1} \tag{7.78}$$

$$\frac{F_A}{\dot{m}_c + \dot{m}_F} = \frac{a_0}{1 + \alpha^*}\left[(1 + f)M_0\frac{u_9}{u_0} - M_0 + \alpha^*\left(M_0\frac{u_{9'}}{u_0} - M_0\right)\right] \quad (7.79)$$

(For British units divide by $g_0 = 32.174$.)

$$S = \frac{f(10^6)}{(1 + \alpha^*)F/(\dot{m}_c + \dot{m}_F)}\left[S = \frac{3600f}{(1 + \alpha^*)F/[g_0(\dot{m}_c + \dot{m}_F)]}\right] \quad (7.80)$$

Optimum Fan Pressure Ratio with α and π_c Prescribed

An equation for the fan stagnation temperature ratio leading to minimum specific fuel consumption for prescribed α follows from $\partial G/\partial \tau_{c'} = 0$. Thus

$$\frac{\partial G}{\partial \tau_{c'}} = \frac{1 + f}{2}\frac{1}{M_0(u_9/u_0)}\frac{\partial[M_0(u_9/u_0)]^2}{\partial \tau_{c'}}$$

$$+ \frac{\alpha}{2}\frac{1}{M_0(u_{9'}/u_0)}\frac{\partial[M_0(u_{9'}/u_0)]^2}{\partial \tau_{c'}} = 0 \quad (7.81)$$

Hence with Eq. (7.68)

$$\left(M_0\frac{u_{9'}}{u_0}\right)^2 = \left\{\frac{\dfrac{2\alpha}{1 + f}\dfrac{\tau_r}{\gamma_c - 1}\left(1 - \dfrac{1 - e_{c'}}{\tau_A}\right)M_0\left(\dfrac{u_9}{u_0}\right)}{\dfrac{\partial[M_0(u_9/u_0)]^2}{\partial \tau_{c'}}}\right\}^2 \quad (7.82)$$

Combination of Equations (7.63), (7.67), and (7.82) then leads to an expression for $\tau_{c'}$ that can be solved for the desired value leading to minimum specific fuel consumption. This particular optimum will not be further pursued here for the reason that such an optimum generally leads to too low a specific thrust. It is of interest, however, to investigate the joint minimum of S with $\tau_{c'}$ and α because, even though the joint minimum generally corresponds to a configuration with a very low $\tau_{c'}$ and very large α (and hence very low specific thrust), the solution does locate the minimum imaginable specific fuel consumption and thus tends to expose the design tradeoffs that must be considered when installation effects are to be accounted for.

Optimum Bypass Ratio with π_c Prescribed

An equation giving the value of $\tau_{c'}$ that leads to the minimum specific fuel consumption for prescribed π_c (only) follows by equating both $\partial G/\partial \alpha$ and

$\partial G/\partial \tau_{c'}$ to zero. Note first that Eq. (7.81) may be written [with Eq. (7.68)],

$$\left(M_0\frac{u_9}{u_0}\right)^2 = \left[\frac{1+f}{\alpha}M_0\frac{u_{9'}}{u_0}\frac{1}{\frac{2\tau_r}{\gamma_c-1}\left(1-\frac{1-e_{c'}}{\tau_A}\right)}\frac{\partial\left(M_0\frac{u_9}{u_0}\right)^2}{\partial\tau_{c'}}\right]^2 \tag{7.83}$$

Thus, combination of Eqs. (7.67), (7.69), (7.70), and (7.83) gives a single equation for $\tau_{c'}$,

$$M_0\frac{u_{9'}}{u_0}\left(M_0\frac{u_{9'}}{u_0}-M_0\right)=\tau_r\frac{\tau_{c'}-1}{\gamma_c-1}\left(1-\frac{1-e_{c'}}{\tau_A}\right) \tag{7.84}$$

Equations (7.63), (7.66), and (7.84) may then be manipulated into the following form, which may be solved by functional iteration with $\tau_{c'_0}=1$,

$$\tau_{c'}=1+2\left[\tau_{c'}\left(1-\frac{1}{\tau_A}\right)\left(\frac{e_{c'}\tau_{c'}}{\tau_A}-\frac{1}{\tau_r}+\frac{1-e_{c'}}{\tau_A}\right)\right.$$

$$\left.+\frac{1-e_{c'}}{4\tau_A}(\tau_{c'}-1)^2\left(2-\frac{1-e_{c'}}{\tau_A}\right)\right]^{\frac{1}{2}} \tag{7.85}$$

It is to be noted that this optimum value of $\tau_{c'}$ is independent of the prescribed value of π_c. Thus, desired performance and design variables can be obtained for this "joint minimum" case by using the value of $\pi_{c'}$ obtained from this calculation as an input in the summary [Eqs. (7.73–7.80)]. A family of solutions can be generated for a range of compressor pressure ratios π_c, and the minimum value of specific fuel consumption obtained (vs π_c) would hence be the minimum conceivable value for the given flight conditions.

Example Results— Turbofan with Losses

As an example study, consider a "core engine" with $\pi_c = 25$ and investigate the effect of the variation in bypass pressure ratio on the optimum bypass ratio α^* (π_c and $\pi_{c'}$ prescribed) and on the performance parameters. In order to emphasize the effects upon the design configuration of changes in component efficiencies, three engines are considered: a perfect engine, an engine with high component efficiencies, and an engine with low component efficiencies. The flight Mach number is $M_0 = 0.9$ and the component efficiencies for the three engines are shown in Table 7.1.

In addition, assume that the exit pressures are matched ($p_{9'} = p_9 = p_0$) and that $h = 4.4194(10^7)$ J/kg, $C_{p_c} = 1004.9$ J/kg·K, $T_0 = 233.3$ K, $\tau_\lambda = 7.71$, and $\gamma_t = 1.35$. The resulting values for optimal bypass ratio α^* vs the bypass pressure ratio $\pi_{c'}$ are shown in Fig. 7.6.

Table 7.1 Component Efficiencies of Three Engines

Engine	Efficiency								
	π_n	$\pi_{n'}$	π_d	π_b	η_b	η_m	e_c	$e_{c'}$	e_t
1	1	1	1	1	1	1	1	1	1
2	0.99	0.99	0.98	0.98	0.98	1	0.93	0.93	0.92
3	0.96	0.96	0.95	0.95	0.96	0.98	0.87	0.86	0.86

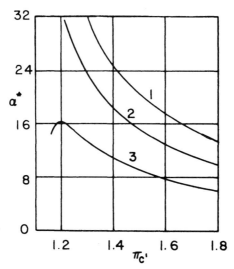

Fig. 7.6 Optimal bypass ratio vs fan pressure ratio.

Several tendencies are notable. It is clear that the optimum bypass ratio for a given $\pi_{c'}$ can change dramatically with changes in component performance, particularly at low values of fan pressure ratio. This again emphasizes the designer's problem in that he must have accurate component performance estimates in order to correctly select his engine configuration. The very large bypass ratios indicated to be optimum (as compared to present-day practice) result because of the very high turbine temperature capability assumed and further because the turbine cooling air penalty has not been included.

Even when the bypass ratio is selected to be optimal for the given component efficiencies, the penalties in performance for the component inefficiencies are substantial, as is evident in Fig. 7.7.

Similar results for higher Mach number flight are easily obtained, although, of course, the optimal bypass ratios are lower and the appropriate fan pressure ratios higher.

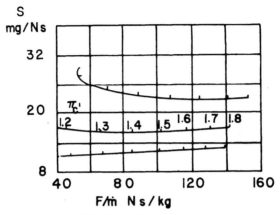

Fig. 7.7 Turbofan performance.

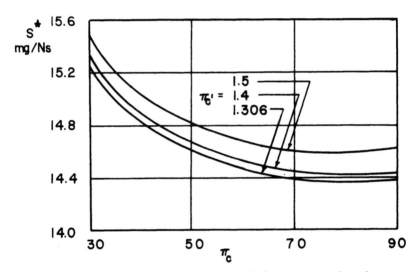

Fig. 7.8 Variation of specific fuel consumption with fan pressure ratio and compressor pressure ratio.

The selection of appropriate compressor and fan pressure ratios is aided by graphs such as Figs. 7.8 and 7.9. The minimum specific fuel consumption attainable for the given conditions (of "engine 2") is indicated in Fig. 7.8 where the minimum value of the "joint minimum" occurs. It is evident however, that the related specific thrust is rather low, the bypass ratio (17.9) rather high, and the compressor pressure ratio (80) very high indeed.

It is readily apparent that installation problems increase for engines with low specific thrust because of the required large cowl diameters. In addition,

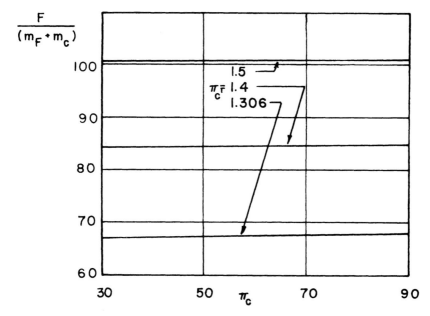

Fig. 7.9   Variation of specific thrust with fan pressure ratio and compressor pressure ratio.

the related bypass ratios are so large that fan-turbine matching becomes a significant problem even if the fan is driven by a separate spool. This in itself results because the fan tip speed limitation brought about by centrifugal stressing requires too low an rpm (and hence tip speed) for the turbine. As a result, it is required either to provide an excessive number of turbine stages to drive the fan or to provide a gearbox to better match the fan and turbine tip speeds. The latter configuration—usually required for bypass ratios in excess of about 10—is referred to as a "geared fan."

The very large compressor pressure ratios identified with minimum specific fuel consumption also introduce severe design problems. Not only does the large compression ratio incur the penalties of large weight and expense, but also, as will be evident in the following chapter, severe off-design penalties are identified with engines having very high-pressure ratios.

A successful design represents an appropriate compromise between the lowest *installed* specific fuel consumption and the cost, weight, off-design behavior, etc., of the engine.

Minimum Specific Fuel Consumption for a Given Specific Thrust

The preceding analyses presented methods of obtaining the minimum specific fuel consumption for an appropriate choice of each of the major individual component design choices, π_c, $\pi_{c'}$, or α. In fact, an optimal design usually involves selection of the best component matching for a prescribed specific thrust. Determination of the combined component configuration is not a difficult task now that high-speed computers (or calculators) are available.

Combination of Eqs. (7.13) and (7.58) gives

$$S = \frac{f(10^6)}{(1+\alpha)\left[F_A/(\dot{m}_c + \dot{m}_F)\right]}, \qquad f = \frac{\tau_\lambda - \tau_r\tau_c}{(h\eta_b/C_{p_c}T_0) - \tau_\lambda} \qquad (7.86)$$

It is evident from Eqs. (7.86) that S will reach a minimum for prescribed π_c and $F_A/(\dot{m}_c + \dot{m}_F)$ when α reaches a maximum *subject to the constrained value of specific thrust.* It can be shown (Problem 7.12) that the expression for the specific thrust itself can be inverted to give an equation of the form

$$\alpha = \alpha\left(\frac{F_A}{\dot{m}_c + \dot{m}_F}, \pi_c, \pi_{c'}\right) \qquad (7.87)$$

Thus, for prescribed values of $F_A/(\dot{m}_c + \dot{m}_F)$ and π_c, the maximum value of α can be obtained numerically by successive calculation with differing values of $\pi_{c'}$. Perusal of Fig. 5.39 makes it evident that the required value of $\pi_{c'}$ will be slightly less than that identified with α^*, as will be the related value of α. A suggested calculation scheme is to estimate the required $\pi_{c'}$ by first calculating that necessary to give the desired specific thrust in the fan stream only (Problem 7.13). Next, initiate the search for α_{max} from Eq. (7.87) with the given or slightly smaller value of $\pi_{c'}$. It is to be noted that the

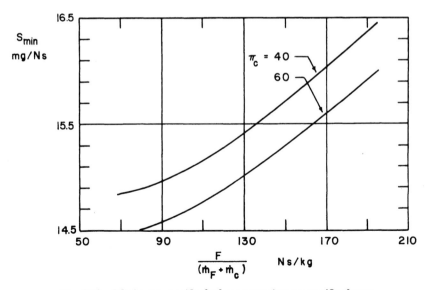

Fig. 7.10   Minimum specific fuel consumption vs specific thrust.

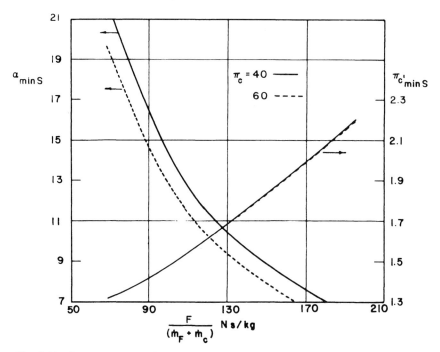

Fig. 7.11 Bypass ratio and fan pressure ratio for minimum specific fuel consumption vs specific thrust.

"true minimum" of S [given $F_A/(\dot{m}_c + \dot{m}_F)$] will involve repetition of this procedure for various values of π_c. Figures 7.10 and 7.11 show the results for engine configuration 2 of the preceding examples. Once again, the values of the compressor pressure ratio corresponding to minimum S are rather large. It is usually the case that off-design and compressor outlet temperature considerations actually determine the final choice of compressor pressure ratio. Example results for $\pi_c = 40$ and 60 are shown in Figs. 7.10 and 7.11.

The Effect of Turbine Cooling

It is readily evident from the preceding calculations that substantial performance benefits can be obtained with increases in the turbine inlet temperature. However, in practice, high turbine inlet temperatures can be achieved only with turbine cooling and the resulting cooling penalties can be substantial. In this section, a simplified approximate analysis of the effects of turbine cooling is developed. The methods are easily extended to more complicated examples, although the algebra can get very tedious.

The configuration to be considered is shown in Fig. 7.12. As indicated in the figure, cooling air is drawn off from the compressor output and injected at the trailing edges of the first-stage nozzle (4a) and first-stage rotor (4b).

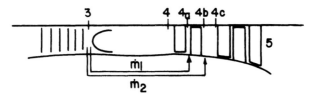

Fig. 7.12 Turbine cooling stations.

Mixing is assumed to occur at each station at the local mainstream stagnation pressure (with $p_{t_{4a}} = p_{t_4}$). Conditions following the mixing behind the rotor are represented with the subscript 4c. Now define,

$$f = \frac{\dot{m}_f}{\dot{m}_c - \dot{m}_1 - \dot{m}_2}, \qquad \varepsilon_1 = \frac{\dot{m}_1}{\dot{m}_c}, \qquad \varepsilon_2 = \frac{\dot{m}_2}{\dot{m}_c}$$

$$\tau_{th} = \frac{T_{t_{4b}}}{T_{t_{4a}}}, \qquad \pi_{th} = \frac{p_{t_{4b}}}{p_{t_{4a}}} = \frac{p_{t_{4b}}}{p_{t_4}}$$

$$\tau_{tL} = \frac{T_{t_5}}{T_{t_{4c}}}, \qquad \pi_{tL} = \frac{p_{t_5}}{p_{t_{4c}}} = \frac{p_{t_5}}{p_{t_{4b}}}$$

Assuming, as previously, that the gas downstream of the burner is calorically perfect with properties γ_t, C_{p_t}, etc., routine but tedious application of the power balance between the compressor, fan, and turbine and of the conservation of stagnation enthalpy in the mixing processes at stations 4a and 4b-c leads to the following equation set for use in determining the engine performance (Problem 7.15).

Additional Inputs to Account for Cooling: π_{th}, ε_1, ε_2, e_{th}, e_{tL}

Equations: Equations (7.47) and (7.48) of the summary [Eqs. (7.38–7.58)] should be replaced by

$$\tau_{th} = \pi_{th}^{(\gamma_t - 1)e_{th}/\gamma_t} \tag{7.88}$$

$$\tau_{tL} = \frac{(\varepsilon_1 + \varepsilon_2)\tau_r\tau_c + (1 - \varepsilon_1 - \varepsilon_2)(1 + f)\tau_\lambda - \dfrac{\tau_r}{\eta_m}\left[\tau_c - 1 + \alpha(\tau_{c'} - 1)\right]}{\left[\varepsilon_1\tau_r\tau_c + (1 - \varepsilon_1 - \varepsilon_2)(1 + f)\tau_\lambda\right]\tau_{th} + \varepsilon_2\tau_r\tau_c} \tag{7.89}$$

$$\pi_{tL} = \tau_{tL}^{\gamma_t/(\gamma_t - 1)e_{tL}} \tag{7.90}$$

$$\pi_t = \pi_{th}\pi_{tL} \tag{7.91}$$

Additionally, in the subsequent equations $\tau_\lambda \tau_t$ and f should be replaced wherever they appear by

$$\tau_\lambda \tau_t \Rightarrow \frac{\tau_{tL}\left\{\left[\varepsilon_1 \tau_r \tau_c + (1 - \varepsilon_1 - \varepsilon_2)(1 + f)\tau_\lambda\right]\tau_{th} + \varepsilon_2 \tau_r \tau_c\right\}}{1 + (1 - \varepsilon_1 - \varepsilon_2)f} \qquad (7.92)$$

$$f \Rightarrow (1 - \varepsilon_1 - \varepsilon_2)f \qquad (7.93)$$

Example calculations (Problem 7.16) indicate that engine performance is very sensitive to the required amount of compressor cooling air.

7.4   The Turboprop or Prop Fan

In recent years renewed interest in highly efficient flight transportation has spurred investigation into "very-high-bypass-ratio fans." Cycle analysis indicates that such bypass ratios (for subsonic flight) could approach those corresponding to the "old" turboprop engines ($\approx 100/1$).

There are several reasons why the turbofan engines became much more popular than the turboprop engines, and it is prudent to review such reasons in order to comprehend why similar concepts are again gaining in popularity. A major reason for the success of the turbofan was its high (subsonic) Mach number capability. In a turboprop, the propeller tip Mach numbers become very large when the flight Mach number approaches about 0.7 and the resultant loss in propeller efficiency limits the turboprop use to $M_0 < 0.7$. With a turbofan, the onset of high Mach number effects is reduced by the diffusion within the inlet duct. In addition, the individual blade loading can be much reduced by utilizing many blades and the cowl much reduces blade tip losses.

A further important benefit of conventional turbofans is that they require no gearbox to reduce the tip speeds of their relatively short blades. (Note, of course, that a turbofan engine usually has multiple spools.) Turboprop gearboxes have to date been heavy and subject to reliability problems.

Finally, the high tip speed of the turboprops led to high noise levels, both in the airport vicinity and within the aircraft at flight speeds.

Recent studies of engines with very high bypass ratios have, however, suggested some compromise designs that show great promise. Thus, if a bypass ratio of (say) 25 is selected, the corresponding cowl could have identified with it weight and drag penalties that are not compensated for by the benefits of the inlet diffusion and the reduction in tip losses. By considering this "in-between" bypass ratio (sometimes termed a "prop fan"), the required shaft speed will be increased with the result that a lighter and simpler gearbox may be utilized. Finally, the effects of tip losses and noise production may be somewhat curtailed by utilizing many (about eight) of the smaller diameter blades and by sweeping the blades to reduce the relative Mach numbers. An additional benefit is available in that the blades may be made variable pitch, which will allow high propeller efficiencies to be maintained over a wide operating range.

It will be recalled (Sec. 5.6) that the propulsive efficiency can be written

$$\eta_p = 2u_0/(u_9 + u_0)$$

where u_0 is the flight speed and u_9 the "jet speed."

This expression is also appropriate for a propeller and serves to emphasize that a large propeller [to reduce u_9 for a given thrust $F = \dot{m}(u_9 - u_0)$] is needed if the propulsive efficiency is to be high. The propulsive efficiency η_p represents the ideal limit of the propeller efficiency defined by

$$\eta_{\text{prop}} = \frac{\text{power to vehicle}}{\text{power to propeller}} \qquad (7.94)$$

Thus,

$$\eta_{\text{prop}} = \frac{Fu_0}{W_{\text{prop}}} = \frac{\dot{m}(u_9 - u_0)u_0}{\frac{1}{2}\dot{m}(u_9^2 - u_0^2)} \frac{\frac{1}{2}\dot{m}(u_9^2 - u_0^2)}{W_{\text{prop}}} \equiv \eta_P \eta_L \qquad (7.95)$$

where W_{prop} is the propeller power input, $\eta_L = \left[\frac{1}{2}\dot{m}(u_9^2 - u_0^2)\right]/W_{\text{prop}}$, and η_L represents the power output of the propeller to the fluid stream (in the "axial" direction) divided by the power input to the propeller.

Thus, the propeller efficiency would be expected to increase with propeller size simply because the ideal propeller efficiency (that is, the propulsive efficiency) increases as more mass is handled. This, of course, relates the propeller efficiency and the "bypass ratio." It should be noted that if the propeller size is increased to the extreme, η_L will begin to decrease because of high Mach number losses in the outer portion of the blades.

Cycle Analysis of the Turboprop

It is appropriate in analyzing the turboprop class of engine to consider the work interaction with the vehicle rather than the thrust. To facilitate this, the "work interaction coefficient" C is introduced, defined by

$$C = \frac{\begin{array}{c}\text{total work interaction with vehicle}/\\ \text{mass of air through core engine}\end{array}}{C_{p_c}T_0} \qquad (7.96)$$

It is usual with turboprop engines to have the core stream exit nozzle unchoked, so the pressure imbalance term will not be included in the expression for the thrust. The numbering of the stations indicated in Fig. 7.13 is used in the following analysis in much the same way as with the previously considered engine types.

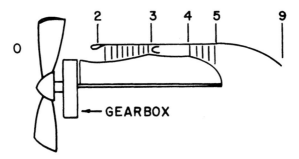

Fig. 7.13 Turboprop reference stations.

Core engine. With

$$P_{t_9} = p_9 \left(1 + \frac{\gamma_t - 1}{2} M_9^2\right)^{\gamma_t/(\gamma_t - 1)} = p_0 \pi_r \pi_d \pi_c \pi_b \pi_t \pi_n \qquad (7.97)$$

$$T_{t_9} = T_9 \left(1 + \frac{\gamma_t - 1}{2} M_9^2\right) = T_0 \tau_\lambda \tau_t \frac{C_{p_c}}{C_{p_t}} \qquad (7.98)$$

and $p_0 = p_9$ and again defining

$$\Pi = \left(\pi_r \pi_d \pi_c \pi_b \pi_n\right)^{(\gamma_t - 1)/\gamma_t} \qquad (7.99)$$

it follows directly that [just as for Eq. (7.64)]

$$\left(M_0 \frac{u_9}{u_0}\right)^2 = \frac{2\tau_\lambda}{\gamma_c - 1}\left(\tau_t - \frac{1}{\Pi}\tau_t^{-[(1-e_t)/e_t]}\right) \qquad (7.100)$$

Thus

$$\frac{F_{\text{core}}}{\dot{m}_c} = a_0 \left[(1 + f) M_0 \frac{u_9}{u_0} - M_0\right] \qquad (7.101)$$

The work interaction coefficient for the core C_c is then given by

$$C_c = \frac{u_0}{C_{p_c} T_0} \frac{F_{\text{core}}}{\dot{m}_c} = \frac{a_0^2 M_0}{\gamma_c R_c T_0} \frac{\gamma_c R_c}{C_{p_c}}\left[(1 + f) M_0 \frac{u_9}{u_0} - M_0\right]$$

or

$$C_c = (\gamma_{c-1}) M_0 \left[(1+f) M_0 \frac{u_9}{u_0} - M_0 \right] \qquad (7.102)$$

Also, it may be noted here that f is still given by Eq. (7.13).

Propeller output. The net work interaction of the turbine with the gearbox input shaft per mass of core engine air may be written

$$(1+f) \eta_m C_{p_t} (T_{t_4} - T_{t_5}) - C_{p_c} (T_{t_3} - T_{t_2})$$

Hence the work interaction with the propeller per mass of core engine air is

$$C_{p_c} T_0 \eta_g \left[\eta_m (1+f) \tau_\lambda (1 - \tau_t) - \tau_r (\tau_c - 1) \right] \qquad (7.103)$$

where η_g is the efficiency of the gearbox.

The work interaction coefficient for the propeller C_{prop} is then given by

$$C_{\text{prop}} = \eta_{\text{prop}} \eta_g \left[\eta_m (1+f) \tau_\lambda (1 - \tau_t) - \tau_r (\tau_c - 1) \right] \qquad (7.104)$$

The total work interaction coefficient C is simply the sum of the core and propeller work interaction coefficients, hence

$$C = C_{\text{prop}} + C_c \qquad (7.105)$$

The work interaction coefficient can then be determined when the "work split" between the core engine and the propeller is decided upon. (This will determine τ_t.) Analogously to the case with the turbofan engine, an optimum value of τ_t may be determined that corresponds to the minimum specific fuel consumption, but before doing so it will be useful to relate the work interaction coefficient to more familiar quantities. Thus, note that the work interaction per second with the flight vehicle is given by the two expressions, Fu_0 and $C_{p_c} T_0 \dot{m}_c C$. Hence, equating these expressions,

$$\text{Specific thrust} = \frac{F}{\dot{m}_c} = C_{p_c} T_0 \frac{C}{u_0} \qquad (7.106)$$

Similarly, the specific fuel consumption follows from

$$S = \frac{f}{F/\dot{m}} (10^6) = \frac{f u_0}{C_{p_c} T_0 C} (10^6) \qquad (7.107)$$

In British units these expressions become

$$\frac{F}{g_0 \dot{m}_c} = 778 \, C_{p_c} T_0 \frac{C}{u_0} \quad \left(778 = \text{ft} \cdot \text{lbf/Btu}, \; C_{p_c} T_0 = \text{Btu/lbm, etc.}\right)$$

$$(7.108)$$

$$S = \frac{3600}{778} \frac{f u_0}{C_{p_c} T_0 C} \frac{\text{lbm fuel/h}}{\text{lbf thrust}} \tag{7.109}$$

A further parameter often used in describing the performance of propeller engines is the power specific fuel consumption S_p, where

$$S_p = \frac{\text{mg fuel/s}}{\text{W to vehicle}} = \frac{f(10^6)}{C_{p_c} T_0 C} \tag{7.110}$$

In the British system, S_p is defined by

$$S_p = \frac{\text{lbm fuel/h}}{\text{hp to vehicle}} = \frac{2545 \, f}{C_{p_c} T_0 C} \tag{7.111}$$

Selection of the optimal turbine temperature ratio. Turbopropeller or prop-fan engines will be designed primarily for low specific fuel consumption. Thus, τ_t is selected to make S a minimum, or equivalently make C a maximum. Thus, from Eqs. (7.102), (7.104), and (7.105)

$$\frac{\partial C}{\partial \tau_t} = \frac{\partial}{\partial \tau_t} \left\{ \eta_{\text{prop}} \eta_g \left[\eta_m (1 + f) \tau_\lambda (1 - \tau_t) - \tau_r (\tau_c - 1) \right] \right.$$

$$\left. + (\gamma_c - 1) M_0 \left[(1 + f) \sqrt{\left(M_0 \frac{u_9}{u_0} \right)^2 - M_0} \right] \right\}$$

$$= (1 + f) \left\{ -\eta_{\text{prop}} \eta_g \eta_m \tau_\lambda + \frac{(\gamma_c - 1) M_0}{2 M_0 (u_9/u_0)} \frac{\partial \left[M_0 (u_9/u_0) \right]^2}{\partial \tau_t} \right\}$$

$$= 0 \qquad \text{when} \qquad \tau_t = \tau_{t*} \tag{7.112}$$

The equation for τ_{t*} may then be written in the form

$$\left(M_0 \frac{u_9}{u_0} \right)^2 = \left\{ \frac{(\gamma_c - 1) M_0}{2 \eta_{\text{prop}} \eta_g \eta_m \tau_\lambda} \frac{\partial \left[M_0 (u_9/u_0) \right]^2}{\partial \tau_t} \right\}^2 \tag{7.113}$$

Equations (7.100) and (7.113) give, after some manipulation

$$\tau_{t^*} = \frac{1}{\Pi} \tau_{t^*}^{-[(1-e_t)/e_t]} + \frac{\gamma_c - 1}{2} \frac{M_0^2}{\tau_\lambda (\eta_{\text{prop}} \eta_g \eta_m)^2} \left[1 + \left(\frac{1 - e_t}{e_t} \right) \frac{1}{\Pi} \tau_{t^*}^{-1/e_t} \right]^2$$

(7.114)

This equation is similar in form to the equivalent equation for the turbofan [Eq. (7.71)] and is also easily solved using functional iteration. A suitable starting guess is obtained by taking $e_t = 1$, to give

$$\tau_{t_0^*} = \frac{1}{\Pi} + \frac{\gamma_c - 1}{2} \frac{M_0^2}{\tau_\lambda (\eta_{\text{prop}} \eta_g \eta_m)^2}$$

(7.115)

It might be noted here that the bypass ratio does not appear in these calculations. This is because the propeller size (and hence the bypass ratio) will be determined once τ_{t^*} and hence the work interaction coefficient C_{prop} is obtained. The propeller will be sized to give the desired propeller efficiency.

Summary of the Equations— Turboprop

Inputs: $T_0 (\text{K})[°\text{R}], \gamma_t, C_{p_t} (\text{J/kg} \cdot \text{K})[\text{Btu/lbm} \cdot °\text{R}],$

$h (\text{J/kg})[\text{Btu/lbm}], \pi_d, \pi_b, \pi_n, \eta_b, \eta_m, \eta_{\text{prop}}, \eta_g, e_c, e_t, \pi_c, \tau_\lambda, M_0$

Note: Standard values for γ_c and C_{p_c} have been incorporated within the summary.

Outputs: $\dfrac{F}{\dot{m}_c} \left(\dfrac{\text{N} \cdot \text{s}}{\text{mg}} \right) \left[\dfrac{F}{g_0 \dot{m}_c} \dfrac{\text{lbf}}{\text{lbm/s}} \right], \quad S \left(\dfrac{\text{mg}}{\text{N} \cdot \text{s}} \right) \left[\dfrac{\text{lbm fuel/h}}{\text{lbf thrust}} \right],$

$S_p \left(\dfrac{\text{mg}}{\text{W} \cdot \text{s}} \right) \left[\dfrac{\text{lbm/h}}{\text{hp}} \right], C, C_c, C_{\text{prop}}, \text{etc.}$

Equations:

$$a_0 = 20.05\sqrt{T_0} \qquad \left[a_0 = 49.02\sqrt{T_0} \text{ ft/s} \right]$$

(7.116)

$$\tau_r = 1 + M_0^2/5$$

(7.117)

$$\pi_r = \tau_r^{3.5}$$

(7.118)

$$\tau_c = \pi_c^{1/3.5 e_c}$$

(7.119)

$$\Pi = (\pi_r \pi_d \pi_c \pi_b \pi_n)^{(\gamma_t - 1)/\gamma_t}$$

(7.120)

If τ_t is provided (nonoptimum case), proceed directly to Eq. (7.123). If the optimum case is desired, calculate τ_t^* from Eq. (7.122) using $\tau_{t_0}^*$ from Eq. (7.121) as the start value.

$$\tau_{t_0}^* = \frac{1}{\Pi} + \frac{M_0^2}{5\tau_\lambda(\eta_{\text{prop}}\eta_g\eta_m)^2} \tag{7.121}$$

$$\tau_t^* = \frac{1}{\Pi}\tau_t^{*-[(1-e_t)/e_t]} + \frac{M_0^2}{5\tau_\lambda(\eta_{\text{prop}}\eta_g\eta_m)^2}\left(1 + \frac{1-e_t}{e_t}\frac{1}{\Pi}\tau_t^{*-1/e_t}\right)^2 \tag{7.122}$$

$$M_0\frac{u_9}{u_0} = \left[5\tau_\lambda\left(\tau_t - \frac{1}{\Pi}\tau_t^{-[(1-e_t)/e_t]}\right)\right]^{\frac{1}{2}} \tag{7.123}$$

$$f = \frac{\tau_\lambda - \tau_r\tau_c}{(h\eta_b/C_{p_c}T_0) - \tau_\lambda} \tag{7.124}$$

$$C_{\text{prop}} = \eta_{\text{prop}}\eta_g\left[\eta_m(1+f)\tau_\lambda(1-\tau_t) - \tau_r(\tau_c - 1)\right] \tag{7.125}$$

$$C_c = 0.4M_0\left[(1+f)M_0(u_9/u_0) - M_0\right] \tag{7.126}$$

$$C = C_{\text{prop}} + C_c \tag{7.127}$$

$$\frac{F}{\dot{m}_c} = 1005\frac{T_0}{a_0M_0}C \qquad \left[\frac{F}{g_0\dot{m}_c} = 186.7\frac{T_0}{a_0M_0}C\right] \tag{7.128}$$

$$S = \frac{f(10^6)}{(F/\dot{m}_c)} \qquad \left[S = \frac{3600f}{(F/g_0\dot{m}_c)}\right] \tag{7.129}$$

$$S_p = \frac{f(10^6)}{1005T_0C} \qquad \left[S_p = \frac{(1.060)(10^4)f}{T_0C}\right] \tag{7.130}$$

Example Results— Turboprop

An engine suitable for use in an 8–10 passenger business aircraft is considered. The parameters assumed are those listed in Table 7.2, which reflect the somewhat modest values it is reasonable to assume for such small, high-reliability engines. The relatively high propeller efficiency has been taken from estimated propeller performance when modern transonic techniques are used in the blade design.

With the values assumed in Table 7.2, a range of compressor pressure ratios is considered, and the thrust and specific fuel consumption corresponding to the minimum fuel consumption at each value of the pressure ratio obtained. Figure 7.14 shows the results. It is interesting to note, also,

Table 7.2   Parameters Assumed for Turboprop Example

$T_0 = 238.9 \text{ K}$	$h = 4.5357 \, (10^7) \text{ J/kg} \cdot \text{K}$	$\eta_b = 0.98$	$\eta_g = 0.99$
[430°R]	[19,500 Btu/lbm]	$\eta_m = 0.95$	$e_c = 0.90$
$\gamma_t = 1.35$	$\pi_d = 0.97$	(power takeoff	$e_t = 0.90$
$C_{p_t} = 1098.2 \text{ J/kg} \cdot \text{K}$	$\pi_b = 0.98$	assumed)	$\tau_\lambda = 6.05$
[0.262 Btu/lbm · °R]	$\pi_n = 0.99$	$\eta_{\text{prop}} = 0.83$	$M_0 = 0.8$

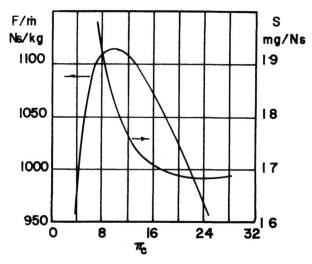

Fig. 7.14 Specific thrust and specific fuel consumption vs compressor pressure ratio for turboprop with optimum turbine temperature ratio.

that at the minimum specific fuel consumption ($\pi_c = 24.5$), the ratio of propeller thrust to core thrust is 8.04.

It is of interest to note that when the turboprop of the above example is compared to a turbofan also optimized to minimum specific fuel consumption (with $\pi_{c'} = 1.4$) and with the same parameters where appropriate, the turboprop shows an approximately 13% improvement in specific fuel consumption.

7.5   The Effects of Nonconstant Specific Heats

The range of accuracy of the cycle analysis equations can be extended substantially by including the variation of the specific heats with temperature. In the range of temperatures found in gas turbines it is an appropriate approximation to retain the assumption of a perfect gas, but to allow specific heat variation. This, in effect, implies that the molecular weight of the air and fuel/air mixture changes very little. As a result of the perfect gas approximation, the convenient relationships of Eqs. (2.42–2.47) remain valid.

For convenience of manipulation, three functions of temperature X_1, X_2, and X_3 are defined. Thus,

$$C_p = RX_1 \tag{7.131}$$

$$h = RTX_2 \tag{7.132}$$

$$X_3 = \int \frac{X_1 - a_1}{T} dT \tag{7.133}$$

The constant a_1 is introduced for convenience. It is to be noted that the function X_1 is of the form

$$X_1 = a_1 + f(T) \tag{7.134}$$

Note that Eq. (2.45) gives

$$X_2 = \frac{1}{T} \int X_1 dT \tag{7.135}$$

In addition,

$$X_1 = \frac{\gamma}{\gamma - 1} \tag{7.136}$$

Stagnation Properties

Application of the flow form of the first law to the imaginary isentropic process connecting the static state to the stagnation state gives

$$h_t = h + (u^2/2)$$

or

$$\frac{h_t}{h} = \frac{T_t X_{2_t}}{TX_2} = 1 + \frac{\gamma RT}{2h} \frac{u^2}{\gamma RT}$$

from which

$$M^2 = 2\frac{X_2}{X_1}(X_1 - 1)\left(\frac{T_t}{T} \frac{X_{2_t}}{X_2} - 1\right) \tag{7.137}$$

In this expression, the additional subscript t, on the symbol X_{2_t}, indicates that the function X_2 is to be evaluated at the stagnation temperature T_t.

The Gibbs equation, with $ds = 0$, gives

$$\frac{dp}{p} = X_1 \frac{dT}{T} = [a_1 + (X_1 - a_1)] \frac{dT}{T}$$

from which

$$\frac{p_t}{p} = \left(\frac{T_t}{T}\right)^{a_1} \exp(X_{3_t} - X_3) \qquad (7.138)$$

and

$$\frac{\rho_t}{\rho} = \left(\frac{T_t}{T}\right)^{(a_1 - 1)} \exp(X_{3_t} - X_3) \qquad (7.139)$$

The Expression for the Mass Flow Rate

Expressions analogous to those obtained in Sec. 2.19 can be obtained directly with the relationships of Eqs. (7.137–7.139). Thus,

$$\dot{m}/A = \rho u = \frac{p_t}{RT_t} \frac{\rho}{\rho_t} \sqrt{\gamma R T} \, M$$

Hence,

$$\dot{m}/A = \sqrt{\frac{2}{R}} \frac{p_t}{\sqrt{T_t}} \left(\frac{T_t}{T}\right)^{-[a_1 - (1/2)]} \sqrt{X_2} \left(\frac{T_t X_{2_t}}{T X_2} - 1\right)^{1/2} \exp\left[-(X_{3_t} - X_3)\right]$$

$$(7.140)$$

It is to be noted that the combination of Eq. (7.137) and (7.140) effectively relates the area variation to the Mach number. Note also that when isentropic flow is considered, by formally taking the derivative of Eq. (7.140) with the temperature T, and equating the result to zero, it follows that $M = 1$ at the throat. This result is also apparent from Eq. (2.76).

Application of the Results to the Turbine Power Balance

As a simple example of the use of the perfect gas results, consider the cycle analysis of a separate stream turbofan engine. It will be assumed that the approximation of a calorically perfect gas in the fan and compressor is still appropriate. In such a case the relationships leading to the fan exit velocity would be those previously obtained, but the expression for the core exit velocity must be rederived. To this end there is defined

$$H = \frac{\tau_r}{\eta_{\dot{m}}(1+f)}\left[\tau_c - 1 + \alpha(\tau_{c'} - 1)\right] \tag{7.141}$$

and

$$\tau_\lambda = \frac{h_{t_4}}{C_{p_c}T_0} \tag{7.142}$$

Note the turbine power balance is simply

$$\frac{h_{t_4} - h_{t_5}}{C_{p_c}T_0} = H \tag{7.143}$$

A further simplifying assumption is now made in which it is assumed that conditions at entry to and exit from the turbine are at effectively zero Mach number, so that the thermodynamic properties are determined in terms of the stagnation temperatures.

The turbine pressure ratio is again related to the temperature ratio by use of the polytropic efficiency. Thus,

$$e_t = \frac{dh_t}{dh_{t_i}} = \frac{C_p}{R}\frac{dT_t/T_t}{dp_t/p_t}$$

So

$$\frac{dp_t}{p_t} = \frac{1}{e_t}X_1\frac{dT_t}{T_t}$$

Hence,

$$\pi_t = \left\{\left(\frac{T_{t_5}}{T_{t_4}}\right)^{a_1}\exp\left[(X_3)_{t_5} - (X_3)_{t_4}\right]\right\}^{1/e_t} \tag{7.144}$$

Finally, the velocity at exit follows from

$$u_9^2 = 2\left(h_{t_5} - h_9\right)$$

$$= 2C_{p_c}T_0\left(\tau_\lambda - H - \frac{h_9}{C_{p_c}T_0}\right)$$

So,

$$\left(\frac{u_9}{u_0}\right)^2 = \frac{1}{\tau_r - 1}\left[\tau_\lambda - H - \frac{R(TX_2)_9}{C_{p_c}T_0}\right] \tag{7.145}$$

With these results, the following summary is obtained.

Summary of the Equations— Variable Specific Heat

Inputs: $M_0, \pi_c, \pi_{c'}, \alpha, C_{p_c}, T_0, \tau_\lambda, \pi_d, \pi_b, \pi_n, e_c, e_{c'}, e_t, \eta_b, \eta_m, h$

Outputs: u_9/u_0

Equations:

$$\tau_c = \pi_c^{(1/3.5e_c)}$$

$$\tau_{c'} = \pi_{c'}^{(1/3.5e_{c'})}$$

$$\tau_r = 1 + \left(M_0^2/5 \right)$$

$$f = \frac{\tau_\lambda - \tau_r \tau_c}{\left(h\eta_b/C_{p_c}T_0 \right) - \tau_\lambda}$$

$$H = \frac{\tau_r}{\eta_m(1+f)} \left[\tau_c - 1 + \alpha(\tau_{c'} - 1) \right]$$

$$h_{t_4} = C_{p_c}T_0\tau_\lambda = R(TX_2)_{t_4} \qquad \left(\text{gives } T_{t_4} \right)$$

$$h_{t_5} = h_{t_4} - C_{p_c}T_0 H = R(TX_2)_{t_5} \qquad \left(\text{gives } T_{t_5} \right)$$

$$\pi_t = \left\{ \left(\frac{T_{t_5}}{T_{t_4}} \right)^{a_1} \exp\left[(X_3)_{t_5} - (X_3)_{t_4} \right] \right\}^{1/e_t}$$

$$\frac{p_{t_9}}{p_0} = \pi_r \pi_d \pi_c \pi_b \pi_t \pi_n$$

$$\frac{p_{t_9}}{p_0} = \left(\frac{T_{t_5}}{T_9} \right)^{a_1} \exp\left[(X_3)_{t_5} - (X_3)_9 \right] \qquad \left(\text{gives } T_9 \right)$$

$$\left(\frac{u_9}{u_0} \right)^2 = \frac{1}{\tau_r - 1} \left[\tau_\lambda - H - \frac{R(TX_2)_9}{C_{p_c}T_0} \right]$$

Effective Ratio of Specific Heats

The preceding equations allow simple determination of the performance of a turbofan. However, the even simpler forms developed in the main text have advantages for rapid calculation, and it would be of use if an "equivalent γ," say γ_e, could be obtained that would lead to the same predicted performance as that obtained for the variable specific heat calculation described above. In fact, by assuming a calorically perfect gas, calculating u_9/u_0, and equating the result to that obtained from Eq.

(7.145), a simple expression for γ_e follows. Thus, given τ_λ, H, C_{p_c}, T_0, and h_9, it follows that

$$\tau_t = 1 - (H/\tau_\lambda)$$

$$\gamma_e = \left\{ 1 + \frac{\ell n \left[\tau_t^{[(1/e_t)-1]} \left(h_9/C_{p_c} T_0 \tau_\lambda \right) \right]}{\ell n \left(\pi_r \pi_d \pi_c \pi_b \pi_n \right)} \right\}^{-1}$$

and

$$T_{t_{4e}} = 3.5 \frac{\gamma_e - 1}{\gamma_e} \tau_\lambda T_0$$

Here, $T_{t_{4e}}$ is the equivalent temperature corresponding to the given τ_λ. It is to be noted that there is no calculational advantage to obtaining γ_e, because it requires evaluation of h_9, which means that the performance of the core stream would already have been determined. The utility in the formulation arises simply because the formulas allow determination of γ_e for a variety of parameter ranges, and hence provides a method of determining appropriate ranges for an effective γ_t.

Example Functional Forms

It is customary to assume simple functional forms for the temperature functions, such as polynomial fits to the experimental data. The coefficients for such polynomial fits are chosen to minimize the least square error of the resultant curve. The reader should do his best to develop his own formulas, given access to experimental data. Failing that, the following forms are recommended from the limited experience of the present author in matching his own performance calculations to the published performance data of the major companies.

Thus, take

$$X_1 = a_1 + a_2 T + a_3 T^2$$

$$X_2 = a_1 + a_2 \frac{T}{2} + a_3 \frac{T^2}{3} + \frac{a_4}{T}$$

$$X_3 = a_2 T + a_3 \frac{T^2}{2}$$

Suggested values for the coefficients (SI system) and R are:

$a_1 = 3.06$ $a_3 = 0.25(10^{-6})$ $R = 287 \text{ m}^2\text{s}^{-2}\text{K}^{-1}$

$a_2 = 1.15(10^{-3})$ $a_4 = 213$

Example calculations were performed with the values $M_0 = 0.85$, $\pi_c = 22$, $\pi_{c'} = 1.5$, $\alpha = 5$, $C_{p_c} = 1004.5$ J·kg^{-1}·K^{-1}, $T_0 = 233$ K, $\pi_d = 0.99$, $\pi_b = 0.97$, $\pi_n = 0.995$, $e_c = e_{c'} = e_t = 0.9$, $\eta_b = 0.995$, $h = 4.42 \, (10^7)$ J/kg^{-1}, $\eta_m = 0.99$, to give

τ_λ	u_9/u_0	γ_e	$T_{t_{4e}}$
6.5	2.45	1.316	1273
7.0	2.80	1.311	1355
7.5	3.11	1.307	1436
8.0	3.38	1.302	1514

7.6   Summary and Conclusions

In the preceding sections, equations capable of describing the expected on-design behavior of several different engine types were developed. The intent was twofold in the sense that it was hoped the reader would gain an understanding of the methodology of cycle analysis, as well as an appreciation for the actual behavior of the several engine types considered. The very simple examples of the analyses considered in this chapter are quite suitable for preliminary design purposes, but it should be realized that more exacting analyses should be utilized if further accuracy is desired. The principal limitations of the analyses considered here arise because of the restriction to calorically perfect gases and because of the lack of inclusion of the effects of power and air takeoffs to operate auxiliary systems. All these effects can be included in a straightforward manner utilizing the same conceptual approaches as already utilized in this chapter, but at the cost of considerably more algebraic complexity.

The very large number of possible input variables in the several example summaries make it difficult indeed to even attempt a comprehensive presentation of the effects of parameter variations. It is of interest, however, to note some of the design trends observable today, the reasons for which are easily shown by utilizing the preceding analyses. It is evident that the industry is spending considerable effort attempting to increase the turbine inlet temperature. The prime benefit for a turbojet resulting from such an increase is in the increased specific thrust. The turbofan engine also benefits from an increase in the turbine inlet temperature because the increased work capability of the turbine causes the optimal bypass ratio (for minimum S) to increase (giving better propulsive efficiency).

Important related changes in the design of other components also occur when an increase in the turbine inlet temperature is attained. Thus, generally, a higher compressor pressure ratio will be utilized to give higher thermal efficiencies. The burner cross section will usually have to be increased because the increased burner outlet temperature will cause increased losses in the stagnation pressure unless the burner inlet Mach number is reduced. As a result of these combined effects, the later stages of even large compressors are becoming excessively small and the burners themselves excessively large. Because of this discrepancy, some modern designs incorporate a single-stage centrifugal compressor following an axial compressor. The centrifugal compressor has the advantage of being rugged and not so subject to such things as tip losses as the several stages of the axial compressor it replaces. The traditional disadvantage of a centrifugal compressor (large cross-sectional area compared to inlet capture area) is not now significant because the compressor is handling high-density air (and hence is relatively small) and is located in front of the necessarily large combustion chambers.

The design of other components is also affected. Thus, if the inlet efficiency as well as the turbine inlet temperature is high, preliminary cycle analysis indicates that the "optimal" bypass ratio will be very large (with a low bypass pressure ratio). Although such large bypass ratio engines look attractive from the point of view of low noise and high propulsive efficiency, the aircraft can be penalized by the requirement of enormous landing gear to accommodate the very-large-diameter engines.

These and similar design interactions must all be considered in a successful aircraft design. If the design is to be successful, accurate estimates of the component efficiencies and an accurate description of the aircraft flight requirements must be available early in the design process.

Problems

7.1   The equation for the ratio of local area to throat area for an isentropic flow, in terms of the Mach number and γ, is given by

$$\frac{A}{A^*} = \left[\frac{2}{\gamma+1}\left(1 + \frac{\gamma-1}{2}M^2\right)\right]^{(\gamma+1)/2(\gamma-1)} \frac{1}{M}$$

(a)   Show by example calculations that when this is written in the form

$$M = \frac{A^*}{A}\left[\frac{2}{\gamma+1}\left(1 + \frac{\gamma-1}{2}M^2\right)\right]^{(\gamma+1)/2(\gamma-1)}$$

functional iteration always gives the subsonic value of M or, for large values of the first guess, diverges.

(b)   Similarly, show that when the equation is written in the form

$$M = \left\{\frac{2}{\gamma-1}\left[\frac{\gamma+1}{2}\left(M\frac{A}{A^*}\right)^{2(\gamma-1)/(\gamma+1)} - 1\right]\right\}^{\frac{1}{2}}$$

functional iteration gives the supersonic value for M, provided that the value of the first guess is not too low.

(c)   Show that when the equation is written in the form

$$F(M) = \frac{1}{M}\left[\frac{2}{\gamma+1}\left(1 + \frac{\gamma-1}{2}M^2\right)\right]^{(\gamma+1)/2(\gamma-1)} - \frac{A}{A^*} = 0$$

ewtonian iteration (using the analytic evaluation of F') usually leads to
e subsonic value for M if the first guess is subsonic and to the supersonic
lue if the first guess is supersonic.

7.2   Consider a family of turbojets with the following parameters:

$T_0 = 220$ K	$\pi_d = 1 - 0.015 M_0^2$	$\eta_{AB} = 0.96$
$\gamma_c = 1.4$	$\pi_b = 0.98$	$\eta_m = 0.99$
$C_{p_c} = 1000$ J/kg $\cdot$ K	$\pi_n = 0.98$ burner off	$e_c = 0.92$
$\gamma_t = 1.3$	$\quad = 0.95$ burner on	$e_t = 0.90$
$C_{p_t} = 1240$ J/kg $\cdot$ K	$\eta_b = 0.99$	$h = 4.5\,(10^7)$ J/kg

   (a)   Plot the specific thrust and specific fuel consumption in the range
$1 \le M_0 \le 3.5$ for the afterburning and nonafterburning cases. Assume $\tau_\lambda = 7$,
$\tau_{\lambda_{AB}} = 8.5$, and $\pi_c = 15$.
   (b)   Plot the specific thrust and specific fuel consumption in the range
$4 \le \pi_c \le$ value giving zero thrust, for the afterburning and nonafterburning
cases. Assume $\tau_\lambda = 7$, $\tau_{\lambda_{AB}} = 8.5$, and $M_0 = 2.2$.

7.3   (a)   Show that the compressor pressure ratio giving maximum
specific thrust when afterburning is present is given by

$$\pi_{c\,\max\,F} = \left[\frac{\eta_m (1+f)\dfrac{\tau_\lambda}{\tau_r} + 1}{\dfrac{\gamma_c - 1}{\gamma_t - 1}\dfrac{\gamma_t}{\gamma_c}\dfrac{1}{e_c e_t} + 1} \right]^{e_c \gamma_c/(\gamma_c - 1)}$$

Note: Ignore the effect of f and f_{AB}, compared to unity, on the magnitude
of the thrust.
   (b)   Plot $\pi_{c\,\max\,F}$ vs M_0 in the range $1 \le M_0 \le 3.5$ for the parameter
values listed in Problem 7.2. Take $\tau_\lambda = 7$ and assume (for this calculation)
that $\eta_m (1+f) = 1$.
   (c)   Plot the related values of specific fuel consumption and specific
thrust for the values of π_c calculated in part (b) for both the afterburning
($\tau_{\lambda_{AB}} = 8.5$) and nonafterburning cases.
   (d)   Obtain and plot $\pi_{c\,\max\,F}$, assuming $\gamma_c = \gamma_t = 1.4$ and $e_c = e_t = 1$,
over the same range as for part (b).

7.4   Investigate the effect of exit pressure mismatch for an engine with
parameters as listed in Problem 7.2 and $M_0 = 2.5$, $\pi_c = 15$, $\tau_\lambda = 7$, and
$\tau_{\lambda_{AB}} = 8.5$.

7.5   Consider a family of nonafterburning turbofan engines with parameters

$$
\begin{array}{lll}
T_0 = 220 \text{ K} & \pi_d = 1 - 0.015\, M_0^2 & e_c = 0.91 \\
\gamma_c = 1.4 & \pi_b = 0.98 & e_{c'} = 0.90 \\
C_{p_c} = 1000 \text{ J/kg} \cdot \text{K} & \pi_n = \pi_{n'} = 0.99 & e_t = 0.89 \\
\gamma_t = 1.32 & \eta_b = 0.99 & h = 4.5\,(10^7)\, \text{J/kg} \\
C_{p_t} = 1200 \text{ J/kg} \cdot \text{K} & \eta_m = 0.99 & p_9 = p_{9'} = p_0
\end{array}
$$

(a)   Plot the bypass ratio α^* identified with minimum specific fuel consumption for prescribed π_c and $\pi_{c'}$ vs $\pi_{c'}$ in the (appropriate) range $1.2 \le \pi_{c'} \le 2.5$ for the case $\tau_\lambda = 7.3$, $M_0 = 2.0$, and $\pi_c = 15$.

(b)   Plot the related values of specific fuel consumption vs specific thrust, as in Fig. 7.7.

(c)   Plot the value of S vs π_c for the "joint minimum" case, and by so doing locate the "absolute minimum" value of S for the prescribed conditions. Obtain the related specific thrust.

7.6   Consider a nonafterburning family of turbofan engines with parameters as listed in Problem 7.5.

(a)   Plot α^* vs M_0 in the range $1.5 \le M_0 \le 3$ for the case $\tau_\lambda = 7.3$ and $\pi_{c'} = 1.6$.

(b)   Plot the related values of specific fuel consumption vs specific thrust similar to those shown in Fig. 7.7. Indicate typical values of M_0 on the curves.

7.7   Consider the effect of afterburning on the family of engines considered in Problem 7.6. For the same Mach number range:

(a)   Obtain S and $F/(\dot{m}_c + \dot{m}_F)$ when $\tau_{\lambda_{AB}} = 8.5$, no fan burning.

(b)   Obtain S and $F/(\dot{m}_c + \dot{m}_F)$ when $\tau_{\lambda_{AB}} = 8.5$, no core burning.

(c)   Obtain S and $F/(\dot{m}_c + \dot{m}_F)$ when $\tau_{\lambda_{AB}} = \tau_{\lambda_{AB'}} = 8.5$.

(d)   Investigate the effect on performance in the given Mach number range of varying α, $\pi_{c'}$, $\tau_{\lambda_{AB}}$, and $\tau_{\lambda_{AB'}}$.

7.8   You are to design an engine for a very-high-performance fighter aircraft. The aircraft, dubbed the "supercruiser," is to be able to cruise at $M_0 = 2$ with no afterburner on, but will then be able to "fight" (maneuver without aircraft energy loss) by utilizing afterburning.

(a)   Develop a preliminary design for a turbojet and give its performance for both afterburning and nonafterburning cases. Investigate also the effects of the changes in your assumed input variables.

(b)   Consider the same "mission" as for part (a), but for a turbofan engine. Take the input parameters to be the same as those of the reference case of part (a) (where appropriate) and compare the performance of candidate turbofans. Include the effects of core and/or duct burning. Discuss the virtues and shortcomings of the various designs.

7.9   Consider two engines with the common parameters

$$M_0 = 0.85 \qquad C_{p_c} = 0.24 \text{ Btu/lbm} \cdot °R \qquad \eta_m = 0.99$$
$$T_0 = 420°R \qquad C_{p_t} = 0.28 \text{ Btu/lbm} \cdot °R \qquad e_t = 0.90$$
$$\gamma_c = 1.4 \qquad h = 19{,}000 \text{ Btu/lbm} \qquad \tau_\lambda = 7.0$$
$$\gamma_t = 1.32 \qquad \pi_b = 0.97 \qquad \tau_c = 36$$
$$\eta_b = 0.97$$

The engines also have:

	π_d	π_n	$\pi_{n'}$	e_c	$e_{c'}$
No. 1	0.990	0.990	0.990	0.90	0.90
No. 2	0.995	0.995	0.995	0.92	0.91

(a) Calculate $\pi_{c'}$, α, $F/g_0(\dot{m}_c + \dot{m}_F)$, and S for each case at the "joint minimum" specific fuel consumption.

(b) For each engine calculate α^*, $F^*/g_0(\dot{m}_c + \dot{m}_F)$, and S^* for $\pi_{c'} = 1.5$.

(c) For each engine calculate $F/g_0(\dot{m}_c + \dot{m}_F)$ and S for $\pi_{c'} = 1.5$ and $\alpha = 0.9\alpha^*$.

7.10   Consider a turbofan engine that has been optimized to have a minimum specific fuel consumption for the case where π_c and $\pi_{c'}$ have been prescribed.

(a) Show that, if all component efficiencies are nonideal *except* that $\eta_m = 1$ and $e_t = 1$, the ratio of the thrust per mass in the primary stream to the thrust per mass in the secondary stream R_A is given by

$$R_A = \tfrac{1}{2} + K$$

where

$$K \equiv \frac{\left[\tau_{c'}^{(1-e_{c'})}(\pi_d \pi_{n'})^{-[(\gamma_c - 1)/\gamma_c]} - 1 \right]}{(\gamma_c - 1)\left[M_0(u_{9'}/u_0) - M_0 \right]^2}$$

(b) Evaluate K for the case $\gamma_c = 1.4$, $e_{c'} = 0.9$, $\pi_{c'} = 1.5$, $M_0 = 0.85$, and $(\pi_d \pi_{n'}) = 0.98$.

7.11   Verify in detail that Eq. (7.85) is correct.

7.12   Equations (7.40–7.57) give the specific thrust of a turbofan engine in terms of prescribed variables. Show that, for the case where no afterburning is present and for which $p_9 = p_{9'} = p_0$, a hierarchy of equations giving the bypass ratio α in terms of prescribed input variables may be obtained in the following form:

Input: $\gamma_t, \tau_\lambda, \mathcal{F} \equiv \dfrac{F}{a_0(\dot{m}_c + \dot{m}_F)}$, $M_0, h, C_{p_c}, T_0, \pi_d, \pi_b, \pi_n, \pi_{n'}, \eta_b,$

$\eta_m, e_{c'}, e_c, e_t, \pi_c, \pi_{c'}$ $(\gamma_c = 1.4)$

Output: α

Equations:

$$\tau_r = 1 + M_0^2/5$$

$$\pi_r = \tau_r^{3.5}$$

$$\tau_c = \pi_c^{1/3.5 e_c}$$

$$\tau_{c'} = \pi_{c'}^{1/3.5 e_{c'}}$$

$$f = \frac{\tau_\lambda - \tau_r \tau_c}{(h\eta_b/C_{p_c} T_0) - \tau_\lambda}$$

$$p_{t_9}/p_9 = \pi_r \pi_d \pi_{c'} \pi_{n'}$$

$$M_0 \frac{u_{9'}}{u_0} = \left\{ 5\tau_r \tau_{c'} \left[1 - \left(\frac{p_{t_9}}{p_9} \right)^{-1/3.5} \right] \right\}^{\frac{1}{2}}$$

$$D_1 = \mathcal{F} + M_0$$

$$D_2 = D_1 - M_0(u_{9'}/u_0)$$

$$D_3 = (\pi_r \pi_d \pi_b \pi_n \pi_c)^{-[(\gamma_t - 1)/\gamma_t]}$$

$$a = D_2^2$$

$$b = 2D_1 D_2 + \frac{5(1+f)}{\eta_m} \tau_r (\tau_{c'} - 1)$$

$$C_0 = D_1^2 - 5(1+f)^2 \tau_\lambda \left[1 - \frac{(\tau_c - 1)}{\eta_m(1+f)} \frac{\tau_r}{\tau_\lambda} - D_3 \right]$$

$$\alpha_0 = \frac{-b + \sqrt{b^2 - 4aC_0}}{2a}$$

$$\tau_{t_j} = 1 - \frac{1}{\eta_m(1+f)} \frac{\tau_r}{\tau_\lambda} \left[\tau_c - 1 + \alpha_j(\tau_{c'} - 1) \right]$$

$$C_{j+1} = C_0 + 5(1+f)^2 \tau_\lambda D_3 \left(\tau_{t_j}^{-[(1-e_t)/e_t]} - 1 \right)$$

$$\alpha_{j+1} = \frac{-b + \sqrt{b^2 - 4aC_{j+1}}}{2a}$$

Calculation note: Particularly with high-bypass ratio turbofans, the parameter a can be very small indeed. Thus note

$$a = D_2^2 = \left\{ \mathscr{F} - \left[M_0(u'_9/u_0) - M_0 \right] \right\}^2$$

The squared quantity is hence the square of the difference of the dimensionless specific thrust of the entire engine less that of the fan stream itself. Straightfoward example calculations indicate that this can be a very small quantity indeed. So small, in fact, that computer accuracy can be lost. In such cases, it is useful to apply an approximate form of the expression for α, which is obtained by binomial expansion of the radical. Thus, write

$$\alpha \approx -\frac{C}{b}\left(1 + \frac{aC}{b^2}\right)$$

7.13   (a) Show that the fan pressure ratio necessary to give a desired specific thrust in the fan stream $F_f/\dot{m}_F$ may be obtained from the expressions (with $\gamma_c = 1.4$):

$$\tau_{c'} = \frac{\left[(F_f/a_0\dot{m}_F) + M_0 \right]^2}{5\tau_r} + \left[\tau_r(\pi_d\pi_{n'})^{1/3.5} \right]^{-1} \tau_c^{(1-e_{c'})}$$

$$\pi_{c'} = \tau_{c'}^{3.5e_{c'}}$$

(b) Plot $\pi_{c'}$ vs $F_f/\dot{m}_F$ N-s kg^{-1} in the range $50 \le F_f/\dot{m}_F \le 500$ for the case $M_0 = 0.85$, $\pi_d\pi_{n'} = 0.99$, $a_0 = 300$ ms^{-1}, and $e_{c'} = 0.9$.

7.14   Utilizing the technique suggested in the text [Eq. (7.87) with Problems 7.12 and 7.13], obtain curves of the bypass ratio, bypass pressure ratio, and specific fuel consumption vs specific thrust in the range $5 \le F/g_0(\dot{m}_c + \dot{m}_F) \le 50$ for the configuration giving minimum specific fuel consumption. Assume the parameters as given for engine 2 of Problem 7.9, including $\pi_c = 36$.

7.15   Derive Eqs. (7.88–7.93).

7.16   (a) Consider a subsonic turbofan engine with parameters as listed for engine 1 of Problem 7.9. Assuming $\pi_{c'} = 1.5$ and for the value of α^* as calculated for Problem 7.9(b), calculate the specific thrust and specific fuel consumption, including the effects of turbine cooling for the cases $\varepsilon_1 = \varepsilon_2 = 0$, 0.01, 0.02, 0.03, 0.04, 0.05 and assuming $\pi_{th} = \frac{1}{2}$ and $\varepsilon_{th} = \varepsilon_{tL} = \varepsilon_t = 0.90$.
     (b) Repeat the calculations of part (a) for the case of $\alpha = 0.9\alpha^*$.

7.17   Consider a family of turboprop engines with the characteristics

$$T_0 = 400°R \qquad\qquad \pi_d = 0.99 \qquad\qquad \eta_g = 0.99$$
$$\gamma_t = 1.3 \qquad\qquad\quad \pi_b = 0.98 \qquad\qquad e_c = 0.93$$
$$\gamma_c = 1.4 \qquad\qquad\quad \pi_n = 0.995 \qquad\qquad e_t = 0.91$$
$$C_{p_t} = 0.27 \text{ Btu/lbm} \cdot °R \qquad \eta_b = 0.99 \qquad\qquad \tau_\lambda = 7.5$$
$$C_{p_c} = 0.24 \qquad\qquad\quad \eta_m = 0.99 \qquad\qquad M_0 = 0.8$$
$$h = 19,000 \text{ Btu/lbm} \qquad \eta_{\text{prop}} = 0.83$$

     (a) Plot the specific fuel consumption and specific thrust vs π_c in the range $15 \leq \pi_c \leq 35$. Consider the "optimal" case, $\tau_t = \tau_{t^*}$.
     (b) For the case $\pi_c = 30$, plot the specific fuel consumption and specific thrust vs τ_t in the range $\tau_{t^*} \leq \tau_t \leq 1.1 \, \tau_{t^*}$.

7.18   Consider a turboprop engine that has the turbine expansion ratio selected to give the minimum specific fuel consumption. It may be assumed that the fuel-to-air ratio may be ignored compared to 1 (i.e., that $1 + f \approx 1$) and that the turbine efficiency may be taken as unity ($e_t = 1$).
     (a) Obtain an expression for the ratio of core engine thrust to propellor thrust in terms of γ_c, τ_λ, M_0, η_{prop}, η_g, η_m, τ_r, τ_c, and $\Pi \equiv (\pi_r \pi_d \pi_c \pi_b \pi_n)^{(\gamma_t - 1)/\gamma_t}$.
     (b) Evaluate the ratio for the example values

$$\gamma_c = 1.4 \qquad \eta_{\text{prop}} = 0.8 \qquad \pi_d = 0.99 \qquad \pi_c = 25$$
$$\tau_\lambda = 7.0 \qquad \eta_g = 0.95 \qquad \pi_b = 0.98 \qquad e_c = 0.92$$
$$M_0 = 0.75 \qquad \eta_m = 0.99 \qquad \pi_n = 1.0 \qquad \gamma_t = 1.35$$

8. ENGINE OFF-DESIGN PERFORMANCE

8.1 Introduction

In the previous chapter, cycle analysis was applied to several example engine types in order to predict the expected performance of such engines as a function of design choices, design limitations, or environmental conditions. The various results obtained are hence to be interpreted as the expected behavior of a family of engines under the various imposed conditions. This chapter considers the related problem of how a given engine (designed for certain prescribed conditions) will behave at conditions other than those for which it was designed.

It is to be noted that in the design process, prescription of the designer variables π_c, $\pi_{c'}$, and α, in fact, actually determines the required turbine expansion ratio because of the required satisfaction of the power balance between the turbine and fan and compressor. The power balance itself is affected by the flight Mach number (through τ_r), the turbine temperature (through τ_λ), and the ambient temperature (also through τ_λ). Thus, when a given engine is operated at other than design conditions, π_c, $\pi_{c'}$, and α may all change, and it is the off-design problem to determine such changes in terms of the imposed changes of other variables. Once such changes have been determined, the new values of π_c, $\pi_{c'}$, and α may then be used in a computational program very similar to those developed in Chap. 7. The most notable change in calculation procedure is evident in the use of component efficiencies rather than the polytropic efficiencies.

Off-design performance analysis can be considered to be of two classes, the first being one where no component performances are available so that the component efficiencies as functions of operating conditions must be estimated, and the second being one where the components have been developed and tested so that the component characteristics are available. The former class of analysis is used in preliminary estimates of engine off-design performance, whereas the second class is used for more exact estimates of the expected performance of an engine that is approaching completion of construction. Both classes of analysis will be considered in the following.

8.2 Off-Design Analysis of the Turbojet

Consider the simple case where both the turbine entrance nozzle and primary nozzle are choked. This puts algebraically simple restrictions on the various relationships and is, in fact, true over a wide operating range for modern turbojets. Further simplifying assumptions consistent with those of Chap. 7 will be made. Thus, the gases will be assumed to be calorically perfect both upstream and downstream of the burner, turbine cooling will be ignored, and no power or air takeoffs will be considered. In addition, the fuel-to-air ratio f compared to unity and the variation of γ_t and C_{p_t} with the power setting will be ignored. These additional assumptions introduce little inaccuracy because it is, in fact, the ratios of the desired quantities to their design values, rather than their absolute values, that are needed.

The station numbering is indicated in Fig. 8.1 with the same notation introduced in Sec. 5.2 and utilized throughout Chaps. 5-7. Equation (2.102), together with the assumption of choked flow at stations 4 and 8, allows

$$\dot{m}_4 = \frac{\Gamma_t}{\sqrt{R}} \frac{A_4 P_{t_4}}{\sqrt{T_{t_4}}} \quad \text{and} \quad \dot{m}_8 = \frac{\Gamma_t}{\sqrt{R}} \frac{A_8 P_{t_8}}{\sqrt{T_{t_8}}} \tag{8.1}$$

where

$$\Gamma \equiv \sqrt{\gamma} \left(\frac{2}{\gamma + 1} \right)^{(\gamma+1)/2(\gamma-1)}$$

Equating $\dot{m}_4$ and $\dot{m}_8$ results in

$$\frac{\tau_t^{\frac{1}{2}}}{\pi_t} = \frac{A_8}{A_4} \frac{\pi_{AB}}{\sqrt{\tau_{AB}}} \tag{8.2}$$

Note that the area ratio A_8/A_4 will be prescribed by the control system and τ_{AB} will be prescribed by the afterburner setting. π_{AB} will change relatively little, so the area ratio and afterburner settings determine the ratio $\tau_t^{\frac{1}{2}}/\pi_t$. Note from Eq. (6.77) that π_t is a unique function of τ_t and the turbine efficiency, so if the turbine efficiency does not change much over the operating range, Eq. (8.2) becomes a single equation for τ_t and hence π_t. As an example, note that for a conventional turbojet without afterburning, A_8 and A_4 remain fixed. (When afterburning is present A_8 is varied so that $A_8(\pi_{AB}/\sqrt{\tau_{AB}})$ remains constant.) For the conventional turbojet, then, the

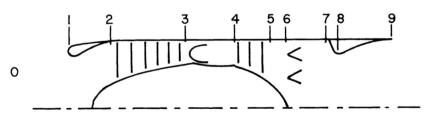

Fig. 8.1 The turbojet.

turbine expansion ratio remains very nearly constant over the entire operating range! (The only change in π_t arises because of changes in $\eta_{t.}$)

With τ_t determined from Eq. (8.2), τ_c may be obtained from the expression for the power balance, Eq. (7.11)

$$\tau_c = 1 + \eta_m(\tau_\lambda/\tau_r)(1 - \tau_t) \tag{8.3}$$

Note here that τ_λ is now not the design limit turbine inlet enthalpy divided by the ambient enthalpy, but rather is the selected turbine inlet enthalpy divided by the ambient enthalpy. Thus, τ_c (and hence π_c) is determined by the throttle setting (T_{t_4}), environment (T_0), and flight condition (τ_r).

The Mass Flow

It will be necessary to determine the variation in the mass flow rate in order to determine the variation in thrust. Thus write (with the assumption $f \ll 1$)

$$\dot{m}_2 = \dot{m}_4 = \frac{\Gamma_t}{\sqrt{R}} \frac{A_4 p_{t_4}}{\sqrt{T_{t_4}}}$$

or

$$\dot{m}_2 = \left(\frac{\Gamma_t}{\sqrt{R}} \sqrt{\frac{C_{p_t}}{C_{p_c}}}\right) \pi_d \pi_b \pi_r \pi_c \frac{p_0}{\sqrt{T_0}} \frac{A_4}{\sqrt{\tau_\lambda}} \tag{8.4}$$

The Corrected Mass Flow

The corrected mass flow $\dot{m}_c$ is defined as the group

$$\dot{m}_c \equiv \dot{m}\sqrt{\theta}/\delta \tag{8.5}$$

where $\dot{m}$ is the actual mass flow at the plane of interest, $\theta = T_t/T_{STP}$, and $\delta = p_t/p_{STP}$. STP refers to standard temperature and pressure so that

$$T_{STP} = 288.33 \text{ K } [519°R]$$

$$p_{STP} = 1.013(10^5) \text{ Pa } [14.69 \text{ lbf/in.}^2]$$

It can be noted from Eq. (2.102) that a given value of corrected mass flow corresponds to a particular value of the Mach number at a given reference area. Thus, a particular corrected mass flow corresponds to a particular engine face Mach number for a given engine.

With Eq. (8.4)

$$\dot{m}_c = \frac{\dot{m}_2\sqrt{\theta_2}}{\delta_2} = \left[p_{STP}\Gamma_t\left(\frac{C_{p_t}}{RC_{p_c}T_{STP}}\right)^{\frac{1}{2}}\right]\pi_b\pi_c A_4\left(\frac{\tau_r}{\tau_\lambda}\right)^{\frac{1}{2}} \tag{8.6}$$

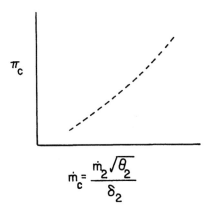

Fig. 8.2 Compressor operating line.

It is to be noted that, when the "schedule" of A_8/A_4 is known, τ_t is fixed by Eq. (8.2) (assuming no afterburning is present). The chosen area ratio A_8/A_4 and τ_c (hence π_c) thus fully determine τ_λ/τ_r, as is apparent from Eq. (8.3). It is then apparent from Eq. (8.6) that the corrected mass flow is a unique function of the compressor pressure ratio and the chosen area scheduling. If the area scheduling is related directly to the compressor pressure ratio, then a unique relationship exists between $\dot{m}_c$ and π_c. (The most obvious example is that occurring in a conventional engine where A_4 and A_8 are fixed.) When the loci of the points defined by the relationship $\dot{m}_c$ vs π_c are plotted on a graph, the resulting line is termed the compressor operating line (i.e., Fig. 8.2).

It is usually most convenient to obtain the off-design behaviors in terms of the ratio of the desired parameter to the value of the parameter at on-design. Denoting the reference, or on-design, quantities by a subscript R, Eq. (8.6) gives

$$\frac{\dot{m}_c}{\dot{m}_{cR}} = \frac{\pi_b \pi_c}{\pi_{bR} \pi_{cR}} \frac{A_4}{A_{4R}} \left(\frac{\tau_{\lambda R}}{\tau_\lambda} \frac{\tau_r}{\tau_{rR}} \right)^{\frac{1}{2}} \tag{8.7}$$

Performance Parameters

Now consider the behavior of the performance parameters when the engine is operated off-design. In order to simplify the equations algebraically it will be assumed that the nozzle exit area is varied so as to keep $p_9 = p_0$ and in addition that $f \ll 1$. Also only nonafterburning turbojets will be considered so that with $\tau_{\lambda_{AB}} = \tau_\lambda \tau_t$ it follows from Eqs. (7.24) and (7.27),

$$p_{t_9}/p_9 = \pi_r \pi_d \pi_c \pi_b \pi_n \pi_t \tag{8.8}$$

$$M_0 \frac{u_9}{u_0} = \left\{ \frac{2}{\gamma_c - 1} \tau_\lambda \tau_t \left[1 - \left(\frac{p_{t_9}}{p_9} \right)^{-[(\gamma_t - 1)/\gamma_t]} \right] \right\}^{\frac{1}{2}} \tag{8.9}$$

The equation for the thrust becomes

$$F = \dot{m}(u_9 - u_0) = \dot{m}a_0 \left[M_0(u_9/u_0) - M_0 \right] \qquad (8.10)$$

from which, with Eq. (8.4)

$$\frac{F}{F_R} = \frac{\pi_d \pi_b \pi_r \pi_c}{(\pi_d \pi_b \pi_r \pi_c)_R} \frac{p_0}{p_{0R}} \frac{A_4}{A_{4R}} \left(\frac{\tau_{\lambda R}}{\tau_\lambda} \right)^{\frac{1}{2}} \frac{\left[M_0(u_9/u_0) - M_0 \right]}{\left[M_0(u_9/u_0) - M_0 \right]_R} \qquad (8.11)$$

With the assumption that $f \ll 1$, or equivalently that $h\eta_b/C_{p_c}T_0 \gg \tau_\lambda$, from Eq. (7.21)

$$f = \frac{C_{p_c}T_0}{h\eta_b} (\tau_\lambda - \tau_r\tau_c) \qquad (8.12)$$

Then from $S = f/(F/\dot{m})$ and Eqs. (8.10) and (8.12)

$$\frac{S}{S_R} = \left(\frac{T_0}{T_{0R}} \right)^{\frac{1}{2}} \frac{\left[M_0(u_9/u_0) - M_0 \right]_R}{\left[M_0(u_9/u_0) - M_0 \right]} \frac{(\tau_\lambda - \tau_r\tau_c)}{(\tau_\lambda - \tau_r\tau_c)_R} \qquad (8.13)$$

Exit Area Variation

It has been assumed that the nozzle exit area will be varied in a manner to keep $p_9 = p_0$. As discussed in Sec. 6.2, it is important to know the exit area variation so that installation penalties (boat-tail drag) can be estimated. Thus write, utilizing Eq. (2.106),

$$\frac{A_9}{A_8} = \Gamma_t \left(\frac{\gamma_t - 1}{2\gamma_t} \right)^{\frac{1}{2}} \frac{\left(p_{t_9}/p_0 \right)^{(\gamma_t + 1)/2\gamma_t}}{\left[\left(p_{t_9}/p_0 \right)^{(\gamma_t - 1)/\gamma_t} - 1 \right]^{\frac{1}{2}}} \frac{\pi_{AB}}{\pi_n} \qquad (8.14)$$

The schedule of A_8 variation is separately prescribed, so that the exit area variation may be obtained from

$$\frac{A_9}{A_{9R}} = \frac{A_9/A_8}{(A_9/A_8)_R} \frac{A_8}{A_{8R}} \qquad (8.15)$$

These equations and appropriate subsidiary equations are summarized in a manner suitable for sequential solution in the following.

Summary of the Equations— Off-Design Turbojet (Nonafterburning)

Inputs: $\gamma_c, \gamma_t, A_8/A_{8R}, p_0/p_{0R}, T_0/T_{0R}, \pi_{cR}$, and both the reference values and off-design values of A_8/A_4, M_0, π_d, π_b, π_{AB}, π_n, τ_λ, η_m, η_c, and η_t

Outputs:
$$\frac{F}{F_R}, \frac{S}{S_R}, \frac{A_9}{A_{9R}}, \frac{\dot{m}_c}{\dot{m}_{cR}}, \frac{\pi_c}{\pi_{cR}}, \frac{\pi_t}{\pi_{tR}}$$

Equations (where appropriate, valid for both design and off-design cases):

$$\frac{\tau_t^{\frac{1}{2}}}{\left[1 - (1/\eta_t)(1 - \tau_t)\right]^{\gamma_t/(\gamma_t-1)}} = \frac{A_8}{A_4} \pi_{AB} \tag{8.16}$$

$$\pi_t = \left[1 - (1/\eta_t)(1 - \tau_t)\right]^{\gamma_t/(\gamma_t-1)} \tag{8.17}$$

$$\tau_r = 1 + \left[(\gamma_c - 1)/2\right] M_0^2 \tag{8.18}$$

$$\pi_r = \tau_r^{\gamma_c/(\gamma_c-1)} \tag{8.19}$$

$$\tau_c = 1 + \eta_m (\tau_\lambda/\tau_r)(1 - \tau_t) \tag{8.20}$$

$$\pi_c = \left[1 + \eta_c(\tau_c - 1)\right]^{\gamma_c/(\gamma_c-1)} \tag{8.21}$$

$$p_{t_9}/p_0 = p_{t_9}/p_9 = \pi_r \pi_d \pi_c \pi_b \pi_n \pi_t \tag{8.22}$$

$$M_0 \frac{u_9}{u_0} = \left\{ \frac{2}{\gamma_c - 1} \tau_\lambda \tau_t \left[1 - \left(\frac{p_{t_9}}{p_9}\right)^{-(\gamma_t-1)/\gamma_t}\right] \right\}^{\frac{1}{2}} \tag{8.23}$$

$$\frac{F}{F_R} = \frac{\pi_d \pi_b \pi_r \pi_c}{(\pi_d \pi_b \pi_r \pi_c)_R} \left(\frac{p_0}{p_{0R}}\right) \frac{A_4}{A_{4R}} \left(\frac{\tau_{\lambda R}}{\tau_\lambda}\right)^{\frac{1}{2}} \frac{\left[M_0(u_9/u_0) - M_0\right]}{\left[M_0(u_9/u_0) - M_0\right]_R} \tag{8.24}$$

$$\frac{S}{S_R} = \left(\frac{T_0}{T_{0R}}\right)^{\frac{1}{2}} \frac{\left[M_0(u_9/u_0) - M_0\right]_R}{\left[M_0(u_9/u_0) - M_0\right]} \frac{(\tau_\lambda - \tau_r \tau_c)}{(\tau_\lambda - \tau_r \tau_c)_R} \tag{8.25}$$

$$\frac{A_9}{A_8} = \Gamma_t \left(\frac{\gamma_t - 1}{2\gamma_t}\right)^{\frac{1}{2}} \frac{(p_{t_9}/p_0)^{(\gamma_t+1)/2\gamma_t}}{\left[(p_{t_9}/p_0)^{(\gamma_t-1)/\gamma_t} - 1\right]^{\frac{1}{2}}} \frac{\pi_{AB}}{\pi_n} \tag{8.26}$$

$$\frac{A_9}{A_{9R}} = \frac{A_9/A_8}{(A_9/A_8)_R} \frac{A_8}{A_{8R}} \tag{8.27}$$

$$\frac{\dot{m}_c}{\dot{m}_{cR}} = \frac{\pi_b \pi_c A_4}{(\pi_b \pi_c A_4)_R} \left(\frac{\tau_{\lambda R}}{\tau_\lambda} \frac{\tau_r}{\tau_{rR}}\right)^{\frac{1}{2}} \tag{8.28}$$

The Fixed-Area Turbojet (FAT)

To date, no main propulsion turbines have been used that incorporate a variable turbine inlet area (A_4). Such "conventional" turbines are also usually coupled with a nozzle of fixed area A_8 except when an afterburner is utilized. However, it is customary to design a primary nozzle variable throat so that, when the afterburner is in operation, A_8 is varied just enough to keep the engine operating at its original setting. It can be seen from Eq. (8.2) that this would require $A_8(\pi_{AB}/\sqrt{\tau_{AB}})$ to remain constant. Now utilize the results of the preceding sections to estimate the off-design behavior of a fixed-area turbojet (FAT) operating without afterburner.

The equations of the preceding sections can be simplified considerably in this special case, for from Eq. (8.2),

$$\tau_t^{\frac{1}{2}}/\pi_t = \text{const} \tag{8.29}$$

Now make the further assumption that (many of) the component efficiencies remain constant in the regime of parameter variation to be considered. This is a very convenient numerical approximation for illustrative purposes, and the results obtained by utilizing this simplification still reveal the principal effects of the off-design behavior that result primarily from changes in the propulsive and thermal efficiencies, rather than from changes in the component efficiencies. If greater accuracy is desired, the more complete equations (8.16–8.28) may be utilized.

With the assumed constant turbine efficiency, the turbine expansion ratio remains fixed. Equation (8.3) then gives

$$\tau_c = 1 + (\tau_{cR} - 1)\frac{\tau_\lambda}{\tau_{\lambda R}}\frac{\tau_{rR}}{\tau_r} \tag{8.30}$$

Utilizing this relationship and further assuming that η_m, η_c, η_t, π_n, π_{AB}, and π_b remain constant, the equations may be simplified and reordered to give the following summary.

Summary of the Equations—Off-Design FAT

Inputs: γ_c, γ_t, p_0/p_{0R}, T_0/T_{0R}, η_m, η_c, η_t, π_{cR}, π_{dR},

$\tau_{\lambda R}$, M_{0R}, π_d, τ_λ, M_0

Outputs: $\dfrac{F}{F_R}$, $\dfrac{S}{S_R}$, $\dfrac{A_9}{A_{9R}}$, $\dfrac{\dot{m}_c}{\dot{m}_{cR}}$, $\dfrac{\pi_c}{\pi_{cR}}$

Equations:

$$\tau_{rR} = 1 + \frac{\gamma_c - 1}{2} M_{0R}^2, \qquad \pi_{rR} = \tau_{rR}^{\gamma_c/(\gamma_c-1)} \tag{8.31}$$

$$\tau_r = 1 + \frac{\gamma_c - 1}{2} M_0^2, \qquad \pi_r = \tau_r^{\gamma_c/(\gamma_c-1)} \tag{8.32}$$

$$\tau_{cR} = 1 + \frac{1}{\eta_c}\left(\pi_{cR}^{(\gamma_c-1)/\gamma_c} - 1\right) \tag{8.33}$$

$$\tau_c = 1 + (\tau_{cR} - 1)\frac{\tau_\lambda}{\tau_{\lambda R}}\frac{\tau_{rR}}{\tau_r}$$

$$\pi_c = \left[1 + \eta_c(\tau_c - 1)\right]^{\gamma_c/(\gamma_c-1)} \tag{8.34}$$

$$\tau_{tR} = 1 - \frac{1}{\eta_m}\frac{\tau_{rR}}{\tau_{\lambda R}}(\tau_{cR} - 1), \qquad \tau_t = \tau_{tR} \tag{8.35}$$

$$\pi_{tR} = \left[1 - (1/\eta_t)(1 - \tau_{tR})\right]^{\gamma_t/(\gamma_t-1)}, \qquad \pi_t = \pi_{tR} \tag{8.36}$$

$$\left(P_{t_9}/P_0\right)_R = \pi_b\pi_t\pi_n\left(\pi_r\pi_d\pi_c\right)_R \tag{8.37}$$

$$\frac{P_{t_9}}{P_9} = \left(\frac{P_{t_9}}{P_0}\right)_R \frac{\pi_r\pi_d\pi_c}{\left(\pi_r\pi_d\pi_c\right)_R} \tag{8.38}$$

$$M_0\frac{u_9}{u_0} = \left\{\frac{2}{\gamma_c - 1}\tau_\lambda\tau_t\left[1 - \left(\frac{P_{t_9}}{P_9}\right)^{-(\gamma_t-1)/\gamma_t}\right]\right\}^{\frac{1}{2}} \tag{8.39}$$

(Note that the formula for $[M_0(u_9/u_0)]_R$ is identical, but R quantities are to be used.)

$$\frac{F}{F_R} = \frac{\pi_r\pi_d\pi_c}{(\pi_r\pi_d\pi_c)_R}\frac{P_0}{P_{0R}}\left(\frac{\tau_{\lambda R}}{\tau_\lambda}\right)^{\frac{1}{2}}\frac{\left[M_0(u_9/u_0) - M_0\right]}{\left[M_0(u_9/u_0) - M_0\right]_R} \tag{8.40}$$

$$\frac{S}{S_R} = \left(\frac{T_0}{T_{0R}}\right)^{\frac{1}{2}}\frac{\left[M_0(u_9/u_0) - M_0\right]_R}{\left[M_0(u_9/u_0) - M_0\right]}\frac{(\tau_\lambda - \tau_r\tau_c)}{(\tau_\lambda - \tau_r\tau_c)_R} \tag{8.41}$$

$$\frac{A_9}{A_{9R}} = \left[\frac{P_{t_9}/P_0}{\left(P_{t_9}/P_0\right)_R}\right]^{(\gamma_t+1)/2\gamma_t}\left[\frac{\left(P_{t_9}/P_0\right)_R^{(\gamma_t-1)/\gamma_t} - 1}{\left(P_{t_9}/P_0\right)^{(\gamma_t-1)/\gamma_t} - 1}\right]^{\frac{1}{2}} \tag{8.42}$$

$$\frac{\dot{m}_c}{\dot{m}_{cR}} = \frac{\pi_c}{\pi_{cR}}\left(\frac{\tau_{cR} - 1}{\tau_c - 1}\right)^{\frac{1}{2}} \tag{8.43}$$

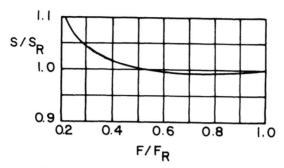

Fig. 8.3 S/S_R vs F/F_R for a fixed-area turbojet.

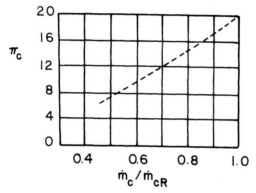

Fig. 8.4 Compressor characteristic, fixed-area turbojet.

As an example calculation, consider an engine held at fixed altitude and Mach number. The parameters assumed are $\gamma_c = 1.4$, $\gamma_t = 1.3$, $\eta_m = 1$, $\eta_c = 0.88$, $\eta_t = 0.9$, $\pi_b = 0.98$, $\pi_n = 0.99$, $\pi_d = 0.97$, $\pi_{cR} = 20$, $\tau_{\lambda R} = 7$, and $M_0 = 0.8$. Figure 8.3 shows the resulting variation in specific fuel consumption with thrust.

It is evident that the specific fuel consumption at first decreases with the decrease in thrust, but later increases at very reduced thrust levels. This behavior is of enormous importance in determining the proper sizing of an engine for use in such vehicles as a high-performance fighter. It is often desirable to have such a fighter have a subsonic "ferry" capability, and in such a case the engine could be required to operate at very low thrust levels. If the engine is very large, such low thrust levels could be well on the "back side of the SFC bucket."

It should be noted that it is the increase in propulsive efficiency which causes the original reduction in specific fuel consumption. At lower thrust levels, however, the decreasing thermal efficiency (caused by decreasing π_c), coupled with the small output compared to the component losses, causes the specific fuel consumption to rise. Figure 8.4 shows the related compressor characteristic as the thrust is decreased

Finally, the substantial contraction required of the exit nozzle to keep the exit pressure balanced is shown in Fig. 8.5.

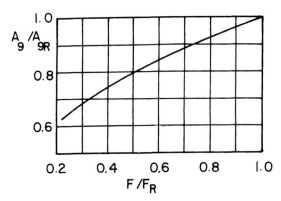

Fig. 8.5   Exit area variation, fixed-area turbojet.

A second example of some interest is that where an aircraft flies at the same Mach number but at two different altitudes. Consider the case where the turbine inlet temperature is held the same so that $\tau_\lambda/\tau_{\lambda R} = T_{0R}/T_0$.

Consider the case in which $T_0/T_{0R} = 0.759$, $p_0/p_{0R} = 0.162$, and all other values are as given in the preceding example. (Note these ratios of T_0/T_{0R} and p_0/p_{0R} correspond approximately to the changes experienced in going from sea level to 40,000 ft altitude.) Straightforward calculation yields $F/F_R = 0.345$, $S/S_R = 0.900$, $\dot{m}_c/\dot{m}_{cR} = 1.57$, $\pi_c/\pi_{cR} = 1.80$, and $A_9/A_{9R} = 1.42$.

At first it is surprising that the thrust decreases as little as indicated, particularly when the large reduction in pressure is considered. The prime reason for the relatively small decrease in thrust is that the reduced compressor inlet temperature reduces the compressor power requirement to sustain a given pressure ratio. Thus, because the turbine inlet temperature is fixed, the turbine has the power capability of providing much higher compression, with a consequent increase in corrected mass flow and hence thrust. There is some question whether, in fact, the compressor could be operated at 1.8 times the sea level value of compressor ratio, and it is possible that the engine would have to be throttled back at altitude. It is, of course, the high compression ratio that is primarily responsible for reducing the specific fuel consumption, even though the propulsive efficiency has decreased.

As a final example of off-design performance of a fixed-area turbojet, consider the problem of designing an engine for an aircraft capable of flying at Mach 3 that is to be able to take off under its own power. Thus, consider an engine with $\pi_{cR} = 9$ at $M_{0R} = 0$ and with $T_0/T_{0R} = 0.759$, $p_0/p_{0R} = 0.162$, and $M_0 = 3$. Other conditions are as in the preceding examples. Straightforward calculation then gives $\pi_c = 3.34$, $\dot{m}_c/\dot{m}_{cR} = 0.541$, $F/F_R = 0.841$, $S/S_R = 1.15$, and $A_9/A_{9R} = 5.04(!)$

It can be seen that when an engine is to be used in an aircraft with such an extreme operating range, the restriction to fixed A_4 and A_8 presents a very serious design problem. Thus, it is hard to imagine how such a huge exit area variation could be accomplished, and it is probable that the engine would have to be operated with substantial overexpansion in the nozzle at takeoff and substantial underexpansion at cruise. The large reduction in compressor pressure ratio, with corresponding reduction in corrected mass flow, would be difficult to achieve and would probably imply the use of compressor bleed at the cruise condition (see Sec. 8.5).

These simple calculations serve to emphasize that the difficult task of designing an engine is further complicated when the engine must operate at more than one "design point" for a substantial length of time. Some of the restrictions are so severe that competing concepts to conventional engines have gained consideration as the demand for multiple-mission aircraft has grown. An example of such a concept is considered in the next section.

Before considering more complicated examples of off-design performance, it is of use to note that very similar methodology can be used to predict the effect of engine redesigns, or of the effect of unexpected operational excursions such as afterburner blowout.

It occurs more often than is desired, that a compressor characteristic map turns out to be other than that assumed in the preliminary design calculations. If, as a result, the engine operating line ends out in too close proximity of the surge line, the components must be rematched. An often used technique to effect such a rematch is that of adjusting either or both of the areas A_4 or A_8. As is to be developed in Problem 8.4, for example, variation of A_4 and A_8 can be used to shift the location of the operating line.

The Variable-Area Turbojet (VAT)

A concept of considerable interest to industry today is that of the variable-area turbojet or VAT. With such a machine, it is planned to make both the turbine inlet area variable (by having movable turbine nozzles for example, Fig. 8.6) and the primary nozzle variable.

In spite of the enormous complexity and difficulty of developing such a concept, the possible performance benefits are sufficiently substantial that considerable research and development effort is presently being devoted to such concepts. To investigate the possible performance benefits, again consider the first example in the preceding section, in which the behavior of

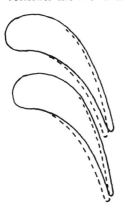

Fig. 8.6 Variable-area turbine nozzles.

an engine at fixed Mach number and altitude was discussed. It was found that the specific fuel consumption rose at low thrust levels, primarily because the compressor pressure ratio decreased with thrust.

Now consider a competitive engine in which the turbine and primary nozzle areas are varied in a manner to keep the compressor pressure ratio and corrected mass flow constant. Note that by so doing not only will the thermodynamic efficiency of the engine be maintained, but the inlet airflow will also be maintained, thereby preventing excessive inlet spillage drag. The required relationship for the area variation follows directly from Eq. (8.7) to give

$$A_4/A_{4R} = (\tau_\lambda/\tau_{\lambda R})^{\frac{1}{2}} \tag{8.44}$$

Equation (8.3) also gives directly

$$\tau_t = 1 - (1 - \tau_{tR})(\tau_{\lambda R}/\tau_\lambda) \tag{8.45}$$

Thus this VAT is quite the opposite to the FAT, in that the VAT has a variable turbine expansion ratio but a fixed compressor pressure ratio, whereas the FAT has a variable compressor pressure ratio but a fixed turbine expansion ratio.

The required primary nozzle area variation follows from Eq. (8.2) to give

$$\frac{A_8}{A_{8R}} = \frac{A_4}{A_{4R}} \left(\frac{\tau_t}{\tau_{tR}} \right)^{\frac{1}{2}} \frac{\pi_{tR}}{\pi_t} = \left(\frac{\tau_\lambda \tau_t}{\tau_{\lambda R} \tau_{tR}} \right)^{\frac{1}{2}} \frac{\pi_{tR}}{\pi_t} \tag{8.46}$$

The remaining equations follow directly from Eqs. (8.16–8.28). In the special case of flight at fixed Mach number and fixed altitude (i.e., fixed M_0, T_0, p_0) for example, the necessary equations may be summarized as follows.

Summary of the Equations—Off-Design VAT

(Note that π_c, M_0, T_0, p_0, and component efficiencies are fixed.)

Inputs: $\gamma_c, \gamma_t, \eta_c, \eta_t, \eta_m, \pi_b, \pi_d, \pi_n, \pi_c, \tau_{\lambda R}, M_0, \tau_\lambda$

Outputs: $\dfrac{F}{F_R}, \dfrac{S}{S_R}, \dfrac{A_4}{A_{4R}}, \dfrac{A_9}{A_{9R}}, \dfrac{1/\pi_t}{1/\pi_{tR}}, \dfrac{A_8}{A_{8R}}$

Equations:

$$\tau_r = 1 + \frac{\gamma_c - 1}{2} M_0^2, \qquad \pi_r = \tau_r^{\gamma_c/(\gamma_c - 1)} \tag{8.47}$$

$$\tau_c = 1 + \frac{1}{\eta_c} \left(\pi_c^{(\gamma_c - 1)/\gamma_c} - 1 \right) \tag{8.48}$$

$$\tau_{tR} = 1 - \frac{1}{\eta_m}\frac{\tau_r}{\tau_{\lambda R}}(\tau_c - 1), \qquad \pi_{tR} = \left[1 - \frac{1}{\eta_t}(1 - \tau_{tR})\right]^{\gamma_t/(\gamma_t - 1)}$$

$$(8.49)$$

$$\tau_t = 1 - (1 - \tau_{tR})\frac{\tau_{\lambda R}}{\tau_\lambda}, \qquad \pi_t = \left[1 - \frac{1}{\eta_t}(1 - \tau_t)\right]^{\gamma_t/(\gamma_t - 1)} \qquad (8.50)$$

$$\left(\frac{P_{t_9}}{P_9}\right)_R = \pi_b\pi_c\pi_n\pi_r\pi_d\pi_{tR}, \qquad \frac{P_{t_9}}{P_9} = \left(\frac{P_{t_9}}{P_9}\right)_R\frac{\pi_t}{\pi_{tR}} \qquad (8.51)$$

$$M_0\frac{u_9}{u_0} = \left\{\frac{2}{\gamma_c - 1}\tau_\lambda\tau_t\left[1 - \left(\frac{P_{t_9}}{P_9}\right)^{-(\gamma_t-1)/\gamma_t}\right]\right\}^{\frac{1}{2}} \qquad (8.52)$$

(Note that the formula for $[M_0(u_9/u_0)]_R$ is identical, but R quantities are to be used.)

$$\frac{F}{F_R} = \frac{M_0(u_9/u_0) - M_0}{[M_0(u_9/u_0) - M_0]_R} \qquad (8.53)$$

$$\frac{S}{S_R} = \frac{\tau_\lambda - \tau_r\tau_c}{(\tau_\lambda - \tau_r\tau_c)_R}\frac{1}{F/F_R} \qquad (8.54)$$

$$A_4/A_{4R} = (\tau_\lambda/\tau_{\lambda R})^{\frac{1}{2}} \qquad (8.55)$$

$$\frac{A_8}{A_{8R}} = \left(\frac{\tau_\lambda\tau_t}{\tau_{\lambda R}\tau_{tR}}\right)^{\frac{1}{2}}\frac{\pi_{tR}}{\pi_t} \qquad (8.56)$$

$$\frac{A_9/A_8}{(A_9/A_8)_R} = \left(\frac{P_{t_9}/P_9}{(P_{t_9}/P_9)_R}\right)^{(\gamma_t+1)/2\gamma_t}\left[\frac{(P_{t_9}/P_9)_R^{(\gamma_t-1)/\gamma_t} - 1}{(P_{t_9}/P_9)^{(\gamma_t-1)/\gamma_t} - 1}\right]^{\frac{1}{2}} \qquad (8.57)$$

$$\frac{A_9}{A_{9R}} = \frac{A_9/A_8}{(A_9/A_8)_R}\frac{A_8}{A_{8R}} \qquad (8.58)$$

As an example calculation, consider an engine with the same on-design characteristics as the engine considered in the first example of Sec. 8.3. Thus, take $\gamma_c = 1.4$, $\gamma_t = 1.3$, $\eta_m = 1$, $\eta_c = 0.88$, $\eta_t = 0.9$, $\pi_b = 0.98$, $\pi_n = 0.99$, $\pi_d = 0.97$, $\pi_c = 20$, $\tau_{\lambda R} = 7$, and $M_0 = 0.8$. Figure 8.7 shows the resulting variation in specific fuel consumption vs thrust. Included for comparison is the FAT result (Fig. 8.3) and also the related turbine enthalpy ratio τ_λ.

The potential operating advantage for a VAT is evident here in that the superior thermal and propulsive efficiencies of the VAT at part-thrust

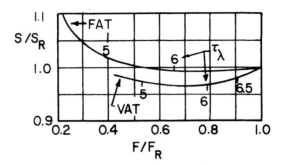

Fig. 8.7 S/S_R vs F/F_R for a VAT and for a FAT.

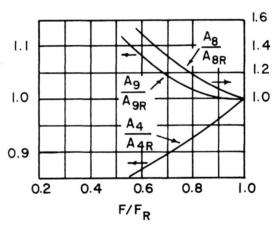

Fig. 8.8 A_4/A_{4R}, A_8/A_{8R}, and A_9/A_{9R} vs F/F_R, VAT.

operation leads to substantial benefits in reduced specific fuel consumption. A further benefit is implied by the behavior of the turbine inlet enthalpy. Thus, the more rapid fall-off of the turbine inlet temperature with thrust for the VAT indicates the possibility that less cooling air will be required from a VAT during part-throttle operation than from a FAT.

The required area variations for the VAT are shown in Fig. 8.8.

It is evident that substantial variations in A_4 and A_8 are required, and there is some question as to whether the more extreme variations could be attained in a working design. Note, however, the very much reduced area variation required of the exit nozzle. (Note that this particular VAT requires an area increase, in contrast to the severe decrease required for the FAT.) This reduced exit area variation could lead to substantial benefits in reduced installation losses.

The required variation in turbine expansion ratio is indicated in Fig. 8.9.

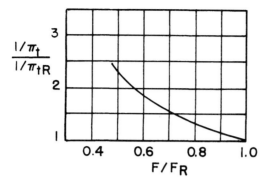

Fig. 8.9   Variation in turbine expansion ratio.

It is clear that at very low thrust levels the turbine expansion ratio compared to that required at the on-design point is very large. This required variation of expansion ratio would seem to imply the need for substantial research and development to extend the capabilities of present-day turbines. It would also seem to imply the use of a turbine with several more stages than would be required for operation at high thrust levels, so that the turbine would be capable of supporting the required large expansion ratios at low thrust levels.

Further potential operating benefits of a VAT become apparent when flight at various altitudes, various Mach numbers, etc., is considered. It is these possible benefits that have stimulated industry interest in the VAT, as well as other variable geometry engines such as the variable bypass ratio turbofan.

Although the analysis of other variable geometry engines will not be included here, a brief description of some possible engine types will be given to illustrate the extent of creative thought that has been directed to the problem of developing engines with efficient multimission capability. In Ref. 1 the annulus inverting valve (AIV) is described and many possible cycles utilizing the valve are considered. The valve has the capability of switching half the flow it encounters from the inside of an annulus to the outside, and vice versa. Alternatively, it can be operated so that the flow passes straight through the valve. This capability offers the opportunity of varying the cycle bypass ratio and bypass pressure ratio to better suit the required operating point. The valve in its most simple utilization is illustrated in Fig. 8.10.

It is evident from Fig. 8.10 that the AIV allows operation of the engine as a turbofan engine with a low compressor pressure ratio, or as a turbojet engine with a high compressor pressure ratio. (In the latter mode, half of the inlet air is bypassed.) The design calculations of Chap. 7 indicated that flight at low Mach numbers is best served by a turbofan, whereas flight at high Mach numbers is best served by a turbojet. Thus, the AIV offers the opportunity of substantially extending an engine's efficient flight envelope.

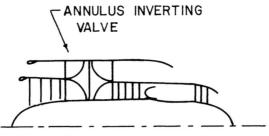

Fig. 8.10 Annulus inverting valve.

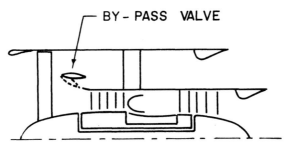

Fig. 8.11 Variable bypass ratio engine.

Many other variable geometry concepts are possible, for example the variable bypass ratio engine illustrated in Fig. 8.11.

In an engine such as that shown in Fig. 8.11, the bypass ratio would be selectively varied by varying the turbine inlet area at the entry to the low-pressure turbine. Use of a relatively simple bypass valve, as indicated, would allow efficient flow adjustment to deliver the required bypass ratio change. The configuration indicated has the advantage that the variable-area nozzles are in a portion of the flowfield where nozzle cooling is not required. As a consequence, the required mechanical complexity of the movable nozzles will be greatly reduced.

The examples of variable geometry engines cited in this section represent only a few of the very large number of possible engine concepts that deserve consideration, and the future should see extensive study of such advanced concepts. The eventual development and production of such engines will depend upon the many technical tradeoffs obviously present, as well as the enormous economic tradeoffs required.

Installation Effects

The example calculations of the preceding sections illustrated the variation of parameters with the "uninstalled" thrust and the specific fuel consumption based upon this value of thrust. Many of the major benefits of variable geometry engines are identified with the reduction of installation effects, however.

When a high Mach number aircraft is flown at low Mach number, the mass of air per second swallowed by the projection of the inlet area decreases, but not, in fact, as rapidly as the engine demand for air decreases. Thus, at low flight Mach numbers the inlet must spill air; and, in the case of a sharp-edged supersonic inlet, substantial spillage losses occur. It can be seen then that the VAT, by maintaining a high airflow demand, will much reduce such spillage losses at low thrust levels.

A further installation benefit arises for the VAT because of the small required exit area variation as the thrust is reduced. Usually, the exit area when no afterburning is present, even with full thrust, is substantially less than the main engine cross-sectional area. This projected area can lead to serious drag penalties and it is important to design the engine aft end so that substantial separation does not occur. When a conventional engine (FAT) is operated at very low thrust levels, the required exit area closure is so substantial that the prevention of separation is usually impossible, and as a result large boat-tail drag penalties are incurred. The VAT, because of its continued handling of a large mass flow rate that is delivered at an ever-decreasing exit velocity, leads to a requirement of very little area change. As a result, the VAT shows promise of leading to much reduced aft end installation losses as well as reduced inlet losses.

In conclusion, it should be pointed out that the effects upon installation losses for a VAT are comparable to the effects upon uninstalled performance. Thus, the VAT, like all competing concepts, should be evaluated in terms of the installed behavior. Of course, this requires knowledge of the inlet and aft end losses, which generally must be obtained from experiment.

8.3 Off-Design Analysis of the Turbofan

When the off-design performance of a turbofan is considered, a slight additional complication arises in that the variation in performance of each separate stream must be determined. In order to determine these separate variations in performance, additional information must be provided to describe the "work split" between the two streams of the turbine output. To illustrate such a procedure, consider a turbofan in which the fan is driven by its own low-pressure turbine, and the high-pressure compressor is driven by the high-pressure turbine. The intermediate location between the two turbines is station $4a$ and the obvious definitions are

$$\tau_{ch} = \frac{T_{t_3}}{T_{t_{3'}}} = \frac{\tau_c}{\tau_{c'}}, \qquad \pi_{ch} = \frac{p_{t_3}}{p_{t_{3'}}} = \frac{\pi_c}{\pi_{c'}}, \qquad \tau_{th} = \frac{T_{t_{4a}}}{T_{t_4}},$$

$$\pi_{th} = \frac{p_{t_{4a}}}{p_{t_4}}, \qquad \tau_{tL} = \frac{T_{t_5}}{T_{t_{4a}}}, \qquad \pi_{tL} = \frac{p_{t_5}}{p_{t_{4a}}} \qquad (8.59)$$

The off-design analysis of the turbofan is much simplified algebraically if, analogously to the assumptions already utilized in the description of the off-design performance of the turbojet, the two exit nozzles and both turbine entrance nozzles may be assumed to be choked. These assumptions prove to be valid over a fairly wide operating range for flight at high Mach numbers, but the exit nozzles tend to unchoke at only slightly reduced power settings at low flight Mach numbers. An algebraic method of handling the unchoked nozzle problem is given in Sec. 8.4 for the turboprop engine and is given as an assignment in Problems 8.6 and 8.7 for the turbofan engine.

For simplicity of presentation, consider the example of fixed turbine inlet areas and fixed exit nozzle throat areas. Analogous to the result of Sec. 8.2, it then follows that τ_{th}, π_{th}, τ_{tL}, and π_{tL} are all constant. A power balance between the high-pressure compressor and the high-pressure turbine then leads to

$$\tau_{ch} = 1 + (\tau_{chR} - 1)\frac{\tau_\lambda/\tau_r\tau_{c'}}{(\tau_\lambda/\tau_r\tau_{c'})_R} \tag{8.60}$$

Then

$$\pi_{ch} = [1 + \eta_{ch}(\tau_{ch} - 1)]^{3.5} \quad \text{and} \quad \tau_{chR} - 1 = \frac{1}{\eta_{chR}}\left[\left(\frac{\pi_{cR}}{\pi_{c'R}}\right)^{1/3.5} - 1\right]$$

give

$$\pi_{ch} = \left\{1 + \frac{\eta_{ch}}{\eta_{chR}}\left[\left(\frac{\pi_{cR}}{\pi_{c'R}}\right)^{1/3.5} - 1\right]\frac{\tau_\lambda/\tau_r\tau_{c'}}{(\tau_\lambda/\tau_r\tau_{c'})_R}\right\}^{3.5} \tag{8.61}$$

Utilizing Eq. (8.4) in both streams leads to

$$\alpha = \alpha_R \frac{\pi_{chR}}{\pi_{ch}}\left[\frac{\tau_\lambda/\tau_r\tau_{c'}}{(\tau_\lambda/\tau_r\tau_{c'})_R}\right]^{\frac{1}{2}} \tag{8.62}$$

Finally, a power balance between the fan and low-pressure turbine gives

$$\tau_{c'} = 1 + (\tau_{c'R} - 1)\frac{\alpha_R + 1}{\alpha + 1}\frac{\tau_\lambda/\tau_r}{(\tau_\lambda/\tau_r)_R} \tag{8.63}$$

These equations can be iterated rapidly to determine the desired values. Thus, for example, the iteration may be started by assuming $\tau_{c'} = \tau_{c'R}$ in Eq. (8.61) to give a first estimate of π_{ch} and thence α from Eq. (8.62). The process is continued until the desired accuracy is obtained.

Example—Turbofan Off-Design

As an example, consider the variation of $\pi_{c'}$, π_c, and α with flight Mach number for an engine with design conditions. $M_{0R} = 2$, $\pi_{c'R} = 1.5$, $\pi_{cR} = 15$, $\alpha_R = 1$, $\eta_{ch} = \eta_{chR}$, and $\eta_{c'} = 0.90$.

Calculation gives

M_0	1.0	1.5	2.0	2.5	3.0
$\pi_{c'}$	2.02	1.73	1.50	1.34	1.24
π_c	35.7	23.4	15.0	9.93	6.94
α	0.661	0.805	1.0	1.23	1.48

It is apparent from these results that variable geometry should be considered for use with turbofan engines. Thus, for example, it is evident that the bypass ratio increases with increase in flight Mach number. This tendency is opposite to that found for best on-design choice of bypass ratio.

The parameters evaluated by the preceding methods can be incorporated in the performance equations to provide the off-design performance. An example summary is provided here (for which it has been assumed that the exit areas have been varied to provide $p_9 = p_{9'} = p_0$)

Summary—Turbofan Off-Design

Inputs:

$$\gamma_t, C_{p_c}, C_{p_t}, \eta_{ch}, \eta_{c'}, \eta_b, \pi_d, \pi_b, \pi_n, \pi_{n'}, T_0, p_0/p_{0R}, h, \tau_\lambda, M_0, M_{0R},$$

$$\tau_{\lambda R}, \pi_{c'R}, \pi_{cR}, \alpha_R, \eta_{chR}, \eta_{c'R}, \eta_{bR}, \pi_{dR}, \pi_{bR}, \pi_{nR}, \pi_{n'R}, T_{0R},$$

$$\left(M_0 \frac{u_9}{u_0}\right)_R, \left(M_0 \frac{u_{9'}}{u_0}\right)_R, \tau_{tR}, \eta_{tR}$$

Outputs: $S, F/F_R$, etc.

Equations:

$$\tau_r = 1 + M_0^2/5 \tag{8.64}$$

$$\pi_r = \tau_r^{3.5} \tag{8.65}$$

$$\tau_{rR} = 1 + M_{0R}^2/5 \tag{8.66}$$

$$a_0 = 20.04\sqrt{T_0} \qquad \left[a_0 = 49.0\sqrt{T_0}\right] \tag{8.67}$$

$$a_{0R} = 20.04\sqrt{T_{0R}} \qquad \left[a_0 = 49.0\sqrt{T_{0R}}\right] \tag{8.68}$$

π_{ch}, α, and $\tau_{c'}$ are obtained from Eqs. (8.61–8.63).

If the turbine entry nozzles and primary and secondary main nozzles are choked, then $\tau_t = \tau_{tR}$ and

$$\pi_t = \pi_{tR} = \left[1 + \frac{1}{\eta_{tR}}(\tau_{tR} - 1)\right]^{\gamma_t/(\gamma_t - 1)} \tag{8.69}$$

(See Problems 8.6 and 8.7 if examples with unchoked nozzles are to be considered.)

$$T_c = T_{c'}T_{ch} = T_{c'}\left[1 + \frac{1}{\eta_{ch}}\left(\pi_{ch}^{1/3.5} - 1\right)\right] \tag{8.70}$$

$$\pi_{c'} = \left[1 + \eta_{c'}(T_{c'} - 1)\right]^{3.5} \tag{8.71}$$

$$f = \frac{T_\lambda - T_r T_c}{\left(h\eta_b/C_{p_c}T_0\right) - T_\lambda} \tag{8.72}$$

$$\frac{p_{t_9}}{p_9} = \pi_r\pi_d\pi_{c'}\pi_{ch}\pi_b\pi_t\pi_n \tag{8.73}$$

$$M_0\left(\frac{u_9}{u_0}\right) = \left\{5T_\lambda T_t\left[1 - \left(\frac{p_{t_9}}{p_9}\right)^{-[(\gamma_t-1)/\gamma_t]}\right]\right\}^{\frac{1}{2}} \tag{8.74}$$

$$\frac{p_{t_{9'}}}{p_9} = \pi_r\pi_d\pi_{c'}\pi_{n'} \tag{8.75}$$

$$M_0\left(\frac{u_{9'}}{u_0}\right) = \left\{5T_rT_{c'}\left[1 - \left(\frac{p_{t_{9'}}}{p_{9'}}\right)^{-1/3.5}\right]\right\}^{\frac{1}{2}} \tag{8.76}$$

$$\frac{F}{\dot{m}_c + \dot{m}_F} = \frac{a_0}{1+\alpha}\left\{(1+f)M_0\frac{u_9}{u_0} - M_0 + \alpha\left[M_0\frac{u_{9'}}{u_0} - M_0\right]\right\} \tag{8.77}$$

$$S = \frac{f}{(1+\alpha)[F/(\dot{m}_c + \dot{m}_F)]} \tag{8.78}$$

$$\frac{F}{F_R} = \frac{\alpha_R}{\alpha}\frac{\pi_r\pi_d\pi_{c'}\pi_{n'}}{(\pi_r\pi_d\pi_{c'}\pi_{n'})_R}\left[\frac{(T_rT_{c'})_R}{T_rT_{c'}}\right]^{\frac{1}{2}}\frac{p_0}{p_{0R}}$$

$$\times \frac{\left[(1+f)M_0\dfrac{u_9}{u_0} - M_0 + \alpha\left(M_0\dfrac{u_{9'}}{u_0} - M_0\right)\right]}{\left[(1+f)M_0\dfrac{u_9}{u_0} - M_0 + \alpha\left(M_0\dfrac{u_{9'}}{u_0} - M_0\right)\right]_R} \tag{8.79}$$

8.4 Off-Design Analysis of the Turboprop

This analysis of the off-design performance of the turboprop begins by again assuming that the entrance areas to both turbines are choked and that the turbine entrance areas and the nozzle exit area are fixed. The turboprop

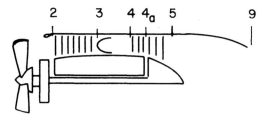

Fig. 8.12 Turboprop station numbering.

station numbering is as shown in Fig. 8.12. The power balance between compressor and high-pressure turbine gives immediately

$$\tau_c - 1 = (\tau_c - 1)_R \frac{\tau_\lambda/\tau_r}{(\tau_\lambda/\tau_r)_R} \tag{8.80}$$

In virtually all operating conditions of a turboprop, the exit nozzle will not be choked, so with Eq. (2.102)

$$\dot{m}_4 = \frac{A_4 P_{t_4}}{\sqrt{T_{t_4}}} \sqrt{\frac{\gamma_t}{R}} \frac{2}{\gamma_t + 1}^{(\gamma_t + 1)/2(\gamma_t - 1)} \tag{8.81}$$

$$\dot{m}_9 = \frac{A_9 P_{t_9}}{\sqrt{T_{t_9}}} \sqrt{\frac{\gamma_t}{R}} M_9 \left[1 + \frac{\gamma_t - 1}{2} M_9^2\right]^{-(\gamma_t + 1)/2(\gamma_t - 1)} \tag{8.82}$$

Equating these two expressions and rearranging slightly, there follows

$$M_9 = \frac{A_4}{A_9} \frac{1}{\pi_n} \frac{\sqrt{\tau_t}}{\pi_t} \left[\frac{2}{\gamma_t + 1}\left(1 + \frac{\gamma_t - 1}{2} M_9^2\right)\right]^{(\gamma_t + 1)/2(\gamma_t - 1)} \tag{8.83}$$

Also

$$\left[1 + \frac{\gamma_t - 1}{2} M_9^2\right]^{\gamma_t/(\gamma_t - 1)} = \frac{P_{t_9}}{P_9} = \frac{p_0}{p_9} \pi_r \pi_d \pi_c \pi_b \pi_t \pi_n \tag{8.84}$$

Thus, noting that $p_0/p_9 = 1$ because the exit is unchoked,

$$\pi_t = \frac{1}{\pi_r \pi_d \pi_c \pi_b \pi_n} \left[1 + \frac{\gamma_t - 1}{2} M_9^2\right]^{\gamma_t/(\gamma_t - 1)} \tag{8.85}$$

Equations (8.83) and (8.85), together with the intermediate equation relating π_t and τ_t, may be easily solved by functional iteration. With the thus determined values of M_9, π_t, and τ_t, the performance variables follow directly. Utilizing the results of the summary of Sec. 7.4, with Eqs. (7.51) and (8.4), the results of the following summary are obtained. Note that the nozzle efficiency varies very little, so it is appropriate to assume $\pi_n = \pi_{nR}$.

Summary of the Equations—Off-Design Turboprop

Inputs: γ_t, η_{prop}, η_g, η_m, η_c, η_t, η_b, π_d, π_b, $\pi_n = \pi_{nR}$, T_0, p_0/p_{0R}, h, τ_λ, M_0,

$\qquad (F/g_0\dot{m})_R$, η_{cR}, η_{tR}, π_{dR}, π_{bR}, T_{0R}, $\tau_{\lambda R}$, M_{0R}, π_{cR}, τ_{tR}, M_{9R}

Outputs: S, F/F_R, $\dot{m}/\dot{m}_R$, $\dot{m}_c/\dot{m}_{cR}$, π_c, etc.

Equations:

$$\tau_r = 1 + M_0^2/5, \qquad \tau_{rR} = 1 + M_{0R}^2/5 \tag{8.86}$$

$$\pi_r = \tau_r^{3.5}, \qquad \pi_{rR} = (\tau_{rR})^{3.5} \tag{8.87}$$

$$\tau_{cR} = 1 + (1/\eta_{cR})(\pi_{cR}^{1/3.5} - 1) \tag{8.88}$$

$$\tau_c = 1 + (\tau_{cR} - 1)\frac{\tau_\lambda/\tau_r}{(\tau_\lambda/\tau_r)_R} \tag{8.89}$$

$$\pi_c = \left[1 + \eta_c(\tau_c - 1)\right]^{3.5} \tag{8.90}$$

$$\pi_{tR} = \left[1 - (1/\eta_{tR})(1 - \tau_{tR})\right]^{\gamma_t/(\gamma_t - 1)} \tag{8.91}$$

$$\pi_t = \frac{1}{\pi_r \pi_d \pi_c \pi_b \pi_n}\left(1 + \frac{\gamma_t - 1}{2}M_9^2\right)^{\gamma_t/(\gamma_t - 1)} \tag{8.92}$$

$$\tau_t = 1 - \eta_t\left(1 - \pi_t^{(\gamma_t - 1)/\gamma_t}\right) \tag{8.93}$$

$$M_9 = M_{9R}\left(\frac{\tau_t}{\tau_{tR}}\right)^{\frac{1}{2}}\frac{\pi_{tR}}{\pi_t}\left(\frac{1 + \frac{\gamma_t - 1}{2}M_9^2}{1 + \frac{\gamma_t - 1}{2}M_{9R}^2}\right)^{(\gamma_t + 1)/2(\gamma_t - 1)} \tag{8.94}$$

The latter three equations are to be iterated for π_t, τ_t, and M_9. Start by assuming $M_9 = M_{9R}$, calculate π_t, τ_t, and M_9. Continue until $(M_{9_{j+1}} - M_{9_j})$ $< 10^{-6}$.

$$M_0 \frac{u_9}{u_0} = \left\{ 5\tau_\lambda \tau_t \left[1 - (\pi_r \pi_d \pi_c \pi_b \pi_t \pi_n)^{-[(\gamma_t - 1)/\gamma_t]} \right] \right\}^{\frac{1}{2}} \tag{8.95}$$

$$f = \frac{\tau_\lambda - \tau_r \tau_c}{(h\eta_b / C_{p_c} T_0) - \tau_\lambda} \tag{8.96}$$

$$C_{prop} = \eta_{prop} \eta_g \left[\eta_m (1 + f) \tau_\lambda (1 - \tau_t) - \tau_r (\tau_c - 1) \right] \tag{8.97}$$

$$C_c = 0.4 M_0 \left[(1 + f) M_0 (u_9 / u_0) - M_0 \right] \tag{8.98}$$

$$\frac{F}{\dot{m}} = 50.1 \frac{\sqrt{T_0} (C_{prop} + C_c)}{M_0}$$

$$\left[\frac{F}{g_0 \dot{m}} = 3.807 \frac{\sqrt{T_0} (C_{prop} + C_c)}{M_0} \right] \tag{8.99}$$

$$S = \frac{f(10^6)}{F/\dot{m}} \qquad \left[S = 3600 \frac{f}{F/g_0 \dot{m}} \right] \tag{8.100}$$

$$\frac{\dot{m}_c}{\dot{m}_{cR}} = \frac{\pi_b \pi_c}{(\pi_b \pi_c)_R} \left[\frac{(\tau_\lambda / \tau_r)_R}{\tau_\lambda / \tau_r} \right]^{\frac{1}{2}} \tag{8.101}$$

$$\frac{\dot{m}}{\dot{m}_R} = \frac{p_0}{p_{0R}} \sqrt{\frac{T_{0R}}{T_0}} \frac{\pi_d}{\pi_{dR}} \left(\frac{\tau_r}{\tau_{rR}} \right)^3 \frac{\dot{m}_c}{\dot{m}_{cR}} \tag{8.102}$$

$$\frac{F}{F_R} = \frac{F/\dot{m}}{(F/\dot{m})_R} \frac{\dot{m}}{\dot{m}_R} \tag{8.103}$$

Sample Calculation— Off-Design Turboprop

As an example calculation, consider the off-design performance of the engine considered in Sec. 7.4. The engine on-design parameters are listed in Table 7.2. Assume the engine is operated at fixed altitude with fixed turbine inlet temperature. The variation of the thrust and specific fuel consumption with the flight Mach number are to be found. As a simple approximation to real propeller efficiency variation with Mach number, η_{prop} is assumed to vary as in Fig. 8.13 and π_c is taken to be 24.

The results of the calculation are shown in Fig. 8.14. The results indicate that the prop fan (or turboprop) represents a very desirable engine for use in low Mach number aircraft ($M_0 < 0.8$), provided that the high propeller efficiencies indicated in Fig. 8.13 (particularly at the higher Mach numbers) can be attained. Thus, as was shown in Sec. 7.4, the prop fan gave substantially improved behavior at the design condition compared to a turbofan engine; it can be seen from Fig 8.14 that the high propulsive efficiency of a prop fan engine leads to very high thrusts and low fuel consumptions when operated at low Mach numbers. These results further

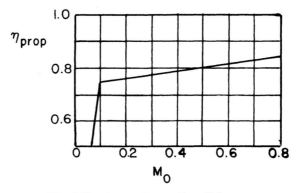

Fig. 8.13   Assumed propeller efficiency.

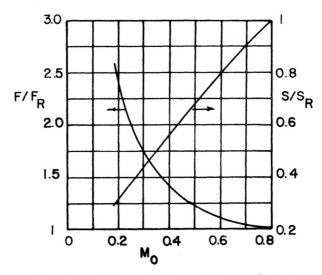

Fig. 8.14 F/F_R and S/S_R vs M_0 for a turboprop.

explain the renewed interest in the development of advanced turboprop engines.

It should be pointed out that the very-high-thrust capability evidenced for flight at low Mach numbers is often far in excess of that which can be usefully exploited. Thus, the related propeller blade loading can be so excessive that blade stalling will occur; and if the required propeller torque is to be provided, the main gearbox must be excessively heavy. As a result of these tendencies, turboprop engines designed for high Mach number capability are usually operated considerably derated in the (low-altitude) takeoff and climb condition. Restricting the engine output in this manner is referred to as "flat rating."

8.5   The Use of Component Characteristics

When the performance characteristics of the various components are available, the combined performance of the compressor, burner, and turbine can be predicted. These combined characteristics, termed the pumping characteristics, can then be utilized to predict the overall engine performance. Once a schedule of turbine area variation and primary nozzle area variation is selected, a unique operating characteristic can be determined.

In what follows, the individual component characteristics will be described, and the method of combining these characteristics to obtain the pumping characteristics and then the operating characteristics will be developed.

The Compressor Characteristics

When a new compressor has been developed, it will be subjected to a compressor rig test to determine its performance capability. It is most efficient to present the results of such a test in terms of dimensionless or pseudodimensionless variables. Routine dimensional analysis reveals that the compressor pressure ratio p_{t_3}/p_{t_2} could be expected to be a function of four dimensionless parameters which could be taken to be the ratio of specific heats γ_c, the Reynolds number R_e, the Mach number at the engine face, and the ratio of the blade (tip) speed to the speed of sound.

Experience has shown that variations in both γ_c and R_e have relatively little effect over much of the operating range of the typical compressor, so it is usual to present the performance in terms of the other dimensionless variables and to provide γ_c and R_e corrections when necessary. (See Sec. 8.6.) It is also usual to utilize variables related to the engine face Mach number and "blade" Mach number, rather than to use those variables directly. Thus, as was shown in Sec. 8.2, a unique value of engine face Mach number corresponds directly to a unique value of corrected mass flow $\dot{m}_c$, and it is usual to utilize $\dot{m}_c$ as a "pseudodimensionless" variable.

When a specific compressor (i.e., given geometry) is to be tested, it is apparent also that a given blade speed occurs for a specific value of rotational frequency, and that the reference speed of sound can be taken to be proportional to the square root of the incoming stagnation temperature T_{t_2}. Thus, it is customary to utilize the corrected speed N_c as the second pseudodimensionless variable, where N_c is defined by

$$N_c = N/\sqrt{\theta} \tag{8.104}$$

where N is the actual rotational speed (in radians/second for the SI system of units, in rpm for the British system) and θ is the dimensionless temperature as already defined in Eq. (8.5).

The method of obtaining a "map" of the compressor characteristics is to set a given corrected speed for the compressor (on a separately driven motor) and to vary the corrected mass flow over the desired range by varying the exit valve opening. By operating at an appropriate number of corrected speed settings and over an appropriate range of corrected mass flows, the operating behavior of the compressor over its entire range can be determined.

A schematic diagram of a typical compressor test facility is shown in Fig. 8.15 and typical results are indicated in Fig. 8.16.

The dotted line indicated in Fig. 8.16 represents the limit of pressure ratio that can be obtained for the given corrected speed. This limit occurs when the pressure rise across the compressor is so extreme that the blade loadings reach levels that cause boundary-layer separation over substantial portions of the blades. In this condition several forms of flow instability can occur, principal among which are rotating stall and compressor surge.

The mechanism of rotating stall is complicated indeed, but some understanding of the phenomenon can be attained from a relatively simple model, as depicted in Fig. 8.17.

Figure 8.17 depicts a packet of fluid that, because of the large imposed pressure gradient, has undergone a severe flow reversal. To the surrounding fluid such a reversal region appears to be a blockage area, and the fluid divides to bypass the area. It can be seen in the figure that the blades on the lower side of the stall packet are thus confronted with a flow of reduced angle of attack, whereas the blades on the upper side of the stall packet are confronted with a flow at increased angle of attack. The result then is that

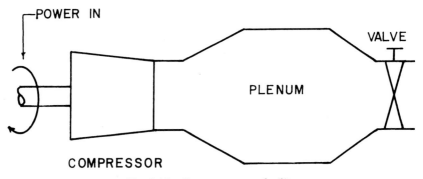

Fig. 8.15   Compressor test facility.

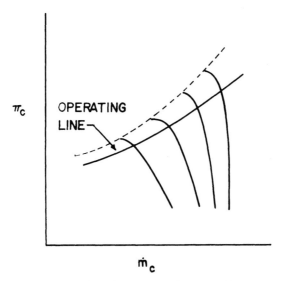

Fig. 8.16 Compressor characteristics.

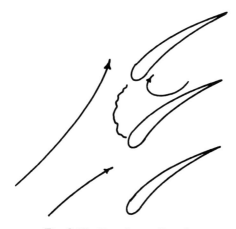

Fig. 8.17 Rotating stall packet.

the lower blades tend to unstall and the upper blades tend to go into stall, leading to a net movement (or rotation) of the stall packet.

Rotating stall, and its inception, are of enormous importance to the industry, because not only does rotating stall limit the attainable pressure ratio in some portions of the compressor characteristics, but the vibratory stresses set up in the blades can lead to very rapid and catastrophic failure of the compressor.

Surge is that phenomenon where a substantial fraction of the compressor blades simultaneously reach the limit of their load-carrying ability. As a result, flow breakdown occurs and the entire compressor loses its capability of supporting the overall pressure rise and massive flow reversal occurs. The frequency of such reversals is related to the storage volume following the compressor as well as to the compressor behavior itself. Compressor surge represents a very serious design limit, because when it occurs in an operating engine, very often engine flameout occurs.

The distance between the operating line and surge line, shown in Fig. 8.16, is referred to as the surge margin. It usually happens that the best engine performance occurs when the compressor operates near the surge line. This introduces a difficult design problem in that "appropriate" precautions must be taken to prevent engine surge due to such things as severe inlet flow distortions occurring (caused by, for example, operation at extreme angle of attack), ingestion of combustibles from gun and rocket firing (leading to added combustion in the burner with consequent pressure rise), burner overpressuring from fuel surges during acceleration, etc. A careful balance must be struck between selecting an overly large surge margin with poor steady-state performance and selecting too small a surge margin with inherent low engine safety.

Compressor Behavior during Starting

Figure 8.18 depicts a typical compressor section. It is apparent that the overall contraction in the annulus area will be selected so that, when the compressor is on its design point, the axial velocity throughout will be appropriate to match the design angles of the many blades. When the compressor is operated at a pressure ratio other than the design one, then, the ratio of exit axial velocity to entrance axial velocity will not be the same as when the compressor is on-design, because the density ratio will be dependent on the pressure ratio.

When the compressor is running at very low rotational speed, as during starting, the first stage will tend to induce a flow at an appropriate angle of attack to the first-stage blades. As the flow proceeds through the compressor, it will tend to accelerate because the low-stage compression ratios identified with the low rotational speed will not introduce a sufficient increase in density to compensate for the annulus area contraction. As a result, the axial velocity can become very large near the back of a high-compression-ratio compressor, leading to "windmilling" of the rear stages and, in severe cases, to choking of the flow. The net result is that during starting the early blade rows operate at high angle of attack (hence tending to stall),

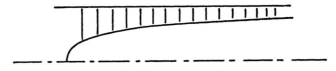

Fig. 8.18   Compressor section.

the mid blade rows operate approximately on-design, and the rearmost blade rows operate at low angle of attack (hence tending to windmill).

There are several design techniques available to relieve these starting problems, some or all of them being utilized in all modern high-compression-ratio compressors. The first technique is to utilize bleed valves. Bleed valves operate by releasing air from an appropriate stage in order to reduce the axial velocity and thereby reduce the tendency to windmill the rear stages.

A second technique to reduce the starting problem is to utilize variable-angle stators. By varying the angle of the stators the angle of attack to the rotor can be changed to more closely approach the design angle and thus the compression per stage can be improved. The effect of variable stators is both to reduce the tendency of the front blades to stall and to reduce the tendency of the rear blades to windmill.

The third technique to reduce the starting problem is to drive the compressor with multiple spools (Fig. 8.19). Thus, the low-pressure portion of the compressor will be connected directly to the low-pressure portion of the turbine and the high-pressure portions of compressor and turbine will be directly connected on another "spool."

It can be noted that the same argument used to explain the stalling and windmilling behavior of the compressor can be used to explain the turbine starting problem. Thus, by providing separate spools, the (for example) high-pressure portion of the compressor will be allowed to match with the high-pressure portion of the turbine, both components in fact operating closer to their design point if allowed to increase speed.

It is obvious that the requirement for these additional complexities increases as the design pressure ratio of the compressor increases, which tends to explain why modern engines usually utilize all three of the techniques described above. Aircraft designed for operation over a wide Mach number range (as considered in the example of Sec. 8.2) are also required to fly with widely different compressor pressure ratios at their various flight conditions, and as a result may even be forced to fly with compressor bleed when cruising at a "design" flight Mach number. These complexities, and the demands that cause them, again serve to emphasize the need for design ingenuity in the development of modern engines.

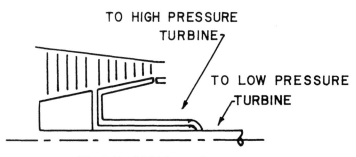

Fig. 8.19 Multiple spool compressor.

The Turbine Characteristics

The same reasoning that lead to the use of the parameters of corrected mass flow and corrected speed for the compressor suggests that the turbine performance characteristics be plotted the same way with the simple exception that the reference inlet and exit quantities are now those at stations 4 and 5. Utilizing this procedure, the results of a turbine rig test would appear, typically, as those shown in Fig. 8.20.

It is evident from Fig. 8.20 that in the case of the turbine an alternative presentation format is desirable because so much of the desired information collapses on the choke limit line, where in fact the turbines usually operate. A simple method for displaying the desired information is to multiply the corrected mass flow by the corrected speed. This has the effect of moving the separate corrected speed lines apart so that the efficiency contours, etc., may be discerned (Fig. 8.21).

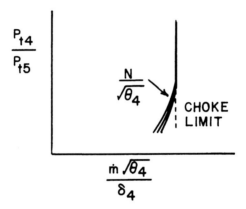

Fig. 8.20   Turbine characteristics.

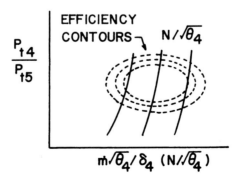

Fig. 8.21   Alternate form of turbine characteristics.

The Pumping Characteristics

The group of the three components (the compressor, burner, and turbine) is termed the gas generator, and the performance of the gas generator is represented by the "pumping characteristics." The pumping characteristics simply give the output variables (station 5) in terms of the input variables (station 2).

To obtain the pumping characteristics from the individual component characteristics, the matching of mass flow, shaft speeds, stagnation pressures, and power requirements are utilized. For simplicity, consider the simple example of a nonafterburning, single-spool turbojet. Then

$$N_t = N_c = N \tag{8.105}$$

$$\dot{m}_4 = (1+f)\dot{m}_2 \tag{8.106}$$

$$p_{t_4} = p_{t_2}\pi_c\pi_b \tag{8.107}$$

$$C_{p_c}(T_{t_3} - T_{t_2}) = \eta_m(1+f)C_{p_t}(T_{t_4} - T_{t_5}) \tag{8.108}$$

From Eq. (8.106) (with $\dot{m}_c = \dot{m}\sqrt{\theta}/\delta$)

$$\dot{m}_{c_2} = \frac{1}{1+f}\dot{m}_{c_4}\frac{p_{t_4}}{p_{t_2}}\left(\frac{T_{t_4}}{T_{t_2}}\right)^{-\frac{1}{2}} = \frac{\dot{m}_{c_4}}{1+f}\pi_c\pi_b\left(\frac{T_{t_4}}{T_{t_2}}\right)^{-\frac{1}{2}} \tag{8.109}$$

In the case where the turbine is choked, the corrected mass flow $\dot{m}_{c_4}$ will be a constant (or proportional to A_4 for a variable-area turbine), and for simplicity it is now assumed that this is the case. (Note that if this is not true, it is simply required to iterate to determine the value of $\dot{m}_{c_4}$.) The burner characteristics will give π_b as a function of $\dot{m}_{c_2}$ and the fuel-

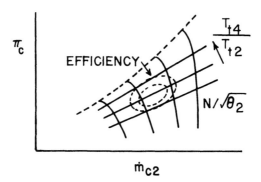

Fig. 8.22   Compressor performance map.

to-air ratio f. Thus, if T_{t_4}/T_{t_2} is specified, $\dot{m}_{c_2}$ is specified as a function of π_c. (Note that the effect of f is small, but to be exact its value will be iterated upon when the actual value of f is determined more exactly later.) The relationship between T_{t_4}/T_{t_2}, π_c, and $\dot{m}_{c_2}$ may be indicated on the compressor performance map as in Fig. 8.22.

Once the compressor performance map has been obtained in the manner indicated in Fig. 8.22, the graph may be utilized to obtain π_c and $\dot{m}_{c_2}$ when $N/\sqrt{\theta_2}$ and T_{t_4}/T_{t_2} are prescribed. Note also that τ_c may be obtained for a given location on the map because both π_c and η_c are provided.

The equation for the power balance [Eq. (8.108)] may be rearranged to give

$$\tau_t = 1 - \frac{1}{\eta_m(1+f)} \frac{C_{p_c}}{C_{p_t}} \left(\frac{T_{t_4}}{T_{t_2}}\right)^{-1}(\tau_c - 1) \qquad (8.110)$$

When T_{t_4}/T_{t_2} and $N/\sqrt{\theta_2}$ are prescribed, π_c, $\dot{m}_{c_2}$, η_c, and hence τ_c can be obtained from Fig. 8.22. Then obtain τ_t from Eq. (8.110) and $N/\sqrt{\theta_4}$ from $N/\sqrt{\theta_4} = N/\sqrt{\theta_2}(T_{t_4}/T_{t_2})^{-1/2}$. This allows locating the position on the turbine performance map (Fig. 8.23), which in turn provides the corresponding value of π_t.

The pumping characteristics (Fig. 8.24) then follow by noting

$$\frac{p_{t_5}}{p_{t_2}} = \pi_t \pi_b \pi_c \qquad \text{and} \qquad \frac{T_{t_5}}{T_{t_2}} = \tau_t \frac{T_{t_4}}{T_{t_2}}$$

Finally, the fuel-to-air ratio is determined from an enthalpy balance across the combustor to give

$$\dot{m}_2 C_{p_c} T_{t_3} + \eta_b h \dot{m}_f = \dot{m}_2 (1+f) C_{p_t} T_{t_4}$$

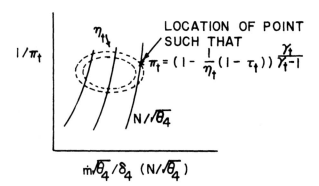

Fig. 8.23 Turbine performance map.

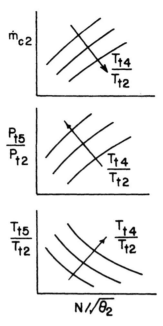

Fig. 8.24 The pumping characteristics.

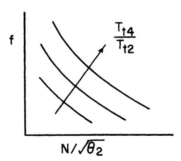

Fig. 8.25 The fuel-to-air ratio.

hence

$$f = \frac{\left(C_{p_t}/C_{p_c}\right)\left(T_{t_4}/T_{t_2}\right) - \tau_c}{\left(\eta_b h/C_{p_c} T_{t_2}\right) - \left(C_{p_t}/C_{p_c}\right)\left(T_{t_4}/T_{t_2}\right)} \qquad (8.111)$$

Thus a relationship of the form of Fig. 8.25 may be obtained.

Note that unlike the results of Fig. 8.24, the results of Fig. 8.25 are valid for only a single specified value of T_{t_2}.

Gas Generator– Nozzle Matching

The pumping characteristics allow determination of conditions at station 5 in terms of those at station 2 when the corrected speed and the ratio T_{t_4}/T_{t_2} are prescribed. When the gas generator is coupled with a nozzle of prescribed throat variation, however, $N/\sqrt{\theta_2}$ and T_{t_4}/T_{t_2} are not separately prescribable. (Recall the results of Sec. 8.2 where it was shown that by prescribing the ratio A_4/A_8 a unique compressor operating line will be determined.) For simplicity again consider the case where the primary nozzle is choked, so that the group $\dot{m}_8\sqrt{\theta_8}/A_8\delta_8$ is a constant. Mass flow continuity then gives

$$\frac{A_8}{A_2} = \left(\frac{\dot{m}_2\sqrt{\theta_2}}{A_2\delta_2}\right)\left(\frac{\dot{m}_8\sqrt{\theta_8}}{A_8\delta_8}\right)^{-1}(1+f)\left(\frac{T_{t_8}}{T_{t_2}}\right)^{\frac{1}{2}}\frac{p_{t_2}}{p_{t_8}} \qquad (8.112)$$

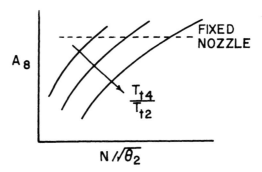

Fig. 8.26   Area variation.

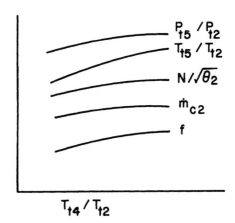

Fig. 8.27   Gas generator-nozzle pumping characteristics.

Noting that for the nonafterburning case $T_{t_8} = T_{t_5}$ and also that p_{t_k}/p_{t_2} $= \pi_n p_{t_5}/p_{t_2}$, this expression may be written

$$\frac{A_8}{A_2} = \left(\frac{T_{t_5}}{T_{t_2}}\right)^{\frac{1}{2}}\left(\frac{p_{t_5}}{p_{t_2}}\right)^{-1}\frac{(1+f)}{\pi_n}\left(\frac{\dot{m}_2\sqrt{\theta_2}}{A_2\delta_2}\right)\left(\frac{\dot{m}_8\sqrt{\theta_8}}{A_8\delta_8}\right)^{-1} \tag{8.113}$$

π_n will be known as a function of the corrected mass flow rate through the tail pipe and nozzle, so that if T_{t_4}/T_{t_2}' and $N/\sqrt{\theta_2}$ are prescribed, then $\dot{m}_2\sqrt{\theta_2}/\delta_2$, p_{t_5}/p_{t_2}, T_{t_5}/T_{t_2}, and f will all be determined from the pumping characteristics. Thus, A_8/A_2 will also be prescribed and a graph of the form of Fig. 8.26 obtained.

When the desired schedule for the nozzle area A_8 is prescribed a unique T_{t_4}/T_{t_2} vs $N/\sqrt{\theta_2}$ curve is established. This allows determination of a unique set of gas generator–nozzle pumping characteristics as indicated in Fig. 8.27.

Performance Prediction with Pumping Characteristics

Once the gas generator-nozzle pumping characteristics are available, the performance parameters follow directly from the equations for the performance variables. Thus, with reference to the summary of Sec. 7.2, the performance variables may be obtained from the pumping characteristic variables from the following equations. (Note that T_{t_4}/T_{t_2} would be assumed and the flight conditions, etc., would be known.)

$$\frac{p_{t_9}}{p_9} = \left(\pi_r\pi_d\pi_n\frac{p_0}{p_9}\right)\frac{p_{t_5}}{p_{t_2}} \tag{8.114}$$

$$\frac{T_9}{T_0} = \tau_r\frac{T_{t_5}}{T_{t_2}}\left(\frac{p_{t_9}}{p_9}\right)^{-(\gamma_t-1)/\gamma_t} \tag{8.115}$$

$$M_0\frac{u_9}{u_0} = \left\{\frac{2}{\gamma_c-1}\frac{C_{p_t}}{C_{p_c}}\tau_r\frac{T_{t_5}}{T_{t_2}}\left[1-\left(\frac{p_{t_9}}{p_9}\right)^{-(\gamma_t-1)/\gamma_t}\right]\right\}^{\frac{1}{2}} \tag{8.116}$$

$$f = \frac{\tau_r\left[(C_{p_t}T_{t_4})/(C_{p_t}T_{t_2})-\tau_c\right]}{(h\eta_b)/(C_{p_c}T_0)-\tau_r(C_{p_t}T_{t_4})/(C_{p_c}T_{t_2})} \tag{8.117}$$

$$\frac{F}{\dot{m}} = a_0\left[(1+f)\left(M_0\frac{u_9}{u_0}\right)-M_0+\frac{(1+f)}{\gamma_c[M_0(u_9/u_0)]}\frac{T_9}{T_0}\left(1-\frac{p_0}{p_9}\right)\right] \tag{8.118}$$

$$S = \frac{f}{F/\dot{m}} \tag{8.119}$$

8.6   Limitations on the Accuracy of Component Characteristics

When utilizing component characteristics to generate pumping characteristics, it is important to be aware of any accuracy limitations. The most obvious accuracy limitations will occur because of the inevitable instrumentation and recording inaccuracies present in any test procedure and, of course, every effort must be made to minimize such sources of error. Other sources of error, however, arise when the effects of changes in Reynolds number and the ratio of specific heats are not included.

In many large engines, the effect of Reynolds number variations are, in fact, very small over virtually the entire operating range of the engine. However, it can happen, particularly in small general aviation jet engines, that Reynolds number effects can become substantial for flight at extreme altitudes. In the compressor, for example, it is entirely possible that the flow over the first blade row will become largely laminar with the consequent onset of early (laminar) separation. Aside from reducing the performance of the first blade row, such separation causes velocity mismatches at all succeeding rows, with consequent substantial change in compressor performance.

In a similar way, operation at extreme altitudes can lead to very low Reynolds numbers at the later turbine stages with a consequent deterioration in performance (unless the blades had been designed "oversize" originally to prevent such deterioration). The point, then, is that if operation over extreme Reynolds number ranges is to be expected, appropriate investigations of the effects of the Reynolds number variations should be included in the engine test program.

An entirely different phenomenon arises when operation at very high humidity is carried out. Several effects arise when large amounts of water vapor are present in the flow, including change in the ratio of specific heats (so that the value of corrected mass flow at a choke condition changes, as does the reference speed of sound in the corrected speed). When condensation (and later evaporation) occurs, substantial effects arise because of the release or absorption of the latent enthalpy of evaporation. As an example, if an inlet is considered, the presence of water vapor causes three major effects to occur:

(1) Mass continuity. The specific density of water is so high that any droplets formed (due to the lowered static temperature as the air accelerates into the inlet) occupy effectively zero volume. The result is that this aspect of condensation allows the inlet to pass a larger mass flow than it can without condensation.

(2) Stagnation enthalpy increase. When droplets form, their latent enthalpy of vaporization is released to the surrounding gas, thereby increasing the stagnation temperature of the gas. This tends to reduce the mass flow handling capability of the inlet (recall $\dot{m} \sim 1/\sqrt{T_t}$).

(3) Stagnation pressure decrease. The latent enthalpy of vaporization, released upon the formation of droplets, reduces the stagnation pressure, as was pointed out in the analysis of Sec. 2.18. This reduction in stagnation pressure tends to reduce the mass flow handling capability of the inlet (recall $\dot{m} \sim p_t$).

The net of these effects is that when condensation occurs, less mass flow can be handled by the inlet than when no condensation occurs. Thus, inlet testing (and engine testing) at ground level on hot humid days can lead to substantial variations in performance (greater than 1% mass flow rate changes) and great care should be exercised in applying data correction procedures.

8.7 Engine Acceleration

In the preceding sections of this chapter, methods for predicting engine performance at various throttle settings or flight conditions were developed. In determining such off-design performance, steady-state operation is assumed, so that appropriate power balances between turbines, fans, and compressors can be applied. When the engine undergoes transient operation, however, the power output of a turbine does not equal the power absorption of its related compression system, but rather a system acceleration exists as a result of such a power imbalance.

The description of the accelerative behavior of the rotating components requires knowledge of the momentary angular velocity of the rotating system, and, as a result, the compressor (and turbine) characteristics must be known. In the following, a simplified representation of the compressor characteristics will be presented, but it is important to note first that to a very high degree of approximation, provided only that the acceleration maneuver does not introduce fluid mechanical instabilities, it is appropriate to assume quasisteady fluid flow throughout the system. This assumption is supported by the observation (and calculation) that engine acceleration transients occur over time periods of several seconds, whereas the residence time of a fluid element convecting through the entire engine is of the order of about 1/100th of a second. As a result, conditions may be assumed quasisteady throughout the transient time of a given fluid element even though conditions do change on a much larger time scale. The importance of such an approximation is that the component performance maps—that are obtained from steady-state tests—may be used to describe the engine behavior during transient operation.

The Compressor Characteristics

A schematic representation of a typical map of compressor characteristics is presented in Fig. 8.16. When transient operation is to be considered, the momentary operating point will depart from a location on the steady-state operating line, and will then tend to return to the operating line at a new steady-state operating condition. The determination of the "path" followed by the operating point requires knowledge of the compressor characteristics, and particularly of the momentary compressor angular velocity. In industrial practice, it is usual to represent the compressor characteristics in great detail within computer codes, so that the history of an operational excursion can be calculated.

In order to facilitate the calculation of example transients, a simplified representation of a compressor map is presented. The resulting simplified representation can be used in two ways. Thus, it may be used as an approximation to an actual compressor characteristic so that simple approximate estimates of transient compressor behavior can be estimated, or alternatively, the representation can be considered a simple example of an actual compressor performance map for interpretive purposes.

For future convenience the compressor characteristic map is represented in nondimensional form in terms of Π, $\dot{M}$, and $\bar{N}$, where by definition

$$\Pi = \frac{\pi_c}{\pi_{c_d}}$$

$$\dot{M} = \frac{\dot{m}_c}{\dot{m}_{c_d}} \tag{8.120}$$

$$\bar{N} = \frac{N_c}{N_{c_d}}$$

Figure 8.28 presents a typical compressor characteristic map, and identifies the stall-surge line (subscript S), the operating line, and seven points that will be used to establish the analytical description of the map. It is now assumed that the dimensionless compressor pressure ratio Π_S, and dimensionless mass flow $\dot{M}_S$, found on the stall-surge line, can be related

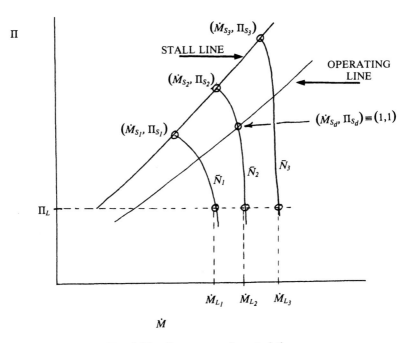

Fig. 8.28   Compressor characteristic map.

quadratically, such that

$$\Pi_S = K_1 + K_2 \dot{M}_S + K_3 M_S^2 \tag{8.121}$$

Similarly, it is assumed that both $\dot{M}_S$ and $\dot{M}_L$ can be related quadratically to the dimensionless speed $\bar{N}$ through the expressions

$$\dot{M}_L = K_4 + K_5 \bar{N} + K_6 \bar{N}^2 \tag{8.122}$$

$$\dot{M}_S = K_7 + K_8 \bar{N} + K_9 \bar{N}^2 \tag{8.123}$$

Note that the value of Π_L has no special physical significance, but rather is chosen as a matter of convenience to locate the outer "edge" of the compressor characteristic that is likely to be included in example calculations.

In practice the values of the constants $K_1 \rightarrow K_9$ are determined by inserting the known values of $\dot{M}_L$, $\dot{M}_S$, and $\bar{N}$ in Eqs. (8.121–8.123) to give after inversion

$$K_1 = \frac{1}{D_M}\left[\dot{M}_{S_3}\dot{M}_{S_2}\left(\dot{M}_{S_3} - \dot{M}_{S_2}\right)\Pi_{S_1} - \dot{M}_{S_3}\dot{M}_{S_1}\left(\dot{M}_{S_3} - \dot{M}_{S_1}\right)\Pi_{S_2} \right.$$

$$\left. + \dot{M}_{S_2}\dot{M}_{S_1}\left(\dot{M}_{S_2} - \dot{M}_{S_1}\right)\Pi_{S_3}\right] \tag{8.124}$$

$$K_2 = \frac{1}{D_M}\left[-\left(\dot{M}_{S_3}^2 - \dot{M}_{S_2}^2\right)\Pi_{S_1} + \left(\dot{M}_{S_3}^2 - \dot{M}_{S_1}^2\right)\Pi_{S_2} - \left(\dot{M}_{S_2}^2 - \dot{M}_{S_1}^2\right)\Pi_{S_3}\right]$$

$$\tag{8.125}$$

$$K_3 = \frac{1}{D_M}\left[\left(\dot{M}_{S_3} - \dot{M}_{S_2}\right)\Pi_{S_1} - \left(\dot{M}_{S_3} - \dot{M}_{S_1}\right)\Pi_{S_2} + \left(\dot{M}_{S_2} - \dot{M}_{S_1}\right)\Pi_{S_3}\right]$$

$$\tag{8.126}$$

where

$$D_M = \left(\dot{M}_{S_3} - \dot{M}_{S_2}\right)\left(\dot{M}_{S_3} - \dot{M}_{S_1}\right)\left(\dot{M}_{S_2} - \dot{M}_{S_1}\right) \tag{8.127}$$

Also, with $J \equiv L$ for $K_4 \rightarrow K_6$, $J \equiv S$ for $K_7 \rightarrow K_9$, have

$$K_4 \text{ or } K_7 = \frac{1}{D_{\bar{N}}}\left[\bar{N}_3\bar{N}_2\left(\bar{N}_3 - \bar{N}_2\right)\dot{M}_{J_1} - \bar{N}_3\bar{N}_1\left(\bar{N}_3 - \bar{N}_1\right)\dot{M}_{J_2} \right.$$

$$\left. + \bar{N}_2\bar{N}_1\left(\bar{N}_2 - \bar{N}_1\right)\dot{M}_{J_3}\right] \tag{8.128}$$

$$K_5 \text{ or } K_8 = \frac{1}{D_{\bar{N}}}\left[-\left(\bar{N}_3^2 - \bar{N}_2^2\right)\dot{M}_{J_1}\right.$$

$$\left. + \left(\bar{N}_3^2 - \bar{N}_1^2\right)\dot{M}_{J_2} - \left(\bar{N}_2^2 - \bar{N}_1^2\right)\dot{M}_{J_3}\right] \qquad (8.129)$$

$$K_6 \text{ or } K_9 = \frac{1}{D_{\bar{N}}}\left[\left(\bar{N}_3 - \bar{N}_2\right)\dot{M}_{J_1} - \left(\bar{N}_3 - \bar{N}_1\right)\dot{M}_{J_2} + \left(\bar{N}_2 - \bar{N}_1\right)\dot{M}_{J_3}\right] \qquad (8.130)$$

where

$$D_{\bar{N}} = \left(\bar{N}_3 - \bar{N}_2\right)\left(\bar{N}_3 - \bar{N}_1\right)\left(\bar{N}_2 - \bar{N}_1\right) \qquad (8.131)$$

Equations (8.121–8.131) describe the relationship between the endpoints of the speed lines and the dimensionless speed $\bar{N}$. To complete the description, an appropriate curve fit for the speed lines themselves is now introduced. Thus, it is assumed that

$$\Pi = \Pi_S - \left(\frac{\dot{M} - \dot{M}_S}{\dot{M}_L - \dot{M}_S}\right)^n \left(\Pi_S - \Pi_L\right) \qquad (8.132)$$

The exponent n is determined by selecting the speed $\bar{N}_2$ as the design speed, that is to say $\bar{N}_2 = 1$. The exponent n is then selected to ensure that for $\bar{N}_2 = 1$, Eq. (8.132) not only passes through the endpoints $(\dot{M}_{S_2}, \Pi_{S_2})$ and $(\dot{M}_{L_2}, \Pi_{L_2})$ (ensured by the assumed form), but also passes through $(\Pi = 1, \dot{M} = 1)$. It follows that

$$n = \ell n\left(\frac{\Pi_{S_2} - 1}{\Pi_{S_2} - \Pi_{L_2}}\right) \bigg/ \ell n\left(\frac{1 - \dot{M}_{S_2}}{\dot{M}_{L_2} - \dot{M}_{S_2}}\right) \qquad (8.133)$$

Recapitulating, it is seen that with $K_1 \rightarrow K_9$ determined from Eqs. (8.124–8.131), then Eqs. (8.121–8.123), (8.132), and (8.133) provide a relationship of the form

$$\Pi = \Pi(\dot{M}, \bar{N}) \qquad (8.134)$$

The remaining task to complete the description of the compressor characteristic map is to provide the equation of the operating line. Here, for simplicity, it is assumed that the entrance and exit areas to the driving turbine are choked and that the areas are fixed. In this simple case, Eq. (8.43) is valid, and may be written in the form

$$\frac{\dot{m}_c}{\dot{m}_{c_d}} = \frac{\pi_c}{\pi_{c_d}}\left[\frac{\eta_c\left(\pi_{c_d}^{(\gamma-1)/\gamma} - 1\right)}{\eta_{c_d}\left(\pi_c^{(\gamma-1)/\gamma} - 1\right)}\right]^{\frac{1}{2}} \qquad (8.135)$$

which becomes in terms of the dimensionless quantities and π_{c_d},

$$\dot{M} = \Pi \left\{ \frac{\eta_c \left(\pi_{c_d}^{(\gamma-1)/\gamma} - 1 \right)}{\eta_{c_d} \left[\left(\Pi \pi_{c_d} \right)^{(\gamma-1)/\gamma} - 1 \right]} \right\}^{\frac{1}{2}} \tag{8.136}$$

In the interest of reducing the complexity of the following analysis, and in the absence of a simple relationship relating η_c/η_{c_d} to $\dot{M}$ and $\bar{N}$, it is now assumed that $\eta_c/\eta_{c_d} \simeq 1$. This assumption is consistent with that introduced in the steady-state off-design examples previously considered. The equation for the operating line then reduces to

$$\dot{M} = \Pi \left[\frac{\pi_{c_d}^{(\gamma-1)/\gamma} - 1}{\left(\Pi \pi_{c_d} \right)^{(\gamma-1)/\gamma} - 1} \right]^{\frac{1}{2}} \tag{8.137}$$

This form may be used directly when a model compressor characteristic map is to be constructed. However, when the compressor characteristic map is intended to approximate an actual (experimentally obtained) map, it is appropriate to determine an "equivalent" design compressor pressure ratio, $(\pi_{c_d})_{eq}$, by selecting $(\pi_{c_d})_{eq}$ so that Eq. (8.137) passes through the design point $(\dot{M}, \Pi) = (1,1)$ and an appropriate reference point $\dot{M}_R, \Pi_R$. In such a case

$$\left(\pi_{c_d} \right)_{eq} = \left[\frac{\left(\dot{M}_R/\Pi_R \right)^2 - 1}{\left(\dot{M}_R/\Pi_R \right)^2 \Pi_R^{(\gamma-1)/\gamma} - 1} \right]^{\gamma/(\gamma-1)} \tag{8.138}$$

Limiting Value for Angular Velocity

This simplified representation of the compressor characteristic map developed in the preceding section allows relatively simple estimation of transient operation. It is important to note, however, that the form assumed becomes inappropriate when the speed line becomes "more than vertical." That is, the limiting value of speed to be considered, $\bar{N}_m$, will be that corresponding to the speed line for which $\dot{M}_S = \dot{M}_L$. This limit occurs [see Eqs. (8.122) and (8.123)] when

$$K_4 + K_5 \bar{N}_m + K_6 \bar{N}_m^2 = K_7 + K_8 \bar{N}_m + K_9 \bar{N}_m^2$$

or

$$\bar{N}_m = \frac{1}{2(K_9 - K_6)} \left\{ K_5 - K_8 + \left[(K_5 - K_8)^2 - 4(K_7 - K_4)(K_9 - K_6) \right]^{\frac{1}{2}} \right\}$$

$$\tag{8.139}$$

In most cases of interest, the value of $\bar{N}_m$ so calculated will correspond to a system speed higher than that allowed from stress limitations.

The Choking Relationship

As previously stated, in the example considered herein, it is assumed that the turbine entrance and exit nozzles are choked, and that the geometry is fixed. It then follows from Eq. (8.7) that

$$\dot{M} = \Pi \left\{ \frac{(\tau_\lambda/\tau_r)_d}{\tau_\lambda/\tau_r} \right\}^{\frac{1}{2}} \qquad (8.140)$$

It is to be noted that this relationship is valid throughout the transient maneuvers. It is only when the system is nonaccelerating that the turbine power output can be equated to the compressor input [as in Eq. (8.3)] to lead to the equation for the operating line, Eq. (8.137).

The Stall Margin

Figure 8.29 illustrates three locations on a compressor characteristic map: the initial location ($\dot{M}_i$, Π_i), the related location of the intersection of the speed line (for $\dot{M}_i$) and the stall line ($\dot{M}_{S_i}$, Π_{S_i}), and the "endpoint" location to be described shortly.

The stall margin SM of the compressor operating at the initial location is defined by

$$SM = \frac{\Pi_{S_i} - \Pi_i}{\Pi_i} \qquad (8.141)$$

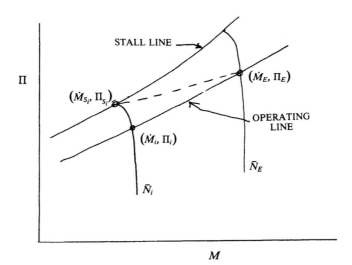

Fig. 8.29 Stall margin.

It subsequently will be found that the acceleration limit is a sensitive function of the stall margin. In fact this can be somewhat foreshadowed by considering an engine acceleration carried out by a sudden ("step") increase of throttle setting (τ_λ/τ_r) that moves the momentary operating point from $(\dot{M}_i, \Pi_i)$ to $(\dot{M}_{S_i}, \Pi_{S_i})$. It is apparent from Eq. (8.140) that the related value of τ_λ/τ_r is given by

$$(\tau_\lambda/\tau_r)_{S_i} = (\tau_\lambda/\tau_r)_d \left(\Pi_{S_i}/\dot{M}_{S_i}\right)^2 \tag{8.142}$$

The endpoint $(\dot{M}_E, \Pi_E)$ is defined as the point on the compressor map that will be reached when the compressor reaches equilibrium at the new operating point without further changes in the ratio (τ_λ/τ_r). Thus with $(\tau_\lambda/\tau_r)_{S_i} = (\tau_\lambda/\tau_r)_E$, Eq. (8.140) gives

$$\dot{M}_E = \Pi_E \frac{\dot{M}_{S_i}}{\Pi_{S_i}} \tag{8.143}$$

The equation for the operating line, Eq. (8.137), may be inverted to give

$$\Pi = \frac{1}{\pi_{c_d}} \left[1 + \left(\pi_{c_d}^{(\gamma-1)/\gamma} - 1 \right) \left(\frac{\Pi}{\dot{M}} \right)^2 \right]^{\gamma/(\gamma-1)} \tag{8.144}$$

With Eq. (8.144) applied at both the initial (i) and end (E) locations, and with Eq. (8.143), it follows that

$$\frac{\Pi_E}{\Pi_i} = \left[\frac{1 + \left(\pi_{c_d}^{(\gamma-1)/\gamma} - 1 \right) \left(\Pi_{S_i}/\dot{M}_{S_i} \right)^2}{1 + \left(\pi_{c_d}^{(\gamma-1)/\gamma} - 1 \right) \left(\Pi_i/\dot{M}_i \right)^2} \right]^{\gamma/(\gamma-1)} \tag{8.145}$$

Example calculations show that Π_E is a sensitive function of the stall margin, with large stall margins leading to large ratios of Π_E/Π_i. The implication of such a result is that care must be taken to restrict the throttle advance allowed (that is, the increase in τ_λ/τ_r) if small stall margins exist and large increases in compressor pressure ratio are desired. The effect of these restrictions is to require automatic control systems to prevent introduction of values of τ_λ/τ_r that will cause compressor stall or surge.

Determination of Starting Conditions

Usually a starting condition will be prescribed in terms of the initial dimensionless pressure ratio Π_i. The related dimensionless mass flow rate $\dot{M}_i$ follows immediately from Eq. (8.137). An equation for the initial rotational speed follows from Eq. (8.132) to give

$$F(N_i) = \Pi_{S_i} - \Pi_i - \left(\frac{\dot{M}_i - \dot{M}_{S_i}}{\dot{M}_{L_i} - \dot{M}_{S_i}} \right)^n \left(\Pi_{S_i} - \Pi_L \right) = 0 \tag{8.146}$$

The subsidiary equations, Eqs. (8.121–8.123), provide the explicit expressions for the $\bar{N}_i$ dependence. This equation may be solved by utilizing Newtonian iteration with

$$F'(\bar{N}_i) = \frac{F(\bar{N}_i + 10^{-6}) - F(\bar{N}_i - 10^{-6})}{2(10^{-6})} \tag{8.147}$$

so that

$$(\bar{N}_i)_{j+1} = \left(\bar{N}_i - \frac{F}{F'}\right)_j \tag{8.148}$$

A simple starting guess is

$$(\bar{N}_i)_0 = \dot{M}_i \tag{8.149}$$

Determination of the System Acceleration

The power extraction by the turbine may be written as the angular frequency of the system ω times the turbine torque T_t. The first law of thermodynamics states that the power extraction is equal to the mass flow rate times the stagnation enthalpy change across the turbine, which leads to

$$T_t = \frac{1}{\omega} \dot{m} c_{p_c} T_0 \tau_\lambda (1 - \tau_t) \tag{8.150}$$

Similarly the power absorption by the compressor gives for the compressor torque T_c,

$$T_c = \frac{1}{\omega} \dot{m} c_{p_c} T_0 \tau_r (\tau_c - 1) \tag{8.151}$$

The angular acceleration times the angular momentum J is equal to the net torque, so that

$$J \frac{d\omega}{dt} = \frac{1}{\omega} \dot{m} c_{p_c} T_0 [\tau_\lambda (1 - \tau_t) - \tau_r (\tau_c - 1)] \tag{8.152}$$

In the case considered here in which the turbine entrance and exit areas remain choked and the geometry remains fixed, Eq. (8.30) may be introduced to give

$$J \frac{d\omega}{dt} = \frac{1}{\omega} \dot{m} c_{p_c} T_0 \tau_r \left[\frac{\tau_\lambda / \tau_r}{(\tau_\lambda / \tau_r)_d} (\tau_{c_d} - 1) - (\tau_c - 1) \right] \tag{8.153}$$

This latter expression is more conveniently written in terms of the corrected variables and the dimensionless forms. Thus write

$$\omega = \frac{\omega / \sqrt{\theta}}{(\omega / \sqrt{\theta})_d} \left(\frac{\omega}{\sqrt{\theta}}\right)_d \sqrt{\theta} = \bar{N} \omega_{c_d} \sqrt{\frac{T_0 \tau_r}{T_{STP}}} \tag{8.154}$$

also

$$\dot{m} = \frac{\dot{m}\sqrt{\theta}}{\delta}\left(\frac{\delta}{\sqrt{\theta}}\right) = \dot{M}\dot{m}_{c_d}\left(\frac{\delta}{\sqrt{\theta}}\right) \tag{8.155}$$

Equations (8.153–8.155) give

$$J\frac{d\left[\overline{N}\sqrt{(T_0\tau_r/T_{STP})}\right]}{dt} = \frac{\dot{M}}{\overline{N}}\frac{\dot{m}_{c_d}}{\omega_{c_d}^2}\sqrt{\frac{T_{STP}}{T_{0\tau_r}}}\frac{\delta}{\sqrt{\theta}}c_{p_c}T_0\tau_r$$
$$\times\left[\frac{\tau_\lambda/\tau_r}{(\tau_\lambda/\tau_r)_d}(\tau_{c_d}-1)-(\tau_c-1)\right] \tag{8.156}$$

This equation can be numerically integrated to determine the angular velocity as a function of time. It is usually true, however, that engine accelerations occur on a somewhat smaller time scale than do airframe accelerations. In such cases it may be assumed that P_0, T_0, $\dot{M}_0$, etc., remain constant throughout the engine transient, and as a result Eq. (8.156) simplifies to the more compact form

$$\frac{d\overline{N}}{dt^*} = \left[\frac{\tau_\lambda/\tau_r}{(\tau_\lambda/\tau_r)_d}(\tau_{c_d}-1)-(\tau_c-1)\right]\frac{\dot{M}}{\overline{N}} \tag{8.157}$$

Here $t^* = t/t_{acc}$ has been introduced, where t_{acc} is the characteristic acceleration time defined by

$$t_{acc} = \frac{J\omega_{c_d}^2}{\dot{m}_{c_d}}\frac{\sqrt{\theta}}{\delta}\frac{1}{c_{p_c}T_{STP}} \tag{8.158}$$

The expression for the characteristic acceleration time reveals some useful physics. Thus, for example, highly energetic systems (large $J\omega_{c_d}^2$) require larger times to acquire increased rotational speeds. Similarly, flight at high altitude (low δ) leads to large acceleration times. This latter effect occurs because the related reduced density leads to reduced torques, and hence lower accelerations. This effect can be dramatic (a factor of more than ten) for an aircraft with high altitude capability.

Equation (8.157) may be integrated numerically by introducing the previously developed analytical representation of the compressor map. It will be assumed that the initial conditions are known, and that the schedule of τ_λ/τ_r as a function of time is prescribed. Integration is carried out by assuming suitably small time steps, δt^*, and calculating the related change in $\overline{N}$. Iteration of the compressor map equations will be required at each time step. A suggested sequence of equations to obtain $\overline{N}_{j+1}$ from $\overline{N}_j$ is

$$\Pi_{j+1} = \dot{M}_j\left[\frac{(\tau_\lambda/\tau_r)_{j+1}}{(\tau_\lambda/\tau_r)_d}\right]^{\frac{1}{2}} \tag{8.159}$$

$$\dot{M}_L = K_4 + K_5 \overline{N}_j + K_6 \overline{N}_j^2 \tag{8.160}$$

$$\dot{M}_S = K_7 + K_8 \overline{N}_j + K_9 \overline{N}_j^2 \tag{8.161}$$

$$\Pi_S = K_1 + K_2 \dot{M}_S + K_3 \dot{M}_S^2 \tag{8.162}$$

$$\dot{M}_{j+1} = \dot{M}_S + \left(\dot{M}_L - \dot{M}_S \right) \left(\frac{\Pi_S - \Pi_{j+1}}{\Pi_S - \Pi_L} \right)^{\frac{1}{n}} \tag{8.163}$$

This equation set is iterated by replacing $\dot{M}_j$ with $\dot{M}_{j+1}$ until the values of Π_{j+1} and $\dot{M}_{j+1}$ stabilize. The new value of $\overline{N}$, $\overline{N}_{j+1}$, then follows from

$$\overline{N}_{j+1} = \overline{N}_j + \frac{\dot{M}_{j+1}}{\overline{N}_j} \left[\frac{(\tau_\lambda/\tau_r)_{j+1}}{(\tau_\lambda/\tau_r)_d} \left(\tau_{c_d} - 1 \right) - \left(\tau_{c_{j+1}} - 1 \right) \right] \delta t^* \tag{8.164}$$

where

$$\tau_{c_{j+1}} - 1 = \frac{\left(\Pi_{j+1} \pi_{c_d} \right)^{(\gamma-1)/\gamma} - 1}{\eta_c} \tag{8.165}$$

Reference

[1]Klees, G. W. and Welliver, A. D., "Variable-Cycle Engines for the Second Generation SST," Society of Automotive Engineers Paper 750630. Air Transportation Meeting, Hartford, Conn., May 1975.

Problems

8.1   Consider the off-design performance of a nonafterburning turbojet. Ideal performance of all components may be assumed ($\pi_d = \pi_n = \eta_c = 1$ and $\gamma_c = \gamma_t = \gamma$, $f \ll 1$, etc.). Both A_4 and A_8 may be considered choked. The engine is flown with fixed τ_λ, but at varying Mach numbers. Reference conditions are $M_{0R} = 1$, $\pi_{cR} = 20$, $\gamma = 1.4$, and $\tau_\lambda = 7$.
   (a)   If A_4 and A_8 are fixed, find π_c when $M_0 = 2$.
   (b)   If the ratio A_4/A_8 is varied in proportion to $1/\tau_r$, find π_c when $M_0 = 2$.

8.2   An ideal, nonafterburning turbojet engine operates with A_4 and A_8 choked. The engine has a variable A_8 that is varied to keep the compressor pressure ratio constant.
   (a)   Obtain an expression for τ_{tR} in terms of τ_c, τ_{rR}, and $\tau_{\lambda R}$.
   (b)   Obtain an expression for τ_t in terms of τ_{tR}, $\tau_{\lambda R}$, τ_{rR}, τ_λ, and τ_r.

(c) Find an expression for A_8/A_{8R} in terms of γ, τ_{tR}, and τ_t.

(d) Given $\pi_c = 15$, $\gamma = 1.4$, $M_{0R} = 2$, $\tau_{\lambda R} = 7.0$, and $M_0 = 2$, $\tau_\lambda = 6$ evaluate τ_{tR}/τ_t and A_8/A_{8R}.

8.3   A concept to improve the off-design performance of a turbojet, which has already seen service, is that A_8 is varied but A_4 remains fixed. Consider a turbojet flying at fixed Mach number, with fixed component efficiencies, fixed γ_c, γ_t, etc., no afterburning, and both A_4 and A_8 choked. A_8 is varied to keep the corrected mass flow constant.

Find and list equations giving π_c, τ_c, π_t, τ_t, and A_8 in terms of the input variables γ_c, γ_t, η_m, η_c, η_t, π_{cR}, $\tau_{\lambda R}$, M_0, and τ_λ.

8.4   "Off-design analysis" can be used also to "redesign" an engine from the reference condition. Thus, for example, we may wish to change the areas A_4 and A_8 to move the operating line further from the surge line.

Consider a turbojet with ideal components, such that $\eta_c = \eta_t = \eta_m = 1$, $f \ll 1$, $\gamma_t = \gamma_c = \gamma$, etc.

(a) Show that the compressor operating line may be determined from the hierarchy of equations:

Inputs: M_0, τ_λ, A_8/A_{8R}, A_4/A_{4R}, M_{0R}, $\tau_{\lambda R}$, π_{cR}, $\dot{m}_{cR}$, $\gamma = 1.4$

Outputs: π_c, $\dot{m}_c$

Equations:

$$\tau_r = 1 + M_0^2/5, \qquad \tau_{rR} = 1 + M_{0R}^2/5$$

$$\tau_{cR} = \pi_{cR}^{1/3.5}$$

$$\tau_{tR} = 1 - \frac{\tau_{cR} - 1}{(\tau_\lambda/\tau_r)_R}$$

$$\tau_t = \tau_{tR}\left[\frac{A_4/A_{4R}}{A_8/A_{8R}}\right]^{\frac{1}{3}}$$

$$\pi_c = \left[1 + \tau_\lambda/\tau_r(1 - \tau_t)\right]^{3.5}$$

$$\dot{m}_c = \dot{m}_{cR}\frac{A_4}{A_{4R}}\frac{\pi_c}{\pi_{cR}}\left[\frac{(\tau_\lambda/\tau_r)_R}{\tau_\lambda/\tau_r}\right]^{\frac{1}{2}}$$

(b) Consider a reference condition with $M_{0R} = 2$, $\tau_{\lambda R} = 7$, $\pi_{cR} = 25$, and $\dot{m}_{cR} = 100$. Calculate and plot π_c vs $\dot{m}_c$ over the range $2.5 \leq \tau_\lambda/\tau_r$

$\leq (\tau_\lambda/\tau_r)_R$ for the reference engine ($A_4 = A_{4R}$, $A_8 = A_{8R}$). Indicate the location of values of τ_λ/τ_r on the curve.

(c) Obtain the compressor operating line over the same range of τ_λ/τ_r for an engine that has been redesigned to have $A_4 = A_{4R}$, $A_8 = 1.2A_{8R}$. Plot on the same graph as in part (b).

(d) Similarly, obtain and plot π_c vs $\dot{m}_c$ for an engine with $A_4/A_8 = (A_4/A_8)_R$, $A_4 = 1.1\,A_{4R}$.

8.5   A designer wishes to design a turbojet engine so that when the flight Mach number changes (at fixed altitude) the inlet will "just swallow its projected image." In order to achieve this objective, he decides to utilize a variable A_8, but retain fixed A_4.

Obtain a series of relationships that would allow the designer to estimate the required variation in A_8/A_{8R} with prescribed variations in M_0 or τ_λ.

8.6   Consider a turbofan engine that operates with the fan stream unchoked (convergent only nozzle), but with the core stream choked at A_4, A_{4a}, and A_8 (see Sec. 8.3 for nomenclature).

Show that the following hierarchy of equations may be used to obtain the off-design performance of the engine:

$$\tau_{ch} = 1 + (\tau_{chR} - 1)\frac{\tau_{c'R}}{\tau_{c'}}\frac{\tau_\lambda/\tau_r}{(\tau_\lambda/\tau_r)_R}$$

$$\pi_{ch} = \left[1 + \eta_{ch}(\tau_{ch} - 1)\right]^{3.5}$$

$$\pi_{c'} = \left[1 + \eta_{c'}(\tau_{c'} - 1)\right]^{3.5}$$

$$M_{9'}^2 = 5\left[(\pi_r\pi_d\pi_{c'}\pi_{n'})^{1/3.5} - 1\right]$$

$$\mathscr{M}_{9'} = \left[\frac{5}{6}\left(1 + \frac{M_{9'}^2}{5}\right)\right]^{-3}M_{9'}$$

$$\frac{\alpha}{\alpha_R} = \frac{\pi_{chR}}{\pi_{ch}}\frac{\mathscr{M}_{9'}}{\mathscr{M}_{9'R}}\left[\frac{\tau_{c'R}}{\tau_{c'}}\frac{\tau_\lambda/\tau_r}{(\tau_\lambda/\tau_r)_R}\right]^{\frac{1}{2}}$$

$$\tau_{c'} = 1 + (\tau_{c'R} - 1)\frac{1 + \alpha_R}{1 + \alpha}\frac{\tau_\lambda/\tau_r}{(\tau_\lambda/\tau_r)_R}$$

8.7   Consider a turbofan engine which operates with both the fan stream and core stream unchoked at the (convergent, $A_{8'} = A_{9'}$ and $A_8 = A_9$) exit nozzles, but with choked flow at A_4 and A_{4a}.

Show that the following hierarchy of equations may be used to obtain the off-design performance of the engine:

$$\tau_{ch} = 1 + (\tau_{chR} - 1)\frac{\tau_\lambda/\tau_r\tau_{c'}}{(\tau_\lambda/\tau_r\tau_{c'})_R}$$

$$\pi_{ch} = [1 + \eta_{ch}(\tau_{ch} - 1)]^{3.5}$$

$$\pi_{c'} = [1 + \eta_{c'}(\tau_{c'} - 1)]^{3.5}$$

$$M_{9'} = \left\{5\left[(\pi_r\pi_d\pi_{c'}\pi_{n'})^{1/3.5} - 1\right]\right\}^{\frac{1}{2}}$$

$$\mathcal{M}_{9'} = \left[\frac{5}{6}\left(1 + \frac{M_{9'}^2}{5}\right)\right]^{-3} M_{9'}$$

$$\frac{\alpha}{\alpha_R} = \frac{\pi_{chR}}{\pi_{ch}}\frac{\mathcal{M}_{9'}}{\mathcal{M}_{9'R}}\left[\frac{\tau_\lambda/\tau_r\tau_{c'}}{(\tau_\lambda/\tau_r\tau_{c'})_R}\right]^{\frac{1}{2}}$$

$$\tau_{c'} = 1 + (\tau_{c'R} - 1)\frac{1 + \alpha_R}{1 + \alpha}\frac{\tau_\lambda/\tau_r}{(\tau_\lambda/\tau_r)_R}\frac{1 - \tau_{tL}}{1 - \tau_{tLR}}$$

$$M_8 = \left\{\frac{2}{\gamma_t - 1}\left[(\pi_r\pi_d\pi_{ch}\pi_{c'}\pi_b\pi_{th}R\pi_{tL}\pi_n)^{(\gamma_t-1)/\gamma_t} - 1\right]\right\}^{\frac{1}{2}}$$

$$\mathcal{M}_8 = \left[\frac{2}{\gamma_t + 1}\left(1 + \frac{\gamma_t - 1}{2}M_8^2\right)\right]^{-[(\gamma_t+1)/2(\gamma_t-1)]} M_8$$

$$\tau_{tL} = 1 - \eta_t\left(1 - \pi_{tL}^{(\gamma_t-1)/\gamma_t}\right)$$

$$\pi_{tL} = \pi_{tLR}\frac{\mathcal{M}_{8R}}{\mathcal{M}_8}\left(\frac{\tau_{tL}}{\tau_{tLR}}\right)^{\frac{1}{2}}$$

8.8   Verify that Eqs. (8.64–8.79) are correct.

8.9   Verify that Eqs. (8.86–8.103) are correct.

8.10   A turboprop engine is flown off-design. It has a very good turbine, so good that you may assume $e_t = \eta_t = 1$.
    (a)   Defining Π as $\Pi \equiv (\pi_r\pi_d\pi_c\pi_b\pi_n)^{(\gamma_t-1)/\gamma_t}$, show that for this case ($\eta_t = 1$)

$$M_9 = M_{9R}\left(\frac{\Pi}{\Pi_R}\right)^{(\gamma_t+1)/2(\gamma_t-1)}$$

$$\frac{\tau_t}{\tau_{tR}} = \frac{\Pi_R}{\Pi} \frac{1 + \frac{\gamma_t - 1}{2} M_9^2}{1 + \frac{\gamma_t - 1}{2} M_{9R}^2}$$

(b) Considering the case where $\pi_d \pi_b \pi_n = (\pi_d \pi_b \pi_n)_R$, $M_0 = M_{0R}$, and $\tau_{cR} = 2.944$, $\eta_c = \eta_{cR} = 0.845$, $M_{9R} = 0.6643$, $\tau_{\lambda R} = 7$, $\tau_\lambda = 6$, $\gamma_c = 1.4$ and $\gamma_t = 1.35$, find π_{cR}, π_c, M_9, and τ_t / τ_{tR}.

8.11   The turboprop engine described in Table 7.2 (with $\pi_{cR} = 24$) is to be flown off-design. The aircraft will be held at constant Mach number and altitude, but the throttle setting will be reduced. It may be assumed that all component efficiencies remain the same except that

$$\eta_{prop} = (\eta_{prop})_R + 0.04(\tau_{\lambda R} - \tau_\lambda)$$

Obtain and plot S/S_R vs F/F_R and π_c / π_{cR} vs $\dot{m}_c / \dot{m}_{cR}$ over the (approximate) range $\frac{1}{2} \le F/F_R \le 1$.

9. ELEMENTARY THEORY OF BLADE AERODYNAMICS

9.1 Introduction

In this chapter the relationships of the desired performance parameters to the related compressor or turbine blade loadings (and resultant fluid flow angles) will be investigated. It is apparent that the flowfield within an actual turbine or compressor is enormously complex; thus, it is desirable to create simplified models of the flowfields if any understanding of the physics of the flow processes is to be attained or if any analytical prediction techniques are to be formulated.

Although several extensive efforts to model the flow through an entire compressor or turbine (including transonic effects, boundary layers, wakes, etc.) have been attempted and have met with some partial preliminary success, it is more common to model the flowfield as a "sum" of less complicated parts. Relatively simple flowfields will be described here and the various separate pieces considered.

In order to analyze what appears at first to be an almost incomprehensibly complicated flowfield, it is customary to model the full three-dimensional flowfield as a compilation of three two-dimensional fields. These fields may be termed the "throughflow field," the "cascade field" (or blade-to-blade field), and the "secondary flowfield." Each of these fields is described in the following sections.

The Throughflow Field

This flowfield is considered to arise because of the influence of all the blades in a row (or rows). The effects of individual blades are not considered, and hence the combined effects of all the blades in the row are obtained by assuming the blade forces to be "smeared out" in the azimuthal direction. Mathematically, this process is accomplished by replacing the blade surface pressures by volumetric (and continuous) body forces. As a result of the simplifications contained in this model, no θ variations occur and the throughflow field is hence two-dimensional with variations occurring in the radial (r) and axial (z) directions.

Figure 9.1 shows the coordinate system and the throughflow representation of the flowfield. It can be noted that the flowfield identified with this approximation would be that approached if the blade row had a very large number of very thin blades.

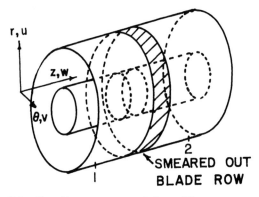

Fig. 9.1   Coordinate system and throughflow representation.

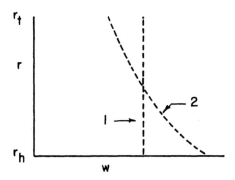

Fig. 9.2   Typical axial velocity profiles.

The solution to such a flowfield would give the axial and tangential velocities throughout. Figure 9.2 shows typical axial velocity profiles that could occur at stations 1 and 2 of Fig. 9.1.

When axial velocity profiles of the form indicated in Fig. 9.2 at station 2 develop, it is apparent from continuity that a downward flow of fluid must occur. Thus, a typical "stream surface" (to be more carefully defined in Sec. 9.4) could appear as depicted in Fig. 9.3.

The throughflow field can be considered the parent flowfield of the cascade and secondary flowfields, and because of this should be calculated with considerable accuracy. For this reason, a somewhat extensive description of throughflow calculation techniques is given in Chap. 10.

The Cascade Field

In order to estimate the flow behavior in the neighborhood of individual blades, a meridional surface such as that indicated in Fig. 9.3 is expanded ("unwrapped") and the individual blade profiles considered (see Fig. 9.4).

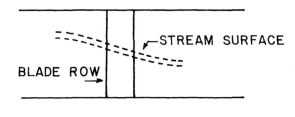

Fig. 9.3   Typical stream surface.

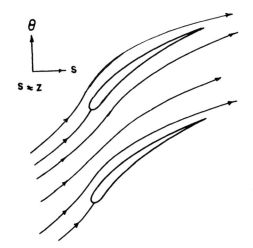

Fig. 9.4   Meridional projections of blade profiles.

When the meridional surface of Fig. 9.3 is unwrapped in this way, a two-dimensional flowfield in the θ and meridional (almost z) coordinates is obtained. If the curvature of the stream surfaces in the throughflow field (Fig. 9.3) is not too great, the pressure gradients across such stream surfaces need not be considered, and hence the individual "strips" may be considered separately (hence, the term "strip theory"). By considering a number of such strips, suitable blade profiles can be determined for a selected number of radial stations on the blade, and the complete blade shape necessary to describe the full three-dimensional blade can be obtained by fairing in the desired profile shape.

Numerical methods may be utilized to calculate the flowfield in this plane (including the boundary layers), but classically the most widely used method is to run cascade tests to obtain the blade performance data. It is to be noted that, in general, the flow between an example pair of stream surfaces in the throughflow field encounters a change in cross-sectional area. As a result, great care must be taken when the corresponding cascade test is conducted to include the effects of the axial variation in flow cross section and its related effect upon the streamwise pressure variation. It is to be further noted that the effect of the cascade sidewall boundary-layer buildup must also be accounted for in order to properly simulate the streamwise pressure variations. The cascade flowfield is more fully described in Chap. 11.

The Secondary Flowfield

The third of the three two-dimensional flowfields considered to comprise the full three-dimensional flowfield of an axial turbomachine is the secondary flowfield. This field exists because the fluid near the solid surfaces will have a lower velocity with respect to those surfaces than does the fluid in the "freestream" (external to the boundary layer). As a result the imposed pressure gradients (created because of the curvature of the freestream) will deflect the fluid within the boundary layers from regions of high pressure to regions of low pressure. Figure 9.5 indicates the possible secondary flows existing within a stator row. (Note that the blade boundary layers on a rotor would tend to be centrifuged outward, whereas the excess pressure existing at the outer annulus tends to deflect the stator blade boundary layer inward.)

Secondary flows are notoriously difficult to analyze; but in spite of this, considerable progress has been made in the analysis of secondary flows in both compressors and turbines. The techniques of analysis for compressors are substantially different than those for turbines. This is because in compressors the adverse pressure gradient leads to low wall shears, so the flows can be analyzed fairly accurately by ignoring the viscous shearing terms. (The presence of the viscous shearing terms is included implicitly in the assumed entrance velocity profile.) The literature of this class of secondary flows is quite extensive, and the interested reader is encouraged to obtain several excellent recent studies.[1-3]

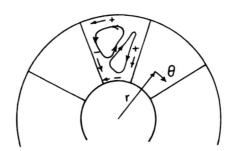

Fig. 9.5 Secondary flows within a stator row.

When secondary flows in turbines are to be considered, the large favorable pressure gradients existing over much of the blade surface and annulus walls leads to thin boundary layers with high shearing rates and consequently high shear stresses. As a result (three-dimensional) boundary-layer calculation techniques must be employed.

9.2   Two-Dimensional Incompressible Flow through Blade Rows

In this section the changes in fluid velocity induced by blade rows are related to the changes in fluid thermodynamic properties. The discussion pertains to the cascade field, but for the present only the far upstream and far downstream conditions will be considered. In this way, description of the details of the flow in the vicinity of the blades will not be necessary, but rather only the changes in fluid properties will be required. The more difficult problem of obtaining the necessary blade geometries to efficiently induce the assumed velocity fields will be addressed in later sections.

The Euler Equation

Now consider the behavior of a single stream tube as it passes through a rotor row. The geometry and nomenclature of the interaction are indicated in Fig. 9.6. For the purposes of this section, it is not necessary to know the details of the interaction within the volume indicated by the dotted lines in the figure. Rather it is required only that stations 1 and 2 be sufficiently far removed from the region of rotor interaction that the flow may be considered time independent. In addition, it is assumed that the entire process is adiabatic.

Δw is defined as the work interaction per unit mass that the stream tube undergoes as it passes through the rotor, and for convenience is defined to be positive for a work "input." The first law of thermodynamics then gives (for this adiabatic flow)

$$\Delta w = h_{t_2} - h_{t_1} \qquad (9.1)$$

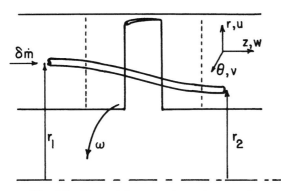

Fig. 9.6   Steam tube geometry and nomenclature.

The work interaction from the rotor is transmitted to the stream tube from the input shaft by the torque of the blades. Denoting that portion of the torque identified with the stream tube as δT and the angular velocity of the shaft as ω, the power input to the stream tube is given by

$$\text{Power input} = \omega\,\delta T \qquad (9.2)$$

The torque on the stream tube itself is equal to the rate of production of angular momentum, so that with the mass flow rate through the stream tube denoted $\delta \dot{m}$,

$$\delta T = \left[(rv)_2 - (rv)_1 \right] \delta \dot{m} \qquad (9.3)$$

The work interaction per unit mass is just the power input divided by the mass flow rate, so from Eqs. (9.2) and (9.3)

$$\Delta w = \omega \frac{\delta T}{\delta \dot{m}} = \omega \left[(rv)_2 - (rv)_1 \right] \qquad (9.4)$$

Combination of Eqs. (9.1) and (9.4) then gives the famous Euler momentum equation,

$$h_{t_2} - h_{t_1} = \omega \left[(rv)_2 - (rv)_1 \right] \qquad (9.5)$$

It is to be noted that no restriction to ideal flow was implied in the development of this equation, the only restriction being that the flow be adiabatic. It would, of course, be hoped that in a compressor the work interaction would occur in a primarily nonviscous manner so that after diffusion (in a stator row) the effects of the work interaction would appear primarily as a pressure increase rather than as a (static) temperature increase. Equation (9.5), however, is applicable no matter what the compressor efficiency.

The Perfect Fluid Approximation

When the flow of a perfect fluid is considered, the stagnation pressure ratio may be related to the stagnation enthalpy ratio because the entropy remains constant. Thus, utilizing Eq. (2.69)

$$\frac{p_{t_2}}{p_{t_1}} = \left(\frac{T_{t_2}}{T_{t_1}} \right)^{\gamma/(\gamma-1)} = \left(\frac{h_{t_2}}{h_{t_1}} \right)^{\gamma/(\gamma-1)} \qquad (9.6)$$

These forms follow because of the assumption of a calorically perfect gas. Thus, with Eqs. (9.5) and (9.6), there follows

$$\frac{p_{t_2}}{p_{t_1}} = \left\{ 1 + \frac{\omega}{h_{t_1}} \left[(rv)_2 - (rv)_1 \right] \right\}^{\gamma/(\gamma-1)} \qquad \text{(perfect fluid)} \qquad (9.7)$$

Now introduce reference quantities denoted by a subscript 0 where such quantities are assumed to be the uniform values of the given quantities found in the approaching flow far upstream. The stagnation enthalpy does not change prior to the blade row, so that

$$h_{t_1} = h_{t_0} = \gamma R T_0 \frac{C_p}{\gamma R} \frac{T_{t_0}}{T_0} = \gamma R T_0 \left(\frac{1}{\gamma - 1} \right) \left(1 + \frac{\gamma - 1}{2} M_0^2 \right)$$

or

$$h_{t_1} = w_0^2 \frac{1}{M_0^2} \left(\frac{1}{\gamma - 1} \right) \left(1 + \frac{\gamma - 1}{2} M_0^2 \right) \tag{9.8}$$

Equations (9.7) and (9.8) then give

$$\frac{p_{t_2}}{p_{t_1}} = \left\{ 1 + \frac{(\gamma - 1) M_0^2}{1 + \frac{\gamma - 1}{2} M_0^2} \frac{\omega}{w_0^2} [(rv)_2 - (rv)_1] \right\}^{\gamma/(\gamma - 1)} \quad \text{(perfect fluid)} \tag{9.9}$$

If attention is now restricted to cases where the amount of turning is small, this expression may be approximated by retaining only the first term in the binomial expansion on the right-hand side. Thus,

$$\frac{p_{t_2}}{p_{t_1}} - 1 \approx \frac{\gamma M_0^2}{1 + \frac{\gamma - 1}{2} M_0^2} \frac{\omega}{w_0^2} [(rv)_2 - (rv)_1] \quad \text{(perfect fluid)} \tag{9.10}$$

It is to be noted that Eq. (9.10) would also be valid in the case of small Mach numbers. In such a case, p_{t_1} becomes an inconvenient reference quantity, and it is more appropriate to reference the "dynamic head," $\rho_0 w_0^2$. Thus, noting

$$\gamma M_0^2 = \frac{\rho_0 w_0^2}{p_0} = \frac{\rho_0 w_0^2}{p_{t_1}} \frac{p_{t_0}}{p_0} = \frac{\rho_0 w_0^2}{p_{t_1}} \left(1 + \frac{\gamma - 1}{2} M_0^2 \right)^{\gamma/(\gamma - 1)}$$

It follows that

$$\frac{p_{t_2} - p_{t_1}}{\rho_0 w_0^2} \approx \left(1 + \frac{\gamma - 1}{2} M_0^2 \right)^{1/(\gamma - 1)} \frac{\omega}{w_0^2} [(rv)_2 - (rv)_1]$$

$$\text{(perfect fluid, small turning)} \tag{9.11}$$

In the limit as M_0 approaches zero, the relationship for incompressible perfect flow is obtained, which may be written

$$\frac{p_{t_2} - p_{t_1}}{\rho_0 w_0^2} = \frac{\omega}{w_0^2} \left[(rv)_2 - (rv)_1 \right] \qquad \text{(perfect incompressible fluid)} \quad (9.12)$$

The expressions represented by Eqs. (9.5–9.12) allow calculation of the changes in stagnation enthalpy or stagnation pressure in terms of the imposed changes in swirl velocities. Each form of the equations should be used only in the regime for which it is valid, as indicated beneath each equation. It is to be noted from Eq. (9.5) that no change in stagnation enthalpy can occur unless the blade row is moving. This situation is quite obvious when it is realized that without a moving surface, no work interaction with the fluid would occur [Eq. (9.4)].

A similar result to that for the stagnation enthalpy occurs in the case of the stagnation pressure of a perfect fluid. When losses are present, however, the accompanying entropy gains will superimpose a stagnation pressure decrease upon the flow processes in the rotor and stator, and the stagnation pressure rise through the rotor will not be as large as it would be if the fluid was perfect. Also, there will be a decrease in stagnation pressure as the fluid passes through the stator.

Velocity Triangles

It was seen in the previous section that desired property changes are brought about by inducing swirl velocities with the rotors and stators. In this section the resulting vector triangles of the fluid velocities are consid-

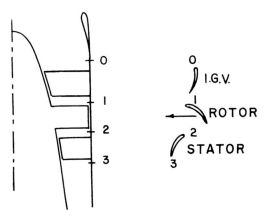

Fig. 9.7   Blade rows and station numbering.

ered in order to gain an appreciation for the fluid field the blades and vanes must create. The blade and vane angles will, of course, be closely related to the fluid flow angles. The flow is considered to be two-dimensional only and is considered to be approaching the first row from afar with conditions denoted by the subscript 0. The flow past three rows are considered: an inlet guide vane row, rotor, and stator. The rows and station numbering are indicated in Fig. 9.7.

Inlet guide vane (IGV).   The flow approaching and following the inlet guide vane (IGV) row is illustrated in Fig. 9.8. The notation for the angle of turning, velocity components, and magnitude of the velocity is indicated in the figure.

An inlet guide vane row tends to be unique among all the rows in a compressor, because in all compressors with IGV rows built to date, the static pressure decreases across the row. The reason for this can be inferred from Fig. 9.8, where it is evident that the magnitude of the velocity has increased upon passage through the row. The IGV row is not moving and hence no increase in stagnation enthalpy occurs and, for an ideal fluid, no change in stagnation pressure occurs. As a result, the increased magnitude of the velocity will have attendant with it a decrease in static pressure. It is conceivable that a severe expansion in the annulus sidewalls could be incorporated, so that the axial velocity could be reduced to a sufficient extent to cause the overall velocity magnitude to reduce, but this has not been incorporated in any design to date. (There would seem to be no reason to do so.) Because of the overall favorable pressure gradient imposed across an IGV row, the tendency of the vane boundary layers to separate is much reduced; as a result, very high turning can be introduced by a single IGV stage.

Fig. 9.8   Flow past an inlet guide vane row.

Rotor. When the air departs the IGV, it has a velocity $V_1 = \sqrt{v_1^2 + w_1^2}$, which is directed at an angle θ_1 from the axial direction. The rotor itself has a velocity $\omega r e_\theta$, and hence sees a relative velocity at inlet given by

$$\mathbf{V}_{\text{rel}_1} = \mathbf{V}_1 - \omega r e_\theta \tag{9.13}$$

The air is then turned within the rotor to the relative velocity $\mathbf{V}_{\text{rel}_2}$, so that the velocity in the absolute frame (the laboratory system of coordinates). $\mathbf{V}_2$ is given by

$$\mathbf{V}_2 = \mathbf{V}_{\text{rel}_2} + \omega r e_\theta \tag{9.14}$$

These velocity relationships are indicated in Fig. 9.9.

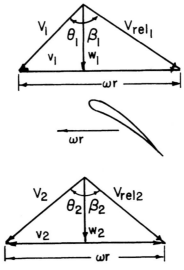

Fig. 9.9   Flow past a rotor row.

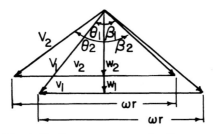

Fig. 9.10   Rotor row composite diagram.

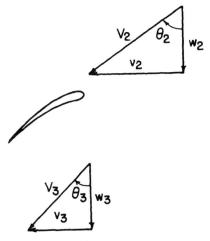

Fig. 9.11   Flow past a stator row.

The information shown in Fig. 9.9 can be more compactly displayed by utilizing a composite diagram wherein both entering and leaving velocity diagrams are superimposed. Such a diagram is indicated in Fig. 9.10.

Stator.   The stator accepts the flow departing from the rotor and turns it to a more axial direction. By reducing the swirl velocity in this way the magnitude of the total velocity is also reduced and the static pressure increases across the vane row. In those designs where the stator returns the velocity vector to that found at entrance to the rotor, the stage is said to be a repeating stage. The stator velocity relationships are indicated in Fig. 9.11.

Pressure Relationships across Blade Rows— The Degree of Reaction and the Diffusion Factor

It is the purpose of a compressor stage to raise the stagnation pressure as much as possible and as efficiently as possible. In addition, it would be desirable to have the mass flow rate per cross-sectional area be as large as possible so as to reduce the required compressor cross-sectional area to the minimum possible. These desired design goals create several conflicting requirements and limitations.

It is evident from Eq. (9.9) that a large stagnation pressure rise would require a large amount of turning in the rotor, and hence also in the stator, to return the velocity vector to (approximately) that entering the rotor. This large turning introduces two related deleterious effects. First, the large stagnation pressure change introduces a related large adverse static pressure gradient. This imposed static pressure gradient enhances the probability of boundary-layer separation on the blade. Second, the large required blade curvature leads to high blade aerodynamic loading with related very low values of minimum static pressure on the blade suction side. As a result, the suction side boundary layer must surmount a locally severe adverse pressure gradient that has already been made worse by the superimposed static pressure gradient due to all other blades in the row, as just described above.

It is apparent [Eq. (9.9)] that the attainment of high pressure ratios will be aided by the use of high blade speeds and high axial Mach numbers, but each of these techniques has its own limitations. It is obvious that the blade speed will be limited by mechanical stressing considerations, although it is notable that great strides in this direction have been made in the last two decades. (Blade speeds have increased from ≈ 300 to ≈ 500 m/s.) Both increasing the blade speed and increasing the axial Mach number lead to large relative Mach numbers on the blades and, if carried to excess, can lead to large shock losses in the stage.

It is the function of a designer to make as optimal a choice as possible of the various conflicting requirements, and some simplified design aids have been developed to help in the various choices. Two coefficients often utilized in the design of axial compressors or turbines, the degree of reaction and the diffusion factor, are defined in the following. Each coefficient is related to the behavior of the static pressure change across the blade rows, so prior to defining the coefficients, the behavior of the static pressure will be investigated. For simplicity, the incompressible perfect fluid case is considered.

From Eq. (9.12), for the two-dimensional case (with $\rho_0 \equiv \rho$),

$$p_2 + \tfrac{1}{2}\rho\left(v_2^2 + w_2^2\right) - \rho\omega\left(rv\right)_2 = p_1 + \tfrac{1}{2}\rho\left(v_1^2 + w_1^2\right) - \rho\omega\left(rv\right)_1 \quad (9.15)$$

Straightforward manipulation of this equation leads to the alternate form

$$p_{t_2\,\text{rel}} - \tfrac{1}{2}\rho\left(\omega r_2\right)^2 = p_{t_1\,\text{rel}} - \tfrac{1}{2}\rho\left(\omega r_1\right)^2 \quad (9.16)$$

Here there has been introduced the relative stagnation pressure as seen by the rotor p_{t_rel}, which is defined by

$$p_{t_\text{rel}} \equiv p + \tfrac{1}{2}\rho\left[w^2 + \left(v - \omega r\right)^2\right]$$

It can be seen that in an axial compressor, where $r_1 \approx r_2$, the relative stagnation pressure across the (ideal) rotor does not change. This, of course, is simply Bernoulli's equation for the relative coordinate system. The terms involving the square of the blade speed are related to the accounting of "energy" stored against the centrifugal forces, and can be of dominant importance in centrifugal compressors with their large change in radius from inlet to outlet. Equation (9.16) is equally valid for stators as well as rotors, where in the case of a stator the equation simply degenerates to the statement that the stagnation pressure does not change across an ideal stator row.

Equation (9.16) is in a useful form for design purposes, because it makes evident that the change in static pressure across an axial compressor rotor row may be estimated simply by observing the fluid behavior in the relative coordinate system (Fig. 9.10). As was evident in the discussion introducing this section, the effect of static pressure increase is of paramount importance upon the operating limit of a blade row, and it is evident that it would be desirable to have such a static pressure increase across a stage nearly evenly divided between the rotor and the stator. A measure of this static pressure split is the degree of reaction $^\circ R$.

Degree of reaction $^\circ R$.   The degree of reaction $^\circ R$ is defined as the static pressure rise across the rotor divided by the static pressure rise across the stage. Thus,

$$^\circ R = \frac{\Delta p_{\text{rotor}}}{\Delta p_{\text{stage}}} = 1 - \frac{\Delta p_{\text{stator}}}{\Delta p_{\text{stage}}} \tag{9.17}$$

In the case of incompressible perfect fluid flow and nearly axial flow, write (utilizing the station numbering of Fig. 9.7)

$$(1/\rho)\Delta p_{\text{stage}} = (1/\rho)\Delta p_{t\,\text{stage}} - \tfrac{1}{2}\Delta V_{\text{stage}}^2$$

or

$$(1/\rho)\Delta p_{\text{stage}} = \omega\left[(rv)_2 - (rv)_1\right] - \tfrac{1}{2}\left(V_3^2 - V_1^2\right) \tag{9.18}$$

where V_i^2 denotes the scalar product $V_i \cdot V_i$ and

$$(1/\rho)\Delta p_{\text{stator}} = \tfrac{1}{2}\left(V_2^2 - V_3^2\right) \tag{9.19}$$

It follows that

$$^\circ R = 1 - \frac{V_2^2 - V_3^2}{V_1^2 - V_3^2 + 2\omega\left[(rv)_2 - (rv)_1\right]}$$

(incompressible, perfect flow, small radius changes)   (9.20)

A repeating stage is one for which, by definition, $V_1 = V_3$, so that with $w_1 = w_2 = w_3$,

$$^\circ R = 1 - \frac{v_2^2 - v_1^2}{2\omega\left[(rv)_2 - (rv)_1\right]}$$

(incompressible, perfect flow, small radius changes,
repeating stages, $w_1 = w_2 = w_3$)   (9.21)

An even simpler form results if the radius change is so small that it does not affect the angular momentum difference. In this case, with $r_1 = r_2 = r$,

$$°R = 1 - \frac{v_2 + v_1}{2\omega r}$$

(incompressible, perfect flow, no radius change, repeating stages) (9.22)

Diffusion factor D.   This factor was developed in an attempt to describe the effects of both the imposed axial pressure gradient and blade turning angle. (It is more fully developed in Ref. 4.) It is intended to be a measure of how highly loaded a single blade row may be considered to be. It is defined by

$$D = \left(1 - \frac{V_e}{V_i}\right) + \frac{|\Delta v|}{2\sigma V_i}$$

(9.23)

where V_e = magnitude of the relative velocity at exit to the blade row, V_i = magnitude of the relative velocity at inlet to the blade row, $|\Delta v|$ = magnitude of the change in tangential component of velocity across the blade row, and σ = solidity $\equiv$ chord/spacing $= c/s$. These relationships are shown in Fig. 9.12.

A typical maximum value of D obtainable for a given family of blade profiles is about 0.6. This value should not be exceeded substantially or it can be expected that flow separation will occur. Such limiting values of the diffusion factor are usually obtained from cascade tests (Chap. 11). Such limiting values can vary substantially in some cases (for blades with ex-

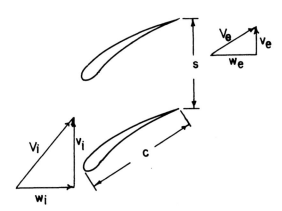

Fig. 9.12   Geometry and velocity changes across blade row.

tremely long chords, for example), but the factor is very convenient to utilize so that possible problem areas can be pinpointed.

Relationships in Terms of Flow Angles

It is often useful to obtain expressions for the pressure change, degree of reaction, and diffusion factor in terms of the fluid flow angles, because the flow angles themselves can be fairly directly related to the blade angles. In particular, the relative velocity departing each row will have a relative angle not far from the blade geometric angle. In fact, relatively simple empirical rules are available to estimate the difference of the relative flow angle and blade angle at the exit to the blade row (Chap. 11), so in the following the desired relationships are obtained in terms of angles θ_1, β_2, and θ_3.

For simplicity of presentation the incompressible perfect fluid case is considered. Thus, Eq. (9.12) may be utilized to give with the relationships apparent from Fig. 9.10, namely

$$v_1 = w_1 \tan \theta_1 \tag{9.24}$$

$$v_2 = \omega r_2 - w_2 \tan \beta_2 \tag{9.25}$$

$$\frac{\Delta p_t}{\rho_0 w_0^2} = \left(\frac{\omega r_2}{w_0} \right)^2 - \left(\frac{\omega r_2}{w_0} \frac{w_2}{w_0} \tan \beta_2 + \frac{\omega r_1}{w_0} \frac{w_1}{w_0} \tan \theta_1 \right)$$

(incompressible, perfect fluid) $\tag{9.26}$

In the special case where no change in axial velocity or radius occurs, this form reduces to

$$\frac{\Delta p_t}{\rho_0 w_0^2} = \left(\frac{\omega r}{w_0} \right)^2 \left[1 - \frac{w_0}{\omega r} (\tan \beta_2 + \tan \theta_1) \right]$$

(incompressible, perfect fluid, $r_1 = r_2 = r$, $w_1 = w_2 = w_0$) $\tag{9.27}$

In Sec. 8.5, it was pointed out that when a compressor is operated at low rotational speeds (during starting, for example) the front blades tend to operate with a low ratio of axial velocity to blade speed, as compared to the design speed ratio. It can be seen from Eq. (9.27) that a further complication will arise, in that the stagnation pressure rise near the tip of a finite length blade will tend to be larger than that near the hub. This is because the flow angles β_2 and θ_1 tend to be relatively insensitive to flow velocity and because β_2 and θ_1, at all radii, would be chosen to be appropriate for the design condition. At off-design when $w_0/\omega r$ decreases, so does the effect of the second term in the brackets of Eq. (9.27). As a result of the larger work interaction at the tip of the blades, the static pressures can become so excessive that a region of reverse flow can possibly exist (Fig. 9.13). The

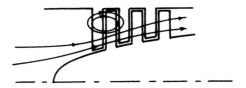

Fig. 9.13   Recirculating region.

fluid contained within such a region of reverse flow can rapidly have its temperature raised (by the work interaction with the blades) to such an extent that blade melting can occur! This is yet another reason why the techniques mentioned in Sec. 8.5 for preventing excessively low axial velocities must be employed.

The following relationships are evident from Fig. 9.10:

$$V_1^2 = w_1^2 \sec^2\theta_1 \tag{9.28}$$

$$V_2^2 = w_2^2 + (\omega r_2 - w_2\tan\beta_2)^2 \tag{9.29}$$

$$V_3^2 = w_3^2 \sec^2\theta_3 \tag{9.30}$$

Equations (9.20), (9.24), (9.25), and (9.28–9.30) may then be combined to give

$$^\circ R = 1 - \left\{ w_2^2\left[1 + \left(\frac{\omega r_2}{w_2} - \tan\beta_2\right)^2\right] - w_3^2\sec^2\theta_3 \right\}$$

$$\div\left\{ w_1^2\sec^2\theta_1 - w_3^2\sec^2\theta_3 + 2w_2^2\left[\left(\frac{\omega r_2}{w_2}\right)^2 - \left(\frac{\omega r_2}{w_2}\tan\beta_2 + \frac{\omega r_1}{w_2}\frac{w_1}{w_2}\tan\theta_1\right)\right]\right\}$$

$$\tag{9.31}$$

In the special case where no change in axial velocity or radius occurs and the stage is repeating, this form reduces to

$$^\circ R = 1 - \frac{(\omega r/w_0) - \tan\beta_2 + \tan\theta_1}{2(\omega r/w_0)}$$

(incompressible perfect flow, small radius changes,

$$r_1 = r_2 = r_3 = r, \; w_1 = w_2 = w_3 = w_0, \; V_1 = V_3) \tag{9.32}$$

Again utilizing Fig. 9.10,

$$(V_e)_{\text{rotor}} = w_2 \sec \beta_2 \tag{9.33}$$

$$(V_i)_{\text{rotor}} = w_1 \left[1 + \left(\frac{\omega r_1}{w_0} - \tan \theta_1 \right)^2 \right]^{\frac{1}{2}} \tag{9.34}$$

$$\Delta v_{\text{rotor}} = \omega r_2 - w_2 \tan \beta_2 - w_1 \tan \theta_1 \tag{9.35}$$

Equations (9.33–9.35) may then be substituted into Eq. (9.23) to give an expression for the rotor diffusion factor. In the special case where no change in axial velocity or radius occurs, there is obtained

$$D_{\text{rotor}} = 1 - \frac{\sec \beta_2 - (1/2\sigma)[(\omega r/w_0) - (\tan \beta_2 + \tan \theta_1)]}{\left\{ 1 + [(\omega r/w_0) - \tan \theta_1]^2 \right\}^{\frac{1}{2}}}$$

(incompressible perfect flow, small radius changes,
$r_1 = r_2 = r$, $w_1 = w_2 = w_0$) \tag{9.36}

Finally, an expression for the stator diffusion factor is obtained by first noting

$$(V_e)_{\text{stator}} = w_3 \sec \theta_3 \tag{9.37}$$

$$(V_i)_{\text{stator}} = w_2 \left[1 + \left(\frac{\omega r_2}{w_2} - \tan \beta_2 \right)^2 \right]^{\frac{1}{2}} \tag{9.38}$$

$$\Delta v_{\text{stator}} = \omega r_2 - w_2 \tan \beta_2 - w_3 \tan \theta_3 \tag{9.39}$$

These equations may then be substituted into Eq. (9.23) to give an expression for the stator diffusion factor. In the special case where no change in axial velocity or radius occurs, there is obtained

$$D_{\text{stator}} = 1 - \frac{\sec \theta_3 - (1/2\sigma)[(\omega r/w_0) - (\tan \beta_2 + \tan \theta_3)]}{\left\{ 1 + [(\omega r/w_0) - \tan \beta_2]^2 \right\}^{\frac{1}{2}}}$$

(incompressible perfect flow, small radius changes $r_1 = r_2 = r$, $w_1 = w_2 = w_0$)

\tag{9.40}

Several tendencies for the degree of reaction and diffusion factor become evident in these expressions. For example, note from Eq. (9.32) that in the case of a repeating stage with "symmetric blading" ($\beta_2 = \theta_1$), the degree of reaction is one-half no matter what the ratio of blade speed to axial velocity. For such a stage, this means that the static pressure rise would remain equally split between the rotor and stator as the compressor was "throttled." When the compressor is throttled (outlet flow decreased, leading to an outlet pressure increase), an increased pressure rise must occur across the stage. The effect of such an increase in pressure rise is seen in D_{rotor} and D_{stator}, which both increase as $\omega r / w_0$ increases. This, of course, means that the compressor is moving toward its design limit.

9.3   Free Vortex Flow

The effects of finite blade height upon the required fluid flow angles is now considered. It is apparent from the preceding sections that the variation in fluid properties across the blade rows is a function of the blade velocity. The blade velocity will change substantially with the radius, so the corresponding effects upon the flowfield should be determined. In most compressor designs, it is desirable to obtain the same stagnation pressure rise across a stage at all radii. (Note that if this is consistently violated in a multistage machine, very large static pressure mismatches can occur with consequent recirculation regimes.)

Consider the case where the change in stagnation enthalpy across the rotor is a constant with radius, so that from Eq. (9.5)

$$h_{t_2} - h_{t_1} = \omega\left[(rv)_2 - (rv)_1\right] = \text{const} \tag{9.41}$$

Thus, a "constant work stage"—that is, one for which the work interaction per mass is independent of radius—requires that the change in angular momentum be the same for all stream tubes. There are many types of flows that could satisfy this requirement, but the simplest such flow would seem to be that for which (for perfect, incompressible flow) the axial velocity remains a constant with radius (a parallel walled annulus is assumed.) It is apparent from the requirement of mass continuity that if no change in axial velocity occurs, no radial flows are induced. The radial momentum equation then reduces to the statement that the radial pressure gradient is balanced by the centrifugal forces. That is,

$$\frac{\partial p}{\partial r} = \rho \frac{v^2}{r} \tag{9.42}$$

The requirement that the stagnation pressure be constant with radius gives

$$p + (\rho/2)(w^2 + v^2) = p_t = \text{const}$$

So that

$$\frac{\partial p}{\partial r} = -\rho v \frac{\partial v}{\partial r} \tag{9.43}$$

Equations (9.42) and (9.43) may be combined to given an equation for v in terms of r. Thus, noting that no changes in properties occur in the axial direction (except within blade rows), it follows that

$$dv/v + dr/r = 0 \qquad \text{or} \qquad rv = \text{const} \qquad (9.44)$$

It can hence be seen that a constant-work machine with a constant axial velocity throughout not only has the jump in angular momentum across the blade rows Δrv equal to a constant, but also must have rv itself a constant in all regimes. Note that the "constant" to which rv is equal changes with passage through each blade row. A machine utilizing blade rows that induce this type of flowfield is termed a "free vortex" machine and is said to be in radial equilibrium throughout. An example free vortex calculation is discussed in the next section and flows of a more general form are considered in Sec. 9.4 and Chap. 10.

Example—Free Vortex Calculation

Several design variables will be prescribed and expressions for hub and tip values of the design limit parameters D_{rotor} and D_{stator}, as well as for $°R$ are to be obtained. In generating these expressions, all of the velocity and relative velocity components necessary to determine the flow vector diagrams will be obtained so that the velocity triangles can be constructed if desired.

Consider a repeating stage and assume the following variables to be prescribed:

$$\omega r_h/w_0, \qquad \Delta p_t/\rho_0 w_0^2, \qquad r_t/r_h, \qquad °R_{r_m}, \qquad \sigma_{r_m} \qquad (9.45)$$

Here $°R_{r_m}$ and σ_{r_m} refer to the degree of reaction and solidity at the "mass average radius" r_m. The mass average radius is that radius which has half the mass flowing within it and the hub and half flowing within it and the tip. In this case of uniform axial velocity,

$$r_m = \sqrt{(r_t^2 + r_h^2)/2} \qquad (9.46)$$

Degree of reaction. Equation (9.22) is valid for this case, so that with Eq. (9.44)

$$°R = 1 - \text{const}/r^2$$

where

$$\text{Const} = r_m^2(1 - °R_{r_m}) = r_h^2\left[\frac{(r_t/r_h)^2 + 1}{2}\right](1 - °R_{r_m})$$

thus

$$^{\circ}R = 1 - \frac{1}{(r/r_h)^2}\left[\frac{(r_t/r_h)^2 + 1}{2}\right](1 - ^{\circ}R_{r_m}) \qquad (9.47)$$

Denoting the value of $^{\circ}R$ at the hub by $^{\circ}R_h$, note also

$$^{\circ}R_h = 1 - \frac{(v_2 + v_1)_h}{2\omega r_h}$$

hence

$$\frac{v_{2_h}}{w_0} + \frac{v_{1_h}}{w_0} = 2\frac{\omega r_h}{w_0}(1 - ^{\circ}R_h) \qquad (9.48)$$

Tangential velocities. An expression for the difference in angular velocities follows from Eq. (9.12) to give

$$\frac{v_{2_h}}{w_0} - \frac{v_{1_h}}{w_0} = \frac{1}{\omega r_h/w_0}\frac{\Delta p_t}{\rho_0 w_0^2} \qquad (9.49)$$

and from Eqs. (9.48) and (9.49) there follows

$$\frac{v_{2_h}}{w_0} = \frac{1}{2}\left[2\frac{\omega r_h}{w_0}(1 - ^{\circ}R_h) + \frac{1}{\omega r_h/w_0}\frac{\Delta p_t}{\rho_0 w_0^2}\right] \qquad (9.50)$$

$$\frac{v_{1_h}}{w_0} = \frac{1}{2}\left[2\frac{\omega r_h}{w_0}(1 - ^{\circ}R_h) - \frac{1}{\omega r_h/w_0}\frac{\Delta p_t}{\rho_0 w_0^2}\right] \qquad (9.51)$$

Equation (9.44) then gives

$$\frac{v_2}{w_0} = \frac{v_{2_h}}{w_0}\frac{r_h}{r} \qquad \text{and} \qquad \frac{v_1}{w_0} = \frac{v_{1_h}}{w_0}\frac{r_h}{r} \qquad (9.52)$$

Diffusion factors. In order to obtain values for the diffusion factors, some assumption must be made with regard to the behavior of the solidity. It is clear that the spacing s will be proportional to the radius. The chord length can be tapered in a variety of ways, but for illustrative purposes it is sufficient to consider the chord length constant. In such a case

$$\frac{\Delta v}{2\sigma V_i} = \frac{\left(\dfrac{v_{2_h}}{w_0} - \dfrac{v_{1_h}}{w_0}\right)\dfrac{r_h}{r}}{2\sigma_{r_m}\dfrac{r_m}{r}\dfrac{V_i}{w_0}} = \frac{\left(\dfrac{v_{2_h}}{w_0} - \dfrac{v_{1_h}}{w_0}\right)}{2\sigma_{r_m}\dfrac{r_m}{r_h}\dfrac{V_i}{w_0}} \qquad \text{(constant chord length)} \quad (9.53)$$

and with Eq. (9.23)

$$D = \left(1 - \frac{V_e}{V_i}\right) + \frac{(v_{2_h}/w_0) - (v_{1_h}/w_0)}{2\sigma_{r_m}(r_m/r_h)(V_i/w_0)} \qquad \text{(constant chord length)} \quad (9.54)$$

The relative velocities for use in this expression follow by reference to Fig. 9.10 and may be written

$$V_{i \text{ rotor}} = \left[(\omega r - v_1)^2 + w_0^2\right]^{\frac{1}{2}} = w_0\left[\left(\frac{\omega r_h}{w_0}\frac{r}{r_h} - \frac{v_{1_h}}{w_0}\frac{r_h}{r}\right)^2 + 1\right]^{\frac{1}{2}}$$

$$V_{e \text{ rotor}} = \left[(\omega r - v_2)^2 + w_0^2\right]^{\frac{1}{2}} = w_0\left[\left(\frac{\omega r_h}{w_0}\frac{r}{r_h} - \frac{v_{2_h}}{w_0}\frac{r_h}{r}\right)^2 + 1\right]^{\frac{1}{2}}$$

$$V_{i \text{ stator}} = \left[w_0^2 + v_2^2\right]^{\frac{1}{2}} = w_0\left[1 + \left(\frac{v_{2_h}}{w_0}\frac{r_h}{r}\right)^2\right]^{\frac{1}{2}}$$

$$V_{e \text{ stator}} = \left[w_0^2 + v_3^2\right]^{\frac{1}{2}} = w_0\left[1 + \left(\frac{v_{1_h}}{w_0}\frac{r_h}{r}\right)^2\right]^{\frac{1}{2}} \qquad (9.55)$$

Discussion. Equations (9.46–9.55) give the desired design limit parameters in terms of the design input variables listed in Eq. (9.45). As an example consider the values

$$\frac{\Delta p_t}{\rho_0 w_0^2} = 0.9, \qquad \frac{r_t}{r_h} = \sqrt{7}, \qquad {}^\circ R_{r_m} = 0.5, \qquad \text{and} \qquad \sigma_{r_m} = 1$$

Consider two values of the blade speed, namely $\omega r_h/w_0 = 0.5$ and 0.7. Direct calculation leads to the results shown in Table 9.1.

Table 9.1   Example Results—Free Vortex Design

Blade speed	$D_{rotor \atop hub}$	$D_{rotor \atop tip}$	$D_{stator \atop hub}$	$D_{stator \atop tip}$
$\omega r_h/w_0 = 0.5$	-0.179	0.558	0.741	0.552
$\omega r_h/w_0 = 0.7$	-0.350	0.381	0.590	0.431
	$^\circ R_t = 5/7, \,^\circ R_h = -1$ in both cases			

Several important trends are evident in this example calculation. Thus, for example, it is apparent that the large value for the diffusion factor found at the hub of the stator for the low blade speed case indicates that such a blade row is too heavily loaded. The designer of such a row would have several options open to him. Perhaps the most obvious change should be to increase the degree of reaction so that the rotor tip will become as highly loaded as the stator hub. A rather peculiar result appears to occur when the degree of reaction $^\circ R_{r_m}$ is raised to 0.7. As expected the stator hub diffusion factor is reduced (to 0.670), but the rotor tip diffusion factor is also reduced (to 0.540!). The explanation for this unexpected behavior lies in the fact that the introduction of a degree of reaction $^\circ R_{r_m}$ (for this geometry) in excess of 0.55 leads to values of v_1 that are negative. As a result, large relative velocities with the rotor occur, so that in spite of the increased static pressure rise, the diffusion factor decreases. This result is not without physical credence, because the larger relative dynamic pressure would help to transfer momentum to the fluid in the boundary layers and hence reduce the tendency to separate. The use of such negative swirl in actual aircraft engines would, however, introduce possible shock losses because of the large relative Mach numbers. It might be noted parenthetically, however, that the use of such negative swirl could hold promise in helium compressors (for use in gas-cooled nuclear reactors) where the Mach numbers are extremely low.

It is obvious that if the designer could utilize materials that allowed a higher blade speed, he could decrease the diffusion factor, as indicated by the result for the higher blade speed case. If this option was not available to him he might choose to reduce the tip-to-hub ratio, thereby preventing the large negative degree of reaction near the hub. This option carries with it the undesirable side effect of reducing the mass flow per cross-sectional area capability of the compressor. Another option available would be to decrease the stagnation pressure rise per stage, but this of course would lead to the requirement of more stages for a given compressor pressure rise.

It is apparent that the free vortex design carries with it some unpleasant design restrictions. Thus, the related rapid change with the radius of the degree of reaction causes the blades to be loaded so that the stator hub requires very high diffusion, as does the rotor tip. Note that the very large turning required in the rotor at the hub leads to such an acceleration that the static pressure actually decreases across the rotor at the hub for this example (note $°R_h = -1$). This means that the stator must undergo an extreme pressure rise in order to return the fluid velocities to those entering the stage (as required for a repeating stage). A design of this sort that places the design limit so strongly at the limits of the blades is quite undesirable because, in effect, the remaining portions of the blades are relatively lightly loaded even when the stator hub and rotor tip are loaded to their limit.

In the following sections alternative flow swirl distributions are considered that reduce the disparity in loading along the blade length, but it should be noted that the tendency for the rotor tip and stator hub to be diffusion limited remains to some extent in all designs. As a result, the allowable stage loading is a function of the tip-to-hub ratio, with the smaller tip-to-hub ratio allowing the attainment of larger stage pressure ratios. A designer, then, can be faced with the choice of providing a large cross-sectional area compressor of fewer stages vs a small cross-sectional compressor of many stages. The appropriate choice often depends on the selected airplane mission.

9.4 Radial Equilibrium Flows

In the preceding section, it was seen that the very simple free vortex theory led to large variations with radius of the degree of reaction. As a result, the free vortex distribution led to blades with excessive diffusion factors at the rotor tip and at the stator hub. Other swirl distributions must hence be investigated and, as a result, the effects of radial flows must be considered.

A very simple limiting case of flows in which radial flows exist is obtained by considering conditions very far from the blade rows. Again assume perfect incompressible flow and consider conditions in a parallel walled annulus, as shown in Fig. 9.14.

The virtue of considering conditions far from the blade rows is that any radial flows will have ceased, because the flow will have become parallel to the containing annulus. In this way ordinary differential equations for the flow properties are obtained. At this stage it is not possible to estimate

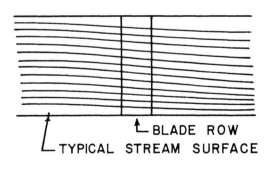

Fig. 9.14 Parallel walled annulus.

how long it takes the flow to approach this condition of "radial equilibrium," but good estimates do follow from the results of the through flow theory, Chap. 10.

The Stream Function

The stream function ψ is introduced here and, for future convenience, the possibility of axial variations in fluid properties is included. Still, consider incompressible flow only, so that the continuity equation may be written

$$\nabla \cdot \mathbf{u} = 0 \qquad (9.56)$$

In cylindrical coordinates this becomes

$$\frac{1}{r}\frac{\partial ru}{\partial r} + \frac{\partial w}{\partial z} = 0 \qquad (9.57)$$

This latter equation is identically satisfied if the stream function ψ is introduced, defined by

$$w = -\frac{1}{r}\frac{\partial \psi}{\partial r} \qquad u = \frac{1}{r}\frac{\partial \psi}{\partial z} \qquad (9.58)$$

An equation for ψ follows by considering the normal derivative of the momentum equation; but before developing such an equation, first consider a physical interpretation of the quantity ψ.

Figure 9.15 shows an annulus bounded by two "meridional surfaces." A meridional surface is defined as a surface through which no fluid passes. Calculate the mass flow rate convected between the two meridional surfaces in either of the alternative ways,

$$\dot{m} = \rho \int_{r_1}^{r_2} w(2\pi r\,dr) = -\rho \int_{\psi_1}^{\psi_2} 2\pi\,d\psi = -2\pi\rho(\psi_2 - \psi_1)$$

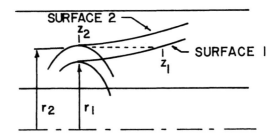

Fig. 9.15   Flow in an annulus.

or

$$\dot{m} = -\rho \int_{z_1}^{z_2} u(2\pi r \, dz) = -2\pi\rho(\psi_2 - \psi_1) \qquad (9.59)$$

Thus, the mass flow rate between the surfaces is proportional to the difference in stream functions of the surfaces. It follows then that if $\psi_2 = \psi_1$, there is no (net) flow through the surface, so that a constant ψ denotes a "stream surface." If the equations of the stream surfaces are known, the velocity components follow from Eq. (9.58).

The Equations for Radial Equilibrium Flows

It is apparent from the Euler equation (9.5) that the stagnation enthalpy does not vary along streamlines that are external to the blade rows. Similarly, it is apparent from Eq. (9.3) that the angular momentum also does not vary along streamlines that are external to blade rows. Because of this, it is convenient (particularly in throughflow theory, Chap. 10) to consider the stagnation enthalpy and angular momentum to be prescribed functions of the stream function ψ. The prescribed function will, of course, be determined by the history of blade loading that the stream tube has encountered.

Note from Eq. (9.58) that for the case of no axial variations,

$$w = -\frac{1}{r}\frac{d\psi}{dr} \qquad \text{or} \qquad \frac{d(\ )}{d\psi} = -\frac{1}{wr}\frac{d(\ )}{dr} \qquad (9.60)$$

This transformation is of particular utility when variations in the stagnation enthalpy and angular momentum are considered because, as previously described, these quantities will be given as functions of the stream function.

For this case of perfect flow, there is no change in entropy throughout, so the Gibbs equation (2.12) gives

$$dh = (1/\rho)\,dp \qquad (9.61)$$

The differential change in stagnation enthalpy may be written

$$dh_t = dh + w\,dw + v\,dv \qquad (9.62)$$

so

$$(1/\rho)\,dp = dh_t - w\,dw - v\,dv \qquad (9.63)$$

Because the flow is in radial equilibrium, the radial momentum equation remains as in Eq. (9.42), so that

$$\frac{1}{\rho}\frac{dp}{dr} = \frac{v^2}{r} = \frac{dh_t}{dr} - w\frac{dw}{dr} - v\frac{dv}{dr}$$

or

$$-\frac{1}{r}\frac{dw}{dr} = -\frac{1}{rw}\frac{dh_t}{dr} - \frac{v}{r}\left(\frac{-1}{rw}\frac{drv}{dr}\right) \qquad (9.64)$$

Utilizing Eq. (9.60), then obtain

$$\frac{1}{r}\frac{d}{dr}\left(\frac{1}{r}\frac{d\psi}{dr}\right) = \frac{dh_t}{d\psi} - \frac{v}{r}\frac{drv}{d\psi} \qquad (9.65)$$

This is a second-order differential equation for the stream function ψ in terms of the prescribed stagnation enthalpy and angular momentum. It is, in general, nonlinear. There are several linear solutions of the equations corresponding to blade loadings of design interest, however, and such special cases will be considered in the next section. It is most convenient to consider a nondimensional form of Eq. (9.65), so introduce (see Fig. 9.16)

$$W = \frac{w}{w_0}, \qquad V = \frac{v}{w_0}, \qquad \Psi = \frac{\psi}{w_0 r_h^2},$$

$$H = \frac{h_t}{w_0^2}, \qquad y = \frac{r}{r_h}, \qquad R = \frac{r_t}{r_h} \qquad (9.66)$$

Routine substitution of the variables of Eq. (9.66) into Eq. (9.65) leads to

$$\frac{1}{y}\frac{d}{dy}\left(\frac{1}{y}\frac{d\Psi}{dy}\right) = \frac{dH}{d\Psi} - \frac{V}{y}\frac{dyV}{d\Psi} \qquad (9.67)$$

The related boundary conditions are obtained by first noting that the stream function may be prescribed within a constant. Thus, for future

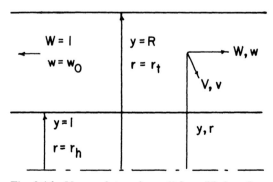

Fig. 9.16 Nomenclature for radial equilibrium flow.

convenience the stream function is assigned on the hub Ψ_h,

$$\Psi_h = -\tfrac{1}{2} \qquad \text{on} \qquad y = 1 \ (r = r_h) \qquad (9.68)$$

Note also

$$W = -\frac{1}{y}\frac{d\Psi}{dy} \qquad (9.69)$$

so obtain (noting that $W = 1$ far upstream)

$$\Psi_t = -\int_1^R y W\, dy - \frac{1}{2} = -\frac{R^2}{2} \qquad \text{on} \qquad y = R \ (r = r_t) \quad (9.70)$$

Equations (9.67–9.70) prescribe the mathematical problem.

Solution of the Radial Equilibrium Equations

Equation (9.67) may be integrated for any arbitrary prescription of H and yV (as a function of Ψ), but particularly simple forms arise with forms for which H and $(yV)^2$ are quadratic in Ψ. Thus, for such flows, Eq. (9.67) reduces to a linear equation in Ψ and a solution can be obtained in terms of known functions. Thus, consider H and $(yV)^2$ to be given by

$$H = A\Psi + H_0 \qquad (9.71)$$

$$(yV)^2 = (-\alpha\Psi + \beta)^2 - \delta\Psi \qquad (9.72)$$

In these expressions, A, H_0, α, β, and δ are constants that may be selected at the designer's discretion. Note

$$\frac{dH}{d\Psi} = A$$

$$V\frac{dyV}{d\Psi} = \frac{1}{2y}\frac{d(yV)^2}{d\Psi} = \frac{1}{y}\left[\alpha^2\Psi - \left(\alpha\beta + \frac{\delta}{2}\right)\right]$$

$$\frac{d}{dy}\left(\frac{1}{y}\frac{d\Psi}{dy}\right) = \frac{d}{dy}\left[\frac{1}{y}\frac{d}{dy}\left(y\frac{\Psi}{y}\right)\right]$$

$$= \frac{d^2(\Psi/y)}{dy^2} + \frac{1}{y}\frac{d(\Psi/y)}{dy} - \frac{1}{y^2}\frac{\Psi}{y}$$

With these expressions, Eq. (9.67) becomes

$$\frac{d^2(\Psi/y)}{dy^2} + \frac{1}{y}\frac{d(\Psi/y)}{dy} + \left(\alpha^2 - \frac{1}{y^2}\right)\left(\frac{\Psi}{y}\right) = yA + \frac{1}{y}\left(\alpha\beta + \frac{\delta}{2}\right) \quad (9.73)$$

The inhomogeneous portion of this equation may be integrated immediately to give

$$\left(\frac{\Psi}{y}\right)_{\text{inhom.}} = \frac{A}{\alpha^2}y + \frac{1}{y}\left(\frac{\beta}{\alpha} + \frac{\delta}{2\alpha^2}\right) \quad (9.74)$$

The homogeneous portion of the solution can be recognized as consisting of a linear combination of Bessel functions of the first and second kinds and of order one. Thus,

$$\left(\frac{\Psi}{y}\right)_{\text{hom.}} = C_1 J_1(\alpha y) + C_2 Y_1(\alpha y) \quad (9.75)$$

and hence

$$\Psi = C_1 y J_1(\alpha y) + C_2 y Y_1(\alpha y) + \frac{A}{\alpha^2}y^2 + \frac{\beta}{\alpha} + \frac{\delta}{2\alpha^2} \quad (9.76)$$

The constants C_1 and C_2 are now determined by applying the boundary conditions [Eqs. (9.68) and (9.70)]. After straightforward but tedious manipulation, there results

$$\Psi = -\left(\frac{A}{\alpha^2} + \frac{\beta}{\alpha} + \frac{\delta}{2\alpha^2} + \frac{1}{2}\right)\frac{yU_1(\alpha y)}{U_1(\alpha)}$$

$$-\left(\frac{A}{\alpha^2}R^2 + \frac{\beta}{\alpha} + \frac{\delta}{2\alpha^2} + \frac{R^2}{2}\right)\frac{yW_1(\alpha y)}{RU_1(\alpha)} + \frac{A}{\alpha^2}y^2 + \frac{\beta}{\alpha} + \frac{\delta}{2\alpha^2} \quad (9.77)$$

where:

$$U_i(\alpha y) \equiv J_i(\alpha y)Y_1(\alpha R) - J_1(\alpha R)Y_i(\alpha y) \quad (9.78)$$

$$W_i(\alpha y) \equiv J_1(\alpha)Y_i(\alpha y) - Y_1(\alpha)J_i(\alpha y) \quad (9.79)$$

where i is an integer.

It is convenient to note that

$$U_1(\alpha R) = 0 = W_1(\alpha) \qquad \text{and} \qquad U_1(\alpha) = W_1(\alpha R) \quad (9.80)$$

With these relationships it can be easily checked that Eq. (9.77) satisfies the boundary conditions, $\Psi = -\frac{1}{2}$ on $y = 1$ and $\Psi = -R^2/2$ on $y = R$.

When the parameter α is equal to zero Eq. (9.77) becomes indeterminate. For this case, rather than taking the mathematical limit of the expression, it is simpler to return to Eq. (9.67), which for the case $\alpha = 0$ becomes

$$\frac{d}{dy}\left(\frac{1}{y}\frac{d\Psi}{dy}\right) = yA + \frac{\delta}{2y} \tag{9.81}$$

Successive integration gives

$$\psi = C_1\frac{y^2}{2} + C_2 + \frac{y^4}{8}A + \frac{\delta}{8}\left(2y^2\ell n y - y^2\right) \tag{9.82}$$

The constants C_1 and C_2 are again determined through application of the boundary conditions [Eqs. (9.68) and (9.70)]. The solution may be written in the form

$$\Psi = -\frac{y^2}{2} - \frac{A}{8}\left(y^2 - 1\right)\left(R^2 - y^2\right) + \frac{\delta}{4(R^2 - 1)}$$

$$\times\left[\left(R^2 - 1\right)y^2\ell n y - \left(y^2 - 1\right)R^2\ell n R\right] \qquad (\alpha = 0) \tag{9.83}$$

Interpretation of the Parameters

Equations (9.77) and (9.83) provide the solution to the equilibrium flow for a rather general swirl and enthalpy distribution leading to linear solutions of the governing equation. It' is of interest to interpret each of the parameters α, β, δ, and A individually, however. To do so, consider the flows prescribed for the cases where all parameters but one are zero.

Free vortex flow (β). When A, α, and δ are all zero, Eq. (9.72) shows that the angular momentum is constant with y, $(yV = \beta)$, which is a free vortex distribution. Equation (9.83) then reduces to

$$\Psi = -y^2/2$$

hence

$$W = -\frac{1}{y}\frac{d\Psi}{dy} = 1 \qquad \text{(free vortex)} \tag{9.84}$$

Thus, for a free vortex distribution, the axial velocity remains unperturbed, just as previously established in Sec. 9.3.

Solid-body-like rotation (α). When A, β, and δ are all zero, Eq. (9.72) reduces to

$$yV = -\alpha\Psi \qquad \text{(solid body like)} \tag{9.85}$$

It is evident from Eq. (9.67) that a perturbation in axial velocity will be present for this case. In the case where very little swirl is introduced (α small), the variation of axial velocity with radius will itself be small. [This may be verified by calculation from Eq. (9.67).] In such a case

$$\Psi = -\int_1^y yW\,\mathrm{d}y - \frac{1}{2} \approx -\frac{y^2}{2} \qquad \text{(small swirl)} \qquad (9.86)$$

then

$$V \approx \alpha y/2 \qquad \text{(small swirl)} \qquad (9.87)$$

Thus, for small swirl this swirl distribution approximates that corresponding to a "solid-body" swirl. Such a distribution is sometimes termed a forced vortex.

Approximately constant swirl (δ). When A, α, and β are zero, Eq. (9.72) reduces to

$$yV = (-\delta\Psi)^{\frac{1}{2}} \qquad \text{(approximately constant swirl)} \qquad (9.88)$$

It can again be argued that for small swirl the perturbation in axial velocity may be expected to be small [see Eq. (9.83)], so that Ψ will be given approximately by $\Psi = -y^2/2$. Hence

$$V \approx \sqrt{\delta/2} \qquad (9.89)$$

Thus for small swirl, this swirl distribution approximates a constant-swirl velocity.

Variable stagnation enthalpy (A). When α, β, and δ are zero, there is no swirl in the flow and because the flow is in radial equilibrium, no static pressure variation with y will be present. From the Gibbs equation, $\mathrm{d}h = T\mathrm{d}s + 1/\rho\,\mathrm{d}p = 0$, and hence

$$\mathrm{d}H = \mathrm{d}(h/w_0^2) + W\mathrm{d}W = W\mathrm{d}W \qquad (9.90)$$

Thus, with $\mathrm{d}H = A\,\mathrm{d}\Psi = -AyW\mathrm{d}y$, it follows that

$$\frac{\mathrm{d}W}{\mathrm{d}y} = -Ay \qquad (9.91)$$

This result is consistent with Eq. (9.83), which simply emphasizes that for this case any change in stagnation enthalpy must be supplied by a variation in the axial velocity.

The stagnation enthalpy will be changed only across the rotor rows, so from the Euler momentum equation (9.5), together with Eq. (9.66) and the definition $\Omega = \omega r_h / w_0$,

$$H_2 - H_1 = \Omega\left[(yV)_2 - (yV)_1\right] \tag{9.92}$$

Thus, in terms of the parameters introduced in Eqs. (9.71) and (9.72),

$$H_2 - H_1 = \Omega\left\{\left[-\alpha_2\Psi + \beta_2\right)^2 - \delta_2\Psi\right]^{\frac{1}{2}}\right.$$
$$\left. -\left[(-\alpha_1\Psi + \beta_1)^2 - \delta_1\Psi\right]^{\frac{1}{2}}\right\} \tag{9.93}$$

It thus follows that if all terms in the expression for angular momentum are to be included when flow across a rotor is considered, the desired linear form for the stagnation enthalpy cannot be utilized. If the "approximately constant swirl" term is excluded, however, it follows that

$$H_2 - H_1 = H_{0_2} - H_{0_1} + (A_2 - A_1)\Psi = \Omega\left[-\alpha_2\Psi + \beta_2 - (-\alpha_1\Psi + \beta_1)\right]$$

$$(\delta = 0) \tag{9.94}$$

and hence

$$H_{0_2} = H_{0_1} + (\beta_2 - \beta_1)\Omega \tag{9.95}$$

and

$$A_2 = A_1 - (\alpha_2 - \alpha_1)\Omega \tag{9.96}$$

In a compressor, little, if any, variation of stagnation enthalpy with radius is desired, so it is usually appropriate to exclude variation of the parameter α across a rotor row. Note, however, that the solid-body-like component of swirl may be introduced by the stator (or inlet guide vane row), which hence gives another parameter to utilize for design purposes. In the next section, a detailed example solution is considered.

Example—Radial Equilibrium Calculation

Consider an example calculation for a rotor-stator pair. The pair will be a repeating stage, with constant-work interaction with radius ($A = 0$) and with no "approximately constant swirl" term ($\delta = 0$). For this special case, the solution of Eq. (9.77) reduces to

$$\Psi = -\left(\beta + \frac{\alpha}{2}\right)\frac{yU_1(\alpha y)}{\alpha U_1(\alpha)} - \left(\frac{\beta}{R} + \frac{\alpha R}{2}\right)\frac{yW_1(\alpha y)}{\alpha U_1(\alpha)} + \frac{\beta}{\alpha} \tag{9.97}$$

The derivatives of Bessel functions may be obtained from any standard mathematics book (for example, Ref. 5). Note here the general relationship

$$\frac{d}{dx}\left\{x\left[C_1 J_1(\alpha x) + C_2 Y_1(\alpha x)\right]\right\} = \alpha x\left[C_1 J_0(\alpha x) + C_2 Y_0(\alpha x)\right] \quad (9.98)$$

Hence,

$$\frac{d}{dy}\left[y U_1(\alpha y)\right] = \alpha y U_0(\alpha y) \qquad \text{and} \qquad \frac{d}{dy}\left[y W_1(\alpha y)\right] = \alpha y W_0(\alpha y)$$

$$(9.99)$$

Thus

$$-\frac{1}{y}\frac{d\Psi}{dy} = W = \left(\beta + \frac{\alpha}{2}\right)\frac{U_0(\alpha y)}{U_1(\alpha)} + \left(\frac{\beta}{R} + \frac{\alpha R}{2}\right)\frac{W_0(\alpha y)}{U_1(\alpha)} \quad (9.100)$$

Now consider $\Delta H = H_2 - H_1$, Ω, β_1, R, and σ_{r_h} to be prescribed and arrange the equations in order to calculate the various velocity components, as well as the diffusion factor at stator hub and rotor tip and the degree of reaction at hub and tip.

Summary—Radial Equilibrium Flows

Inputs: $\qquad\qquad \Delta H, \Omega, \alpha, \beta_1, R, \sigma_{r_h}, y$

Outputs: $\quad V_1(y), V_2(y), W_1(y), W_2(y), D_{\substack{\text{stator}\\\text{hub}}}, D_{\substack{\text{rotor}\\\text{tip}}}, {}^\circ R_h, {}^\circ R_t$

Equations:

$$\beta_2 = \beta_1 + \Delta H/\Omega \qquad\qquad (9.101)$$

$$\Psi = -\left(\beta + \frac{\alpha}{2}\right)\frac{y U_1(\alpha y)}{\alpha U_1(\alpha)} - \left(\frac{\beta}{R} + \frac{\alpha R}{2}\right)\frac{y W_1(\alpha y)}{\alpha U_1(\alpha)} + \frac{\beta}{\alpha} \quad (9.102)$$

$$W = \left(\beta + \frac{\alpha}{2}\right)\frac{U_0(\alpha y)}{U_1(\alpha)} + \left(\frac{\beta}{R} + \frac{\alpha R}{2}\right)\frac{W_0(\alpha y)}{U_1(\alpha)} \qquad (9.103)$$

$$V = (1/y)(-\alpha\Psi + \beta) \qquad\qquad (9.104)$$

[Note that Eqs. (9.102–9.104) may be used for any value of y and stations 1 and 2 (with $\beta = \beta_1$ or β_2)]. It is convenient also to employ the relationships of Eq. (9.80) as well as the relationships[5]

$$U_0(\alpha R) = -\frac{2}{\pi \alpha R}, \qquad W_0(\alpha) = \frac{2}{\pi \alpha} \tag{9.105}$$

$$V_{i \text{ stator} \atop \text{hub}} = \left(W_{h_2}^2 + V_{h_2}^2 \right)^{\frac{1}{2}} \tag{9.106}$$

$$V_{e \text{ stator} \atop \text{hub}} = \left(W_{h_1}^2 + V_{h_1}^2 \right)^{\frac{1}{2}} \tag{9.107}$$

$$V_{i \text{ rotor} \atop \text{tip}} = \left[W_{t_1}^2 + \left(\Omega R - V_{t_1} \right)^2 \right]^{\frac{1}{2}} \tag{9.108}$$

$$V_{e \text{ rotor} \atop \text{tip}} = \left[W_{t_2}^2 + \left(\Omega R - V_{t_2} \right)^2 \right]^{\frac{1}{2}} \tag{9.109}$$

$$D = \left(1 - \frac{V_e}{V_i} \right) + \frac{\Delta H / \Omega}{2 \sigma_{r_h} V_i} \tag{9.110}$$

[Note that Eq. (9.110) may be used for both stator hub and rotor tip. The appropriate values for V_e and V_i follow from Eqs. (9.106–9.109). Note, also, it has again been assumed that the solidity is inversely proportional to the radius.]

$$^\circ R_h = 1 + \frac{1}{2\Delta H} \left(V_{e \text{ stator} \atop \text{hub}}^2 - V_{i \text{ stator} \atop \text{hub}}^2 \right) \tag{9.111}$$

$$^\circ R_t = 1 + \frac{1}{2\Delta H} \left(W_{t_1}^2 + V_{t_1}^2 - W_{t_2}^2 - V_{t_2}^2 \right) \tag{9.112}$$

As an example calculation, the case $\Delta H = 0.9$, $\Omega = 0.5$, $\alpha = 0.15$, $\beta_1 = -0.4$, $R = \sqrt{7}$, $\sigma_{r_h} = 2$ was considered. This case is comparable to the first case considered in Sec. 9.3. Thus, the geometry and "stage loading" considered are the same and the tip swirl velocity at the inlet to the rotor is very nearly the same, 0.0472 here compared to 0.0378. This configuration then does not acquire an advantage because of increased relative velocity at the rotor tip (where Mach losses could be important). Straightforward calcula-

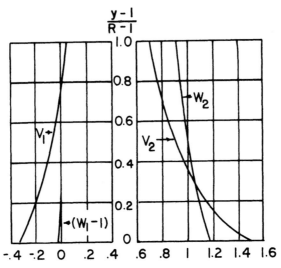

Fig. 9.17   Radial equilibrium velocity profiles.

tion gives:

$$D_{\substack{\text{stator} \\ \text{hub}}} = 0.689\,(0.741) \qquad {}^\circ R_h = -0.352\,(-1)$$

$$D_{\substack{\text{rotor} \\ \text{tip}}} = 0.606\,(0.558) \qquad {}^\circ R_t = 0.803\,(0.714)$$

The comparative values for the free vortex case are shown in parentheses. It can be seen that the introduction of a rather small amount of solid-body-like swirl has helped to reduce the stator hub diffusion factor substantially (even though the value is still rather large). Figure 9.17 shows the related equilibrium velocity profiles.

These results give a relatively quick estimate of the flow profiles to be expected between the blade rows and may be utilized for simple approximate design calculations. The behavior of the throughflow flowfield (which includes the axial variation of the fluid properties) will be investigated in Chap. 10.

9.5   The Effects of Compressibility

In modern axial flow turbomachinery, the pressure ratios found across typical stages, particularly across turbine stages, are so large that the effects of variation in fluid density cannot be ignored. In order to consider such compressibility effects, again consider the pseudo-two-dimensional flow of a perfect gas across a blade row. The Euler equation (9.5) relates the change in the stagnation enthalpy across a rotor row to the blade angular velocity and the change in the fluid angular momentum. This change in stagnation enthalpy is, of course, the total change in stagnation enthalpy for the stage, so from Eqs. (6.58), (6.80), and (9.5), with the stage efficiency written η_{c_j} for the compressor and η_{t_j} for the turbine,

$$\frac{p_{t_2}}{p_{t_1}} = \left\{1 + \frac{\eta_{c_j}\omega}{h_{t_1}}\left[(rv)_2 - (rv)_1\right]\right\}^{\gamma_c/(\gamma_c - 1)} \qquad \text{(compressor)} \quad (9.113)$$

$$\frac{p_{t_2}}{p_{t_1}} = \left\{1 - \frac{\omega}{\eta_{t_j}h_{t_1}}\left[(rv)_1 - (rv)_2\right]\right\}^{\gamma_t/(\gamma_t - 1)} \qquad \text{(turbine)} \quad (9.114)$$

In these expressions, in each case the conditions at 1 are those at the entry to the rotor and conditions at 2 are those at the exit from the rotor. Equation (9.113) is different from the previously obtained Eq. (9.7) only in that the effect of the stage efficiency has been included. Again introducing the upstream reference quantities denoted by a subscript 0, in a similar manner to Sec. 9.2, leads to

$$\frac{p_{t_2}}{p_{t_1}} = \left\{1 + \frac{(\gamma_c - 1)M_0^2\eta_{c_j}}{1 + \frac{\gamma - 1}{2}M_0^2}\frac{\omega}{w_0^2}\left[(rv)_2 - (rv)_1\right]\right\}^{\gamma_c/(\gamma_c - 1)} \qquad \text{(compressor)}$$

$$(9.115)$$

$$\frac{p_{t_2}}{p_{t_1}} = \left\{1 - \frac{(\gamma_t - 1)M_0^2}{\eta_{t_j}\left(1 + \frac{\gamma_t - 1}{2}M_0^2\right)}\frac{\omega}{w_0^2}\left[(rv)_1 - (rv)_2\right]\right\}^{\gamma_t/(\gamma_t - 1)} \qquad \text{(turbine)}$$

$$(9.116)$$

Simple approximate forms follow from these equations for the case of small turning or small Mach number by simply expanding the bracketed terms to only the first term in their binomial expansions.

Turbine Aerodynamics

The aerodynamic and engineering limitations of turbines are of a substantially different nature than those of compressors. First, the extremely favorable pressure gradients allow very high blade loadings before the local adverse pressure gradients on the blades approach values leading to boundary-layer separation. The expansion ratio can, in fact, be limited by choking of the downstream flow. The materials problem will clearly be aggravated as the blades are submerged in a high-temperature corrosive environment. Modern blades are cooled, which in itself adds greatly to the complexity, from both the gasdynamic and stress points of view.

In turbines it is possible to be confronted with a design tradeoff between mass flow capability and power output per stage. The mass flow per cross-sectional area increases as the approach axial Mach number nears unity, and the work interaction per mass increases as the nozzle outlet swirl velocity increases. It will be shown shortly that the maximum work interaction for a single stage occurs when the entire static pressure drop occurs across the stator, the function of the rotor then being to remove the kinetic energy identified with the swirl component of velocity with no further static pressure drop. There tend to be two restrictions on obtaining high swirl velocities. Thus, if the flow Mach number becomes too large severe shock losses can occur, and it also happens that a maximum possible swirl occurs because the flow will choke. This latter restriction is a function of the approach Mach number and conflicts with the desired high mass flow capability.

To analyze the effect of this compressible limitation on the maximum attainable swirl, consider the flow of a perfect fluid through a turbine nozzle row (Fig. 9.18). The approaching flow is assumed to be uniform and swirl free and the cross-sectional areas at inlet and exit, respectively, are denoted A_0 and A_1. Then have

Continuity: $$\rho_0 w_0 A_0 = \rho_1 V_1 A_1 \cos \alpha \qquad (9.117)$$

Isentropic: $$\frac{\rho_1}{\rho_0} = \left(\frac{T_1}{T_0}\right)^{1/(\gamma-1)} \qquad (9.118)$$

Enthalpy: $$T_0\left(1 + \frac{\gamma-1}{2}M_0^2\right) = T_1\left(1 + \frac{\gamma-1}{2}M_1^2\right) \qquad (9.119)$$

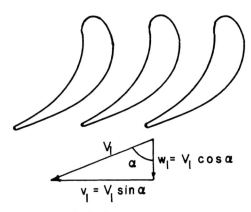

Fig. 9.18 Turbine nozzle row.

These equations may be combined to give

$$\frac{M_1}{M_0} = \frac{V_1}{w_0}\sqrt{\frac{T_0}{T_1}} = \left(\frac{T_0}{T_1}\right)^{(\gamma+1)/2(\gamma-1)}\frac{A_0}{A_1}\frac{1}{\cos\alpha}$$

hence

$$\frac{M_1}{M_0} = \left(\frac{1+\dfrac{\gamma-1}{2}M_1^2}{1+\dfrac{\gamma-1}{2}M_0^2}\right)^{(\gamma+1)/2(\gamma-1)}\frac{A_0}{A_1}\frac{1}{\cos\alpha} \qquad (9.120)$$

It is desired to find the maximum value of the swirl velocity for a given approach Mach number M_0. Thus noting

$$\left(\frac{V_1\sin\alpha}{a_0}\right)^2 = \frac{T_1}{T_0}M_1^2(1-\cos^2\alpha)$$

find

$$\left(\frac{V_1\sin\alpha}{a_0}\right)^2 = \frac{1+\dfrac{\gamma-1}{2}M_0^2}{1+\dfrac{\gamma-1}{2}M_1^2}M_1^2 - \left(\frac{1+\dfrac{\gamma-1}{2}M_1^2}{1+\dfrac{\gamma-1}{2}M_0^2}\right)^{2/(\gamma-1)}M_0^2\left(\frac{A_0}{A_1}\right)^2$$

$$(9.121)$$

The derivative of this equation with respect to M_1^2 may now be taken and equated to zero. After some manipulation, it follows that

$$M_0^2\left(\frac{A_0}{A_1}\right)^2\left(\frac{1+\dfrac{\gamma-1}{2}M_1^2}{1+\dfrac{\gamma-1}{2}M_0^2}\right)^{(\gamma+1)/(\gamma-1)} = 1 \qquad (9.122)$$

Equations (9.120) and (9.122) then give immediately that the maximum possible swirl occurs when

$$\cos\alpha = 1/M_1 \qquad \text{(max swirl)} \qquad (9.123)$$

It is clear then that the maximum swirl occurs when the downstream axial Mach number is unity, or in other words, when the flow chokes. The maximum value of the swirl velocity follows from Eqs. (9.121) and (9.122) to give

$$\left(\frac{V_1 \sin \alpha}{a_0}\right)^2_{\max} = \frac{2}{\gamma - 1}\left[1 + \frac{\gamma - 1}{2}M_0^2 - \frac{\gamma + 1}{2}\left(M_0 \frac{A_0}{A_1}\right)^{2(\gamma - 1)/(\gamma + 1)}\right]$$

(9.124)

Figure 9.19 shows the dimensionless kinetic energy of the swirl component plotted vs incoming Mach number ($A_0 = A_1$, $\gamma = 1.4$).

The related stage pressure ratio identified with a nozzle row producing this maximum swirl may be estimated by noting that the work interaction per mass for the stage will be equal to the kinetic energy identified with the swirl. (Note that the rotor simply removes this kinetic energy with no further pressure drop.) Thus, with

$$a_0^2 = \gamma R T_0 = \frac{\gamma R}{C_p}\frac{T_0}{T_{t_0}}C_p T_{t_0} = \frac{\gamma - 1}{1 + \frac{\gamma - 1}{2}M_0^2}h_{t_0}$$

(9.125)

Eq. (9.124) gives

$$\left(h_{t_0} - h_{t_2}\right)_{\max} = \frac{1}{2}(V_1 \sin \alpha)^2_{\max}$$

$$= h_{t_0}\left[1 - \frac{(\gamma + 1)\left(M_0 \frac{A_0}{A_1}\right)^{2(\gamma - 1)/(\gamma + 1)}}{2 + (\gamma - 1)M_0^2}\right]$$

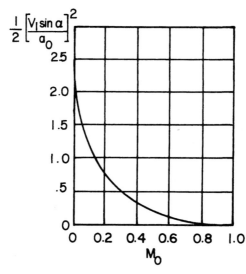

Fig. 9.19 Maximum swirl kinetic energy.

hence

$$\left(\frac{h_{t_2}}{h_{t_0}}\right)_{\max} = \frac{(\gamma+1)\left(M_0\dfrac{A_0}{A_1}\right)^{2(\gamma-1)/(\gamma+1)}}{2+(\gamma-1)M_0^2} \tag{9.126}$$

If the efficiency of the turbine stage η_{t_j} can be estimated, the related stage pressure ratio can be approximated by

$$\left(\frac{p_{t_2}}{p_{t_0}}\right)_{\max} = \left\{1 - \frac{1}{\eta_{t_j}}\left[1-\left(\frac{h_{t_2}}{h_{t_0}}\right)_{\max}\right]\right\}^{\gamma/(\gamma-1)} \tag{9.127}$$

Example results from the case for $A_0/A_1 = 1$, $\eta_{t_j} = 1$, and $\gamma = 1.3$ are shown in Fig. 9.20.

It is apparent that the power extraction from the fluid will soon drive the axial Mach number to unity unless an axial area change is incorporated; and, of course, turbines usually have increases in cross-sectional areas in the axial direction. A measure of the effectiveness of incorporating an axial area variation can be obtained by noting the increase in maximum swirl velocity attainable and the decrease in enthalpy ratio identified with an axial area change. Thus, choosing $M_0 = 0.5$ and $\gamma = 1.3$, the results of Fig. 9.21 are obtained from Eqs. (9.124) and (9.126).

The Impulse Turbine

An impulse turbine stage is defined as a turbine stage in which the entire static pressure drop occurs across the stator. Equivalently, of course, the impulse turbine stage is a stage with a degree of reaction of zero. Because no

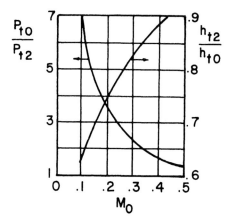

Fig. 9.20 Stage pressure ratio vs axial Mach number.

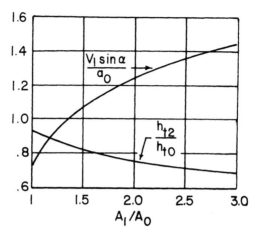

Fig. 9.21   Enthalpy ratio and maximum swirl velocity vs area ratio.

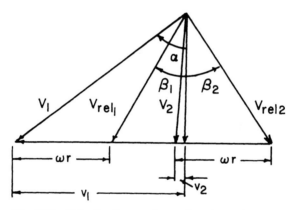

Fig. 9.22   Velocity diagram for an impulse turbine stage.

pressure drop occurs in the rotor, the relative velocity in the rotor will not change if the fluid is assumed perfect. A typical velocity diagram would hence be as indicated in Fig. 9.22.

The Euler momentum equation gives

$$C_p \left(T_{t_1} - T_{t_2} \right) = \omega r \left(v_1 - v_2 \right)$$

It is apparent from Fig. 9.22 that, for this case where the axial velocity has not changed across the rotor, $\beta_1 = -\beta_2$ and hence

$$v_1 - \omega r = \omega r - v_2$$

and

$$C_p(T_{t_1} - T_{t_2}) = 2\omega r(v_1 - \omega r) \qquad (9.128)$$

It is clear that the maximum work interaction per mass for the given pressure drop will occur when the rotor leaves no residual swirl ($v_2 = 0$). At this condition, $v_1 = 2\omega r$ and from Eq. (9.128)

$$1 - \frac{T_{t_2}}{T_{t_1}} = \frac{2(\omega r)^2}{C_p T_{t_1}} \qquad \text{(max work impulse turbine)} \qquad (9.129)$$

The Relative Stagnation Temperature

The relative stagnation temperature, or that temperature the gas would achieve if brought to rest adiabatically on the rotor, is of great importance because it determines what the heat-transfer loading to the blade will be. Defining T_{t_r} as the relative stagnation temperature as seen by the rotor, write

$$C_p T_{t_r} = C_p T_1 + \tfrac{1}{2}\left[w_1^2 + (v_1 - \omega r)^2 \right]$$

$$= C_p T_1 + \tfrac{1}{2}\left(w_1^2 + v_1^2 \right) + \tfrac{1}{2}\left[(v_1 - \omega r)^2 - v_1^2 \right]$$

hence

$$\frac{T_{t_r}}{T_{t_1}} = 1 - \frac{1}{2 C_p T_{t_1}}\left[v_1^2 - (v_1 - \omega r)^2 \right] \qquad (9.130)$$

In the maximum work impulse turbine case, $v_1 = 2\omega r$, so that

$$\frac{T_{t_r}}{T_{t_1}} = 1 - \frac{3(\omega r)^2}{2 C_p T_{t_1}} \qquad \text{(max work impulse turbine)} \qquad (9.131)$$

It can be seen that because the blades are retreating from the flow, the relative stagnation temperature is reduced. This effect can be quite significant and allows the rotors to operate at higher stress levels than might at first be expected. For example, with $T_{t_1} = 1600$ K, $\omega r = 400$ m/s, $C_p = 1250$ J/kg · K, it follows that $T_{t_r}/T_{t_1} = 0.88$. Hence, the effective stagnation temperature is reduced 192 K because of the blade movement.

The Reaction Turbine

The reaction turbine stage is defined simply as a turbine stage in which the degree of reaction is other than zero. As an illustrative example, consider a stage for which the velocity triangles are those that would give $^\circ R = 0.5$ if the flow was incompressible and perfect. It is to be noted that, when losses are present and the Mach numbers are finite, the static pressure behavior

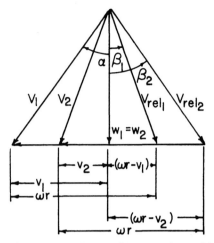

Fig. 9.23 Velocity diagram for a reaction turbine stage.

will not strictly correspond to $°R = 0.5$. As a further simplification, consider no change in axial velocity and assume a symmetric diagram as illustrated in Fig. 9.23.

Because of the assumed symmetry of the velocity diagram, it follows that $v_2 = \omega r - v_1$, so that the Euler momentum equation becomes

$$C_p\left(T_{t_1} - T_{t_2}\right) = \omega r(v_1 - v_2) = (\omega r)^2\left(2\frac{v_1}{\omega r} - 1\right)$$

Again restricting attention to the case where the maximum work interaction occurs, which as before corresponds to no residual swirl remaining in the flow, note that $v_1 = \omega r$ and hence

$$1 - \frac{T_{t_2}}{T_{t_1}} = \frac{(\omega r)^2}{C_p T_{t_1}} \qquad \text{(max work "50\%" reaction turbine)} \quad (9.132)$$

Equation (9.130) is valid generally and hence here reduces to

$$\frac{T_{t_r}}{T_{t_1}} = 1 - \frac{(\omega r)^2}{2C_p T_{t_1}} \qquad \text{(max work "50\%" reaction turbine)} \quad (9.133)$$

Comparison of the Impulse Turbine and "50%" Reaction Turbine

It is of interest to compare the behavior of the impulse turbine and 50% reaction turbine. The behavior of important parameters is compared in Table 9.2.

Table 9.2   Comparative Turbine Behavior

Parameter	Item	Impulse	50% reaction	Impulse	50% reaction
$1 - \dfrac{T_{t_2}}{T_{t_1}}$	Work interaction	$\dfrac{2(\omega r)^2}{C_p T_{t_1}}$	$\dfrac{(\omega r)^2}{C_p T_{t_1}}$	High	Low
$\dfrac{T_{t_r}}{T_{t_1}} - 1$	Increment in rotor temperature	$-\dfrac{3(\omega r)^2}{2C_p T_{t_1}}$	$-\dfrac{(\omega r)^2}{2C_p T_{t_1}}$	High	Low
	Efficiency			Low	High

It can be seen from the summary of Table 9.2 that, for a given wheel speed, the impulse turbine has a larger work interaction and experiences a lower relative stagnation temperature on the rotor than does the 50% reaction turbine. (Note that the lower relative stagnation temperature could allow operation at a slightly higher wheel speed.) These advantages for the impulse turbine do not come without penalty, however, as the stage efficiencies tend to be lower than those of the reaction turbines. This is because the Mach numbers (and hence frictional and shock losses) tend to be large in impulse turbines, and also because the rotors operate without the benefit of an ambient favorable pressure gradient.

In practice, impulse turbines are often used in very high thrust-to-weight engines where their enormous work capability is of direct benefit in reducing the required number of stages and hence weight. In some cases, the first few stages of a turbine will have impulse blading so that the number of stages requiring cooling will be reduced. Transport aircraft, which require highly efficient engines, will usually have blading with 30–50% reactions.

References

[1] Horlock, J. H. and Lakshminarayana, B., "Secondary Flows," *Annual Review of Fluid Mechanics*, Vol. 5, 1973, pp. 247–279.

[2] Hawthorne, W. R., "The Applicability of Secondary Flow Analysis to the Solution of Internal Flow Problems," *Fluid Mechanics of Internal Flow*, edited by G. Sovran, Elsevier, Amsterdam, 1967, pp. 238–269.

[3] Hawthorne, W. R., "Secondary Vorticity in Stratified Compressible Fluids in Rotating Systems," Dept. of Engineering, University of Cambridge, England, Rept. CUED/A-TURBO/TR63, 1974.

[4] Johnsen, I. A. and Bullock, R. O., eds., "Aerodynamic Design of Axial Flow Compressors," NASA SP-36, 1965.

[5] Gray, A. and Mathews, G. B., *A Treatise on Bessel Functions and Their Applications to Physics*, Dover Publications, New York, 1966.

Problems

9.1   Consider the perfect flow of an incompressible fluid through a compressor stage. The stage is of "constant-work" design (Δp_t constant with radius) and has purely axial flow at inlet and exit from the rotor-stator pair.

(a) Show that $°R = 1 - \Delta p_t / 2\rho(\omega r)^2$.

(b) For the case where $r_t/r_h = 3$, $\omega r_h/w = 1$, and the turning angle in the rotor at the hub is 30 deg, find $\Delta p_t/\rho w^2$ and $°R$ at $r = r_h$ and $r = r_t$.

9.2   Consider the ideal flow of an incompressible fluid through a free vortex compressor stage. For the example values $\omega r_h/w = 0.8$, $\Delta p_t/\rho w^2 = 1.0$, $r_t/r_h = 2.8$, $\sigma_h = 1.9$, and $°R_h = -0.65$, (constant chord),

(a) Find the diffusion factor at stator hub and rotor tip.

(b) Find the degree of reaction at the tip.

(c) Draw accurately the combined velocity triangles at the tip and at the hub.

9.3   Consider the ideal flow of an incompressible fluid through a free vortex compressor stage. You are given $\Delta p_t/\rho w^2 = 1.0$, $\theta_1 = \theta_3$, $r_t/r_h = 2.6$, $°R_h = -0.6$, $\sigma_h = 1.7$ $[\sigma_r = \sigma_h(r_h/r)]$, and $\omega r_h/w = 0.7$ and 0.8.

(a) Calculate v_1/w, v_2/w, $(V_e/w)_R$, $(V_i/w)_R$, $(V_e/w)_S$, $(V_i/w)_S$, D_R, and D_S at the hub and at the tip.

(b) Draw accurately the combined velocity triangles at the tip and at the hub for both cases $\omega r_h/w = 0.7$ and 0.8.

9.4   Consider the ideal flow of an incompressible fluid through a rotor-stator pair. There are no inlet guide vanes, so the approaching flow is purely axial. The stage is a free vortex stage, with constant axial velocity, purely axial flow at the exit, and blades of constant chord.

(a) Show that

$$\frac{\Delta p_t}{\rho(\omega r_t)^2} = \frac{w}{\omega r_t}\frac{1}{R}\frac{1}{(1/2\sigma_t^2 R^2) - 2(1 - D)^2}$$

$$\times \left\{ \left[\frac{1}{\sigma_t R} - 2(1 - D) \right] \left[\frac{1}{4\sigma_t^2 R^2} + 1 - (1 - D)^2 \right]^{\frac{1}{2}} \right\}$$

where D is the stator hub diffusion factor.

(b) For the case $\sigma_t = 0.4$, $w/\omega r_t = \frac{1}{3}$, $D = 0.6$ plot $\Delta p_t/\rho(\omega r_t)^2$ vs R for $2 \le R \le 3$.

(c) Repeat (b) with $w/\omega r_t = \frac{1}{4}$.

(d) Repeat (b) with $\sigma_t = 0.5$.

(e) Repeat (b) with $D = 0.5$.

9.5   Consider the ideal compressible flow of a calorically perfect gas through a fan stage. The entering and departing flows as well as the relative velocity at station 2 are all axial. The annulus cross section is varied so as to keep the axial velocity at the (single) radius considered constant.

(a) Find an expression for p_{t_2}/p_{t_1} in terms of γ, M_1, and $\omega r/w$.
(b) Find an expression for M_2 in terms of γ, M_1, and $\omega r/w$.
(c) Find an expression for M_3 in terms of γ, M_1, and $\omega r/w$.
(d) Show that the degree of reaction is given by

$$°R = \frac{\left[1 + \frac{\gamma - 1}{2} M_1^2 \left(\frac{\omega r}{w}\right)^2\right]^{\gamma/(\gamma-1)} - 1}{\left[1 + (\gamma - 1) M_1^2 \left(\frac{\omega r}{w}\right)^2\right]^{\gamma/(\gamma-1)} - 1}$$

(e) Evaluate $°R$ for $\gamma = 1.4$, $\omega r/w = 1$ and 1.5, and $M_1 = 0$, 0.2, 0.4, and 0.6.

9.6   A uniform flow of incompressible fluid passes through a stator row which imparts a swirl such that $(yV)^2 = -\delta\Psi$. The flow process can be considered ideal.

(a) Find expressions for the dimensionless velocities V and W that will exist far downstream in terms of y, δ, and R.
(b) Find an expression, in terms of R, for the limiting value of dimensionless swirl velocity at the tip, V_{tL}, that just leads to $W_t = 0$.
(c) For the case $R = 3$ and $\delta = 4$, plot W and V vs $(y - 1)/(R - 1)$.

9.7   A single-stage fan has a rotor followed by a stator. The flow may be approximated as incompressible and the annulus as parallel walled. The rotor is not a free vortex rotor. The stator removes all the swirl. Given that the departing stagnation enthalpy is given by

$$H = \text{const} - \Omega\Psi$$

find an expression for the far downstream stream function in terms of y, Ω, and R.

9.8   Consider the ideal flow of a calorically perfect gas through a single-stage turbine. The stage is an impulse stage, and the entering and departing flows are purely axial. Entering conditions are $\gamma_t = 1.3$, $M_0 = 0.4$, $C_{p_t} = 7500$ ft^2/s$^2 \cdot °R$, $T_{t_0} = T_{t_1} = 2600$ $°R$, $\omega r = 1300$ ft/s. Given that $A_1/A_0 = 1.25$,

(a) Find the stagnation pressure ratio p_{t_2}/p_{t_1} and stagnation temperature ratio T_{t_2}/T_{t_1}.
(b) Check that this stagnation temperature is allowable in this impulse turbine.

(c) If from part (b) it is found that such a stagnation temperature change is possible, find the Mach number M_1 of the flow departing the nozzle and find the flow angle α.

(d) Repeat parts (a–c) given that $A_1/A_0 = 1.5$.

9.9   "Ducted windmills" are being considered for energy production. It is apparent that such windmills can be viewed as "full-reaction turbines" ($^\circ R = 1$) because there is no pressure drop across the [nonexistent (!)] stator.

(a) Sketch the velocity triangles for such a (single radius) rotor.

(b) If the rotor turns the flow by an angle $\Delta\beta$, show that ideally

$$\frac{T_{t_2}}{T_{t_1}} = 1 - \frac{(\omega r)^2}{C_p T_{t_1}} \left\{ \frac{w}{\omega r} \tan\left[\left(\tan^{-1}\frac{\omega r}{w} \right) + \Delta\beta \right] - 1 \right\}$$

(c) Evaluate T_{t_2}/T_{t_1} and p_{t_2}/p_{t_1} to six significant figures for the conditions $T_{t_1} = 500\ ^\circ R$, $C_p = 6000\ \text{ft}^2/\text{s}^2 \cdot {}^\circ R$, $\omega r = w = 36$ fps, $\Delta\beta = 10$ deg.

9.10   An ideal impulse turbine is to be designed to operate with the maximum possible work interaction per mass for given inlet conditions and axial area ratio A_1/A_0. Given $\gamma = 1.32$, $M_0 = 0.3$, $C_{p_t} = 7400\ \text{ft}^2/\text{s}^2 \cdot {}^\circ R$, $T_{t_0} = 2500\ ^\circ R$, and $A_1/A_0 = 1.3$, find the required blade speed ωr.

10. THROUGHFLOW THEORY

10.1 Introduction

In an early paper,[1] Wu formulated the basic concept of representing the inviscid three-dimensional flowfield as the sum of two separate two-dimensional flowfields. These two separate flowfields were composed of surfaces located in the blade-to-blade direction (s_1 surfaces) and surfaces lying in the hub-to-tip direction (s_2 surfaces). The solution to the flowfield composed of these combined two-dimensional fields would formally require an iterative solution, since the solution for either surface requires a knowledge of the shape of the other surface.

In practice,[2-4] the problem is simplified by taking appropriate averages for the blade-to-blade direction and then assuming the flow to be axisymmetric. In this way, the s_2 planes become meridional surfaces and the expressions for the derivatives along and normal to such surfaces are easily represented in terms of the derivatives in the radial and axial directions.

When the complete throughflow equations are considered, the variations of the fluid properties in the axial direction are included. It is to be expected then that the radial equilibrium solutions obtained in Chap. 9 will again appear as the asymptotic limits of the far upstream and far downstream flows.

The equations will be formulated in a manner that allows the inclusion of compressibility and entropy variation effects. However, the solution of the resulting rather general equation must rely upon the application of advanced computer techniques that cannot reasonably be reported in detail in this book because of space limitations. Section 10.6 briefly describes some of the various calculational methods available and provides references for further reading.

10.2 The Throughflow Equations

In the following, equations appropriate for the description of the (axisymmetric) throughflow are developed. The viscous stresses in the fluid will not be included explicitly, but a general "body force" $\mathbf{F}$ will be included. It is possible to utilize this general body force to artifically introduce the effects of viscosity,[2] but in any case the flowfield considered is primarily external to the blade rows, and it is consistent with throughflow theory to consider the viscous effects to be negligible in this region. The effects of viscous stresses within the blade row will appear implicitly in the variation of entropy in the direction normal to the stream surfaces and also in the effective body force of the related blade forces. (See Secs. 10.3 and 10.4.)

The Conservation Equations

When vector notation is utilized, the continuity, momentum, and Gibb's equations may be written

$$\frac{\partial \rho}{\partial t} + \nabla \cdot (\rho \mathbf{u}) = 0 \tag{10.1}$$

$$\frac{D\mathbf{u}}{Dt} = \mathbf{F} - \frac{1}{\rho} \nabla p \tag{10.2}$$

$$\nabla h - \frac{1}{\rho} \nabla p = T \nabla s \tag{10.3}$$

also

$$\frac{Dh}{Dt} - \frac{1}{\rho} \frac{Dp}{Dt} = T \frac{Ds}{Dt} \tag{10.4}$$

The Convective Derivative

The convective derivative, represented by the operator $(\mathbf{u} \cdot \nabla)$, may be expanded as

$$(\mathbf{u} \cdot \nabla) = u \frac{\partial}{\partial r} + \frac{v}{r} \frac{\partial}{\partial \theta} + w \frac{\partial}{\partial z} \tag{10.5}$$

It is important here to retain the partial derivative with respect to θ because, even though the fluid properties have no θ variation (because of the assumption of axial symmetry), the coordinate directions themselves change with θ. Thus note in particular

$$\frac{\partial \mathbf{e}_r}{\partial \theta} = \mathbf{e}_\theta \tag{10.6}$$

$$\frac{\partial \mathbf{e}_\theta}{\partial \theta} = -\mathbf{e}_r \tag{10.7}$$

Utilizing Eqs. (10.5–10.7), the convective derivative of the velocity vector may be expanded into the components,

$$(\mathbf{u} \cdot \nabla)\mathbf{u} = \left[(\mathbf{u} \cdot \nabla) u - \frac{v^2}{r} \right] \mathbf{e}_r + \left[\frac{1}{r} (\mathbf{u} \cdot \nabla)(rv) \right] \mathbf{e}_\theta + [(\mathbf{u} \cdot \nabla) w] \mathbf{e}_z$$

$$\tag{10.8}$$

The Equation for the Tangential Momentum

The equation for the tangential momentum follows immediately by taking the scalar product of Eq. (10.2) with $\mathbf{e}_\theta$ and utilizing Eq. (10.8) to give

$$\frac{Drv}{Dt} = r\left(f_\theta - \frac{1}{\rho}\frac{1}{r}\frac{\partial p}{\partial \theta} \right) \tag{10.9}$$

It is to be noted that the tangential body force existing in this expression may include a viscous as well as a nonviscous term. The equation may be interpreted to explain the mechanism by which angular momentum is introduced into the flow. Thus, if perfect flow through an actual blade row is considered, no body forces exist and the angular momentum is imparted by the pressure forces of the blade row. Clearly, if the contribution of such pressure forces to a control volume between repetitive streamlines of a cascade external to the cascade is considered (Fig. 10.1, case A), the stream surfaces cannot support a pressure change and the (net) angular momentum cannot change. Once within the blade row, however, the blade surfaces can support the pressure change and the angular momentum changes (Fig. 10.1, case B).

It can be noted here that the formal mechanism for obtaining the throughflow form of the equations is to replace the term

$$\frac{1}{\rho}\frac{1}{r}\frac{\partial p}{\partial \theta}$$

in Eq. (10.9) by an equivalent (artificial) body force f_θ. In this way the θ dependence of the properties is removed. The relationship between this artificial body force and the torque and forces existing on the blade is further developed in Sec. 10.4.

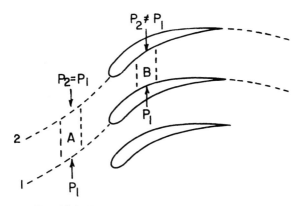

Fig. 10.1 Pressure stresses on stream surfaces.

The Euler Turbine Equation

The vector form of the Euler turbine equation (see also Sec. 9.2) may be obtained directly by taking the scalar product of Eq. (10.2) with $\mathbf{u}$ and adding the result to Eq. (10.4), to give

$$\frac{Dh_t}{Dt} = \frac{1}{\rho}\frac{\partial p}{\partial t} + \mathbf{u}\cdot\mathbf{F} + T\frac{Ds}{Dt} \qquad (10.10)$$

The velocity may be written as the sum of the velocity relative to the blade $\mathbf{u}_{rel}$ and the blade velocity $\omega r e_\theta$ to give

$$\mathbf{u}\cdot\mathbf{F} = (\mathbf{u}_{rel} + \omega r e_\theta)\cdot\mathbf{F} = \mathbf{u}_{rel}\cdot\mathbf{F} + \omega r f_\theta \qquad (10.11)$$

Combination of Eqs. (10.9–10.11) then gives

$$\frac{Dh_t}{Dt} = \frac{1}{\rho}\left[\frac{\partial p}{\partial t} + \omega r\left(\frac{1}{r}\frac{\partial p}{\partial \theta}\right)\right] + \left(\mathbf{u}_{rel}\cdot\mathbf{F} + T\frac{Ds}{Dt}\right) + \omega\frac{Drv}{Dt} \qquad (10.12)$$

It can be noted that the expression within the first pair of brackets represents the pressure perturbation as viewed by an observer fixed to the rotating blade. Under normal circumstances where the pressure perturbations are created by the blades themselves, this group of terms is zero. When considering the terms in the second pair of brackets, note that because the body force term is identified with the blade forces, the product $\mathbf{u}_{rel}\cdot\mathbf{F}$ represents the effect of the frictional forces alone. (Note that if there were no frictional stress, $\mathbf{F}$ would be perpendicular to the relative velocity giving $\mathbf{u}_{rel}\cdot\mathbf{F} = 0$.) The term $\mathbf{u}_{rel}\cdot\mathbf{F}$ is hence the sole source of entropy generation, so as with Eq. (2.63), the terms in the second pair of brackets also cancel. There is thus obtained the vector form of the Euler turbine equation,

$$\frac{Dh_t}{Dt} = \omega\frac{Drv}{Dt} \qquad (10.13)$$

The Compressible Form of the Stream Function

When the throughflow limit of the equations is taken, the replacement of the group

$$\frac{1}{\rho}\frac{1}{r}\frac{\partial p}{\partial \theta}$$

by an equivalent body force not only reduces the equations to axisymmetric form, but, in addition, removes the time dependency. Thus, the continuity equation may be written

$$\frac{1}{r}\frac{\partial(\rho r u)}{\partial r} + \frac{\partial(\rho w)}{\partial z} = 0 \qquad (10.14)$$

This equation is identically satisfied with introduction of the compressible form of the stream function ψ, defined by

$$\rho u = \rho_0 \frac{1}{r} \frac{\partial \psi}{\partial z}, \qquad \rho w = -\rho_0 \frac{1}{r} \frac{\partial \psi}{\partial r} \qquad (10.15)$$

In these expressions ρ_0 is a suitable reference density. It follows, just as in the development in Sec. 9.4 of the incompressible form of the stream function, that the mass flow rate between two stream surfaces $\dot{m}$ is related to the stream function by

$$\dot{m} = -2\pi\rho_0(\psi_2 - \psi_1) \qquad (10.16)$$

The Natural Coordinate System

The stream tube surfaces form natural coordinates for throughflow studies and lead to the introduction of a coordinate system with directions $\mathbf{n}$ (normal to stream tube surfaces), $\mathbf{e}_\ell$ (meridional component), and $\mathbf{e}_\theta$ (tangential). This coordinate system is depicted in Fig. 10.2.

The velocity vector in the natural coordinate system has components given by

$$\mathbf{u} = (0, v, v_\ell) \qquad (10.17)$$

The meridional velocity v_ℓ is related to the cylindrical velocity components by

$$v_\ell^2 = u^2 + w^2 \qquad (10.18)$$

and the unit vectors are given in terms of the cylindrical coordinate quantities by

$$\mathbf{e}_\ell = \frac{1}{v_\ell}(u\mathbf{e}_r + w\mathbf{e}_z) \qquad (10.19)$$

$$\mathbf{n} = \mathbf{e}_\theta \times \mathbf{e}_\ell = \frac{1}{v_\ell}(\mathbf{e}_\theta \times \mathbf{u}) \qquad (10.20)$$

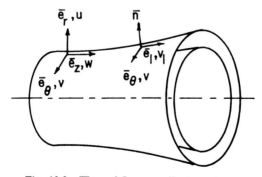

Fig. 10.2 Throughflow coordinate systems.

The scalar magnitude of the differential difference dn occurring for a differential displacement dr is given by

$$dn = \mathbf{n} \cdot \mathbf{dr} = \frac{w}{v_\ell} dr - \frac{u}{v_\ell} dz \qquad (10.21)$$

This may be related to the corresponding change in the stream function by utilizing Eqs. (10.15) to give

$$dn = -\frac{1}{rv_\ell} \frac{\rho_0}{\rho} \left(\frac{\partial \psi}{\partial r} dr + \frac{\partial \psi}{\partial z} dz \right) = -\frac{1}{rv_\ell} \frac{\rho_0}{\rho} d\psi \qquad (10.22)$$

The unit derivative in the direction normal to the stream surfaces $(\mathbf{n} \cdot \nabla)$ is thus given in terms of the stream function as

$$(\mathbf{n} \cdot \nabla) \equiv \frac{\partial}{\partial n} = \frac{\partial \psi}{\partial n} \frac{\partial}{\partial \psi} = -\frac{\rho r v_\ell}{\rho_0} \frac{\partial}{\partial \psi} \qquad (10.23)$$

This same derivative in cylindrical coordinates follows directly from Eq. (10.20) and the definition of the operator ∇ (in cylindrical coordinates) to give

$$(\mathbf{n} \cdot \nabla) = \frac{1}{v_\ell} (\mathbf{e}_\theta \times \mathbf{u}) \cdot \left(\mathbf{e}_r \frac{\partial}{\partial r} + \mathbf{e}_\theta \frac{1}{r} \frac{\partial}{\partial \theta} + \mathbf{e}_z \frac{\partial}{\partial z} \right)$$

or

$$(\mathbf{n} \cdot \nabla) = \frac{w}{v_\ell} \frac{\partial}{\partial r} - \frac{u}{v_\ell} \frac{\partial}{\partial z} \qquad (10.24)$$

Equation for the Tangential Vorticity

The Euler momentum equation and the conservation of angular momentum equation describe the behavior of the angular momentum and stagnation enthalpy along stream surfaces. It can be expected that the essence of the two-dimensional problem lies in the variation of properties across stream surfaces, so to this end the rate of change of properties normal to stream surfaces is investigated.

The steady-state form of the momentum equation (10.2) may be added to Eq. (10.3) to give

$$\nabla h_t + \omega \times \mathbf{u} = T \nabla s + \mathbf{F} \qquad (10.25)$$

In obtaining this expression the vector identity

$$(\mathbf{u} \cdot \nabla)\mathbf{u} = \nabla \frac{u^2}{2} + \omega \times \mathbf{u}$$

has been utilized. This in turn introduces the vorticity ω, defined by

$$\omega = \xi \mathbf{e}_r + \eta \mathbf{e}_\theta + \zeta \mathbf{e}_z \equiv \nabla \times \mathbf{u} \tag{10.26}$$

Utilizing Eq. (10.20) it follows from vector expansion that

$$\mathbf{n} \cdot (\omega \times \mathbf{u}) = \frac{1}{v_\ell}[(\mathbf{e}_\theta \cdot \omega)(\mathbf{u} \cdot \mathbf{u}) - (\mathbf{e}_\theta \cdot \mathbf{u})(\mathbf{u} \cdot \omega)]$$

Thus expanding the vorticity into its components there is obtained

$$\mathbf{n} \cdot (\omega \times \mathbf{u}) = \eta v_\ell - \frac{v}{r}(\mathbf{n} \cdot \nabla)(rv) \tag{10.27}$$

An equation for the tangential vorticity η follows by taking the scalar product of Eq. (10.25) with $\mathbf{n}$ and utilizing Eq. (10.27). Thus,

$$\eta = \frac{1}{v_\ell}\left[\mathbf{n} \cdot \mathbf{F} - (\mathbf{n} \cdot \nabla)h_t + T(\mathbf{n} \cdot \nabla)s + \frac{v}{r}(\mathbf{n} \cdot \nabla)rv\right] \tag{10.28}$$

This expression may be written as an equation for the stream function ψ, density ratio ρ/ρ_0, and prescribed variables $(h_t, s, rv, \mathbf{F})$ by utilizing

$$\eta = \frac{\partial u}{\partial z} - \frac{\partial w}{\partial r}$$

and Eqs. (10.15) and (10.23) to give

$$\frac{\partial}{\partial z}\left(\frac{\rho_0}{\rho}\frac{1}{r}\frac{\partial \psi}{\partial z}\right) + \frac{\partial}{\partial r}\left(\frac{\rho_0}{\rho}\frac{1}{r}\frac{\partial \psi}{\partial r}\right) = \frac{1}{v_\ell}\mathbf{n} \cdot \mathbf{F} + \frac{\rho}{\rho_0}r\left(\frac{\partial h_t}{\partial \psi} - T\frac{\partial s}{\partial \psi} - \frac{v}{r}\frac{\partial rv}{\partial \psi}\right) \tag{10.29}$$

The density ratio follows from Eq. (2.57) and the perfect gas equation of state to give

$$\frac{\rho}{\rho_0} = \left(\frac{T}{T_0}\right)^{1/(\gamma-1)}e^{-[(s-s_0)/R]} = \left[\frac{h_t - \frac{1}{2}(w^2 + u^2 + v^2)}{h_{t0} - \frac{1}{2}(w^2 + u^2 + v^2)_0}\right]^{1/(\gamma-1)}e^{-[(s-s_0)/R]}$$

Utilizing Eq. (10.15) this expression may be rearranged to give

$$\left(\frac{\rho_0}{\rho}\right)^2\frac{1}{2}\left[\left(\frac{1}{r}\frac{\partial \psi}{\partial r}\right)^2 + \left(\frac{1}{r}\frac{\partial \psi}{\partial z}\right)^2\right]$$
$$+ \left(\frac{\rho}{\rho_0}\right)^{\gamma-1}\left[h_{t0} - \frac{1}{2}(w^2 + u^2 + v^2)_0\right]e^{(\gamma/c_p)(s-s_0)} - \left[h_t - \frac{1}{2r^2}(rv)^2\right] = 0 \tag{10.30}$$

Equations (10.29) and (10.30) constitute two coupled, highly nonlinear equations for the stream function in terms of the reference 0 conditions and the prescribed quantities $\mathbf{F}$, h_t, s, and rv.

10.3   The Actuator Disk

The presence of the body force term $\mathbf{F}$ in Eq. (10.29) complicates the solution for the throughflow considerably. A useful approximate solution can be obtained by considering all the forces to be concentrated in an infinitesimally thin disk—the actuator disk. The forces are chosen to have the same integral effect upon the tangential momentum, stagnation enthalpy, and entropy as would the axially distributed forces of the actual blade rows (see Sec. 10.4). The result is then that the body forces themselves do not appear in the resulting equations, but the effects of the blade forces appear as matching conditions across the (infinitely thin) blade row. The approximation can be made as accurate as desired by taking a number of actuator disks to simulate a single blade row.

The equation for the stream function (10.29) thus reduces to

$$\frac{\partial}{\partial z}\left(\frac{\rho_0}{\rho}\frac{1}{r}\frac{\partial \psi}{\partial z}\right) + \frac{\partial}{\partial r}\left(\frac{\rho_0}{\rho}\frac{1}{r}\frac{\partial \psi}{\partial r}\right) = \frac{\rho}{\rho_0}r\left(\frac{\partial h_t}{\partial \psi} - T\frac{\partial s}{\partial \psi} - \frac{v}{r}\frac{\partial rv}{\partial \psi}\right)$$

$$(10.31)$$

Because in the actuator disk approximation the blade row is taken to be infinitely thin, the continuity of mass ensures that the value of the stream function on one side of the disk is identical to that at the same radial location on the other side. This simple relationship may be written

$$[\psi]_d = 0 \qquad (10.32)$$

where the notation [ ] refers to the jump in the value of the quantity and the subscript d refers to conditions at the disk.

A second matching condition may be obtained by considering the radial momentum equation. Thus, noting that by the assumption of axisymmetry no θ derivatives of fluid properties are present and combining Eqs. (10.2) and (10.8), there is obtained

$$\rho u \frac{\partial u}{\partial r} + \rho w \frac{\partial u}{\partial z} - \rho \frac{v^2}{r} + \frac{\partial p}{\partial r} = \rho f_r \qquad (10.33)$$

This equation is now integrated from an infinitesimal distance ε upstream of the disk to an infinitesimal distance downstream of the disk. It is to be noted that no terms on the left side of the equation can become infinite except, possibly, those involving an axial derivative. Thus, integration over the infinitesimal distance (noting ρw is constant) gives

$$\int_{-\epsilon}^{\epsilon} \frac{\partial u}{\partial z} \, dz \equiv [u]_d = \frac{1}{\rho w} \int_{-\epsilon}^{\epsilon} \rho f_r \, dz$$

Utilizing Eq. (10.15) there is then obtained

$$\left[\frac{\rho_0}{\rho} \frac{1}{r} \frac{\partial \psi}{\partial z} \right]_d = \frac{1}{\rho w} \int_{-\epsilon}^{\epsilon} \rho f_r \, dz \qquad (10.34)$$

The integral on the right side represents the effect of the total radial force of the blade row at the given radius of the actuator disk. In many applications, the radial force of the almost radial blades is very small and Eq. (10.34) is often approximated by

$$\left[\frac{1}{\rho} \frac{\partial \psi}{\partial z} \right]_d = 0 \qquad (10.35)$$

Equations (10.30–10.32) and (10.35), together with the boundary conditions that the stream function is prescribed on the containing walls (or that the pressure be constant on the bounding streamline in the case of a free streamline), constitute the mathematical statement of the problem. In Sec. 10.5 an example solution for incompressible flow will be presented and methods for calculating compressibility effects will be described in Sec. 10.6. Before proceeding to the calculational examples, however, the relationships of the overall torque applied to the blade row and the overall axial force on the blade row to the resulting changes in fluid properties are considered.

10.4 Integral Relationships

An equation for the tangential momentum was developed in Sec. 10.2, and the equivalence of the body force field utilized in the throughflow approximation and the pressure field existing on the actual blade surfaces was discussed. The relationship between the equivalent force field and the torque on the blade (taken about the axis) may be determined by noting that the differential contribution to the torque of an annular volume $2\pi r \, dr \, dz$ is given by

$$d\tau = 2\pi \rho f_\theta r^2 \, dr \, dz \qquad (10.36)$$

The torque upon the entire blade row is then obtained by integrating from the blade leading edge to the trailing edge (z_1 to z_2) and from the hub radius to the tip radius (r_h to r_t). Hence,

$$\tau_{1-2} = 2\pi \int_{z_1}^{z_2} \int_{r_h}^{r_t} \rho r^2 f_\theta \, dr \, dz \qquad (10.37)$$

The throughflow form of the tangential momentum equation (10.9) gives

$$rf_\theta = u\frac{\partial rv}{\partial r} + w\frac{\partial rv}{\partial z} \tag{10.38}$$

By adding rv times the continuity equation (10.14) to this expression, there is obtained

$$\rho rf_\theta = \frac{1}{r}\left[\frac{\partial(\rho ur)(rv)}{\partial r} + \frac{\partial(\rho rw)(rv)}{\partial z}\right] \tag{10.39}$$

When Eq. (10.39) is substituted into Eq. (10.37), the resulting expression may be integrated immediately to give with Eq. (10.15)

$$\tau_{1-2} = 2\pi\rho_0\int_{\psi(z_1)}^{\psi(z_2)}\left[(rv)_t - (rv)_h\right]d\psi - 2\pi\rho_0\int_{\psi_h}^{\psi_t}\left[(rv)_{z_2} - (rv)_{z_1}\right]d\psi$$

In this expression the first integral vanishes, because the value of the stream function does not change along the hub or tip (i.e., $d\psi = 0$ on r_t or r_h). There is thus obtained a relationship that is itself obvious from first principles when the relationship between the stream function and the mass flow [(Eq. (10.16)] is recognized. Thus

$$\tau_{1-2} = -2\pi\rho_0\int_{\psi_h}^{\psi_t}\left[(rv)_{z_2} - (rv)_{z_1}\right]d\psi \tag{10.40}$$

Then noting from Eq. (10.16) that $-2\pi\rho_0\,d\psi = d\dot{m}$,

$$\tau_{1-2} = \int_{hub}^{tip}\left[(rv)_{z_2} - (rv)_{z_1}\right]d\dot{m} \tag{10.41}$$

When the desired angular momentum distribution with stream function is prescribed, the resulting torque on the blade row can be immediately determined from Eq. (10.40). It is to be noted that there is no need to obtain any of the detailed flow information through the blade row. The overall torque depends only upon the overall change in angular momentum through the blade row.

The axial force on the blade row and containing annulus walls may be obtained by considering the control volume shown in Fig. 10.3.

Equating the rate of production of momentum to the force applied to the fluid, it follows that the upstream directed force on the blade row and annulus wall F_A is given by

$$F_A = \left[\int_{r_h}^{r_t}(p + \rho w^2)2\pi r\,dr\right]_2 - \left[\int_{r_h}^{r_t}(p + \rho w^2)2\pi r\,dr\right]_1 \tag{10.42}$$

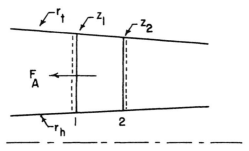

Fig. 10.3 Axial forces.

When the actuator disk limit is considered, F_A becomes the force on the blade row alone. It is convenient to write this expression in the form

$$F_A = 2\pi\rho_0[M]$$ (10.43)

where M is defined by

$$M = \int_{r_h}^{r_t}\left(\frac{p}{\rho_0} + \frac{\rho}{\rho_0}w^2\right)r\,dr$$ (10.44)

It is clear from the way M has been defined that M will be constant with axial position if the walls are parallel and no blades are present. (Recall that the flow was assumed to be nonviscous outside the blade rows.) The quantity M is most easily evaluated by considering the radial equilibrium form of the solutions, which is possible because M will stay constant in a parallel-walled annulus until all radial flows vanish. The expression for M can be manipulated in the following manner (noting $u = 0$ in radial equilibrium):

$$M = \int_{r_h}^{r_t}\left[\frac{p}{\rho_0} + \frac{1}{2}\frac{\rho}{\rho_0}u^2 - \frac{1}{2}\frac{\rho}{\rho_0}(v^2 - w^2)\right]r\,dr$$

$$= \int_{r_h}^{r_t}\left[r\left(\frac{p}{\rho_0} + \frac{1}{2}\frac{\rho}{\rho_0}u^2\right) - \frac{\rho}{\rho_0}\frac{(rv)^2}{2r} + \frac{1}{2r}\frac{\rho_0}{\rho}\left(\frac{\partial\psi}{\partial r}\right)^2\right]dr$$ (10.45)

where as before u^2 denotes the scalar product $\mathbf{u}\cdot\mathbf{u}$. Noting

$$\frac{d}{dr}\left(\frac{\rho_0}{\rho}\frac{\psi}{r}\frac{d\psi}{dr}\right) = \psi\frac{d}{dr}\left(\frac{\rho_0}{\rho}\frac{1}{r}\frac{d\psi}{dr}\right) + \frac{\rho_0}{\rho}\frac{1}{r}\left(\frac{d\psi}{dr}\right)^2$$

and utilizing the radial equilibrium form of Eq. (10.31), Eq. (10.45) may be written

$$M = \int_{r_h}^{r_t} \left[r \left(\frac{p}{\rho_0} + \frac{1}{2} \frac{\rho}{\rho_0} \mathbf{u}^2 - \frac{\rho}{\rho_0} \frac{\psi}{2} \frac{dh_t}{d\psi} \right) - \frac{\rho}{\rho_0} \frac{(rv)}{2r} \left(rv - \psi \frac{drv}{d\psi} \right) \right.$$

$$\left. + \frac{1}{2} \frac{d}{dr} \left(\frac{\rho_0}{\rho} \frac{\psi}{r} \frac{d\psi}{dr} \right) + \frac{\rho}{\rho_0} \frac{\psi}{2} T \frac{ds}{d\psi} \right] dr \qquad (10.46)$$

When the stagnation enthalpy, entropy, and angular momentum are prescribed, all terms in this integral may be evaluated [with the solution of Eqs. (10.30) and (10.31)] and the integral obtained. In the following section simple explicit solutions are obtained for special forms of the blade loading and for these cases evaluation of the integral is very straightforward. It may be noted here that when the flow considered is perfect and incompressible, the group

$$\frac{p}{\rho_0} + \frac{1}{2} \frac{\rho}{\rho_0} \mathbf{u}^2$$

is identical to the stagnation enthalpy h_t.

10.5 Example Solutions

As an example set of solutions consider the simple case of flow of a perfect, incompressible fluid in a parallel-walled annulus. As indicated in Fig. 10.4, there are N actuator disks located at $x = x_j$ $(1 < j < N)$ and $N + 1$ regions of flow in regions $i (0 \le i \le N)$.

It is most convenient to consider the nondimensional form of the equations so again utilize the dimensionless variables introduced in Eq. (9.66). Thus, with the addition of a dimensionless axial variable x, define

$$W = \frac{w}{w_0}, \qquad V = \frac{v}{w_0}, \qquad \Psi = \frac{\psi}{w_0 r_h^2}, \qquad H = \frac{h_t}{w_0^2}$$

$$y = \frac{r}{r_h}, \qquad x = \frac{z}{r_h}, \qquad R = \frac{r_t}{r_h} \qquad (10.47)$$

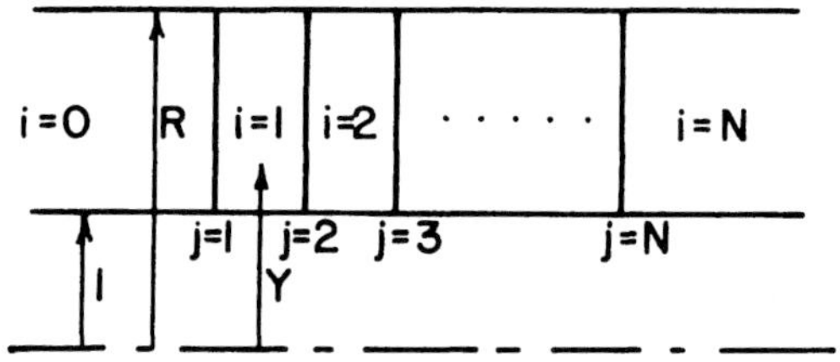

Fig. 10.4　Actuator disks and nomenclature.

With the restrictions listed above and with the introduction of these new variables, Eq. (10.31) reduces to

$$\frac{1}{y}\frac{\partial^2 \Psi}{\partial x^2} + \frac{\partial}{\partial y}\left(\frac{1}{y}\frac{\partial \Psi}{\partial y}\right) = y\frac{\partial H}{\partial \Psi} - V\frac{\partial yV}{\partial \Psi} \qquad (10.48)$$

The boundary conditions are

$$\Psi = -\tfrac{1}{2} \quad \text{on} \quad y = 1 \quad \text{and} \quad \Psi = -R^2/2 \quad \text{on} \quad y = R \qquad (10.49)$$

The matching conditions are

$$[\Psi]_j = 0 \qquad \text{and} \qquad \left[\frac{\partial \Psi}{\partial x}\right]_j = 0 \qquad (10.50)$$

The mathematical statement of the problem is completed by prescribing the desired distribution of the stagnation enthalpy and angular momentum with the stream function. A particularly simple and instructive set of solutions is obtained when forms of H and yV leading to linear forms of the equations are considered. Thus, consider (as in Sec. 9.4) the special forms given by

$$H_i = H_0 - \omega_i \Psi + \theta_i \qquad \text{and} \qquad (yV)_i = -\alpha_i \Psi + \beta_i \quad (10.51)$$

It follows from the Euler momentum equation that

$$\omega_i = \omega_{i-1} + \Omega_i(\alpha_i - \alpha_{i-1}) \qquad (10.52)$$

$$\theta_i = \theta_{i-1} + \Omega_i(\beta_i - \beta_{i-1}) \qquad (10.53)$$

Here Ω_i is the nondimensional angular velocity $(\omega r_h/w_0)$ of the rotor row located at x_j where $j = i$.

Equations (10.48–10.53) constitute the mathematical statement of the problem. Begin the solution by incorporating Eq. (10.51) into Eq. (10.48) to give

$$\frac{\partial^2 \Psi}{\partial x^2} + y\frac{\partial}{\partial y}\left(\frac{1}{y}\frac{\partial \Psi}{\partial y}\right) + \alpha_i^2 \Psi = \alpha_i \beta_i - \omega_i y^2 \qquad (10.54)$$

Solution of the Homogenous Equation—The Natural Eigenfunctions

Now consider solution of the homogeneous form of Eq. (10.54) by separation of variables. Thus substitute

$$\Psi = yE(y)\Phi(x)$$

to obtain the two ordinary differential equations

$$\frac{d^2\Phi}{dx^2} - \left(\lambda^2 - \alpha_i^2\right)\Phi = 0 \tag{10.55}$$

$$\frac{d^2E}{dy^2} + \frac{1}{y}\frac{dE}{dy} + \left(\lambda^2 - \frac{1}{y^2}\right)E = 0 \tag{10.56}$$

where λ^2 is the separation constant.

The solution of Eq. (10.56) is obtained in terms of Bessel functions of the first order and may be written

$$E = C_1 J_1(\lambda y) + C_2 Y_1(\lambda y) \tag{10.57}$$

In order to satisfy the boundary conditions that $\partial\Psi/\partial x$ be zero on the hub, one of the constants may be chosen such that

$$E = C_3 \left[J_1(\lambda y) Y_1(\lambda) - J_1(\lambda) Y_1(\lambda y) \right] \tag{10.58}$$

In order to satisfy the remaining boundary condition that $\partial\Psi/\partial x$ be zero on the tip, only selected values—the eigenvalues—of the separation constant λ can be allowed. Thus a family of solutions is obtained, each solution having a corresponding eigenvalue λ_n. These eigenvalues follow from solution of the equation

$$J_1(\lambda_n R) Y_1(\lambda_n) - J_1(\lambda_n) Y_1(\lambda_n R) = 0 \tag{10.59}$$

A series approximation to the value of λ_n is given on p. 261 of Ref. 5 and may be written

$$\lambda_n = \frac{n\pi}{R-1}\left\{ 1 + \left(\frac{R-1}{n\pi}\right)^2 \frac{3}{8R} - \left(\frac{R-1}{n\pi}\right)^4 \right.$$

$$\left. \times \left[\frac{21(R^3-1)}{128R^3(R-1)} + \frac{9}{64}\frac{1}{R^2} \right] + \cdots \right\} \tag{10.60}$$

This series is quite accurate, the third term reaching a value of only 0.015 for the case $R = 3$, $n = 1$. This set of values corresponds to about the largest correction expected in practice. If a higher accuracy is desired for exceptional cases, Eq. (10.59) is easily solved by iteration.

The constant C remains arbitrary because the functions E have yet to be multiplied by the still to be determined functions Φ. Because there is an infinite set of the functions Φ and E, the solution for Ψ is written in the form

$$\Psi = \sum_{n=1}^{\infty} \Phi_n(x)\, yE_{1n}(y) \tag{10.61}$$

here the subscript n denotes the eigenfunction corresponding to the eigenvalue λ_n, and the added subscript 1 in the function E_{1n} has been introduced to indicate that the Bessel functions with argument $\lambda_n y$ contained in E_{1n} are of first order.

It is a very useful result that the functions $E_{1n}(y)$ are orthogonal, which can be shown directly from the two integral formulas (Ref. 6, p. 146)

$$\int y Z_p(\alpha y)\bar{Z}_p(\beta y)\,\mathrm{d}y$$

$$= \frac{1}{\alpha^2 - \beta^2}\left[\beta y Z_p(\alpha y)\bar{Z}_{p-1}(\beta y) - \alpha y Z_{p-1}(\alpha y)\bar{Z}_p(\beta y)\right] \tag{10.62}$$

$$\int y\left[Z_p(\alpha y)\right]^2\mathrm{d}y = \frac{y^2}{2}\left[Z_p^2(\alpha y) - \frac{2p}{y}Z_{p-1}(\alpha y)Z_p(\alpha y) + Z_{p-1}^2(\alpha y)\right] \tag{10.63}$$

In these expressions $Z_p(\alpha y)$ and $\bar{Z}_p(\beta y)$ refer to any groups of the form $C_1 J_p(\alpha y) + C_2 Y_p(\alpha y)$ or $C_3 J_p(\beta y) + C_4 Y_p(\beta y)$. It then follows directly, with Eqs. (10.58) and (10.59), that

$$\int_1^R yE_{1n}(\lambda_n y)E_{1m}(\lambda_m y)\,\mathrm{d}y = 0 \qquad m \neq n \tag{10.64}$$

$$\int_1^R y\left[E_{1n}(\lambda_n y)\right]^2\mathrm{d}y = C_3^2\left\{\frac{R^2}{2}\left[J_0(\lambda_n R)Y_1(\lambda_n) - J_1(\lambda_n)Y_0(\lambda_n R)\right]^2\right.$$

$$\left. - \tfrac{1}{2}\left[J_0(\lambda_n)Y_1(\lambda_n) - J_1(\lambda_n)Y_0(\lambda_n)\right]^2\right\} \tag{10.65}$$

Equation (10.65) may be greatly simplified by noting from Eq. (10.59)

$$\frac{Y_1(\lambda_n)}{Y_1(\lambda_n R)} = \frac{J_1(\lambda_n)}{J_1(\lambda_n R)} \tag{10.66}$$

and the relationship (Ref. 6, p. 144)

$$Y_{p-1}(x)J_p(x) - Y_p(x)J_{p-1}(x) = 2/\pi x \tag{10.67}$$

to obtain

$$\int_1^R y[E_{1n}(\lambda_n y)]^2 \, dy = \frac{2C_3^2}{(\pi\lambda_n)^2}\left\{\left[\frac{J_1(\lambda_n)}{J_1(\lambda_n R)}\right]^2 - 1\right\} \qquad (10.68)$$

This suggests the convenient choice for the constant C_3 as that value which will render the functions orthonormal. Thus, define the orthonormal set of functions $E_{1n}(\lambda_n y)$ by

$$E_{1n}(y) = \lambda_n \frac{\pi}{\sqrt{2}} \frac{J_1(\lambda_n y)Y_1(\lambda_n) - J_1(\lambda_n)Y_1(\lambda_n y)}{\left\{\left[\dfrac{J_1(\lambda_n)}{J_1(\lambda_n R)}\right]^2 - 1\right\}^{\frac{1}{2}}} \qquad (10.69)$$

These functions have the property that

$$\int_1^R yE_{1n}(y)E_{1m}(y)\,dy = 0 \qquad n \neq m$$

$$= 1 \qquad n = m \qquad (10.70)$$

In addition note from Eq. (10.56)

$$\frac{d}{dy}\left\{\frac{1}{y}\frac{d}{dy}[yE_{1n}(y)]\right\} = -\lambda_n^2 E_{1n}(y) \qquad (10.71)$$

Also, it may be noted that

$$\frac{1}{y}\frac{d}{dy}[yE_{1n}(y)] = \lambda_n E_{0n}(y) \qquad (10.72)$$

Solution of the Inhomogeneous Form of the Equation

The emergence of the orthonormal functions $E_{1n}(y)$ from solution of the homogeneous equation suggests consideration of $\Phi_n(x)$ as a transformed variable defined by

$$\Phi_n(x) = \int_1^R \Psi(y, x)E_{1n}(y)\,dy \qquad (10.73)$$

Obviously, the inverse of the transformation is

$$\Psi(y,x) = \sum_{n=1}^{\infty} \Phi_n(x)\, y E_{1n}(y) \tag{10.74}$$

The transformation of Eq. (10.54) is now taken by multiplying the equation by $E_{1n}(y)$ and integrating between $y = 1$ and $y = R$. The several terms appearing are evaluated as follows.

Define

$$A_n = \int_1^R y^2 E_{1n}(y)\, dy = \frac{1}{\lambda_n}\left[y^2 \left(\frac{2}{\lambda_n y} E_{1n} - E_{0n} \right) \right]_1^R$$

or

$$A_n = \frac{1}{\lambda_n}\left[E_{0n}(1) - R^2 E_{0n}(R) \right] \tag{10.75}$$

With the definition of $E_{0n}(y)$ and with Eqs. (10.66) and (10.67), this expression may be rearranged to give

$$A_n = \frac{\sqrt{2}}{\lambda_n} \frac{R J_1(\lambda_n) - J_1(\lambda_n R)}{\left[J_1^2(\lambda_n) - J_1^2(\lambda_n R) \right]^{\frac{1}{2}}} \tag{10.76}$$

Now define

$$B_n = \int_1^B E_{1n}(y)\, dy = \frac{1}{\lambda_n}\left[E_{0n}(1) - E_{0n}(R) \right]$$

With Eqs. (10.66) and (10.67) this becomes

$$B_n = \frac{\sqrt{2}}{\lambda_n} \frac{(1/R) J_1(\lambda_n) - J_1(\lambda_n R)}{\left[J_1^2(\lambda_n) - J_1^2(\lambda_n R) \right]^{\frac{1}{2}}} \tag{10.77}$$

Then

$$\int_1^R \frac{\partial^2 \Psi}{\partial x^2} E_{1n}(y)\, dy = \frac{\partial^2}{\partial x^2} \int_1^R \Psi E_{1n}(y)\, dy = \frac{d^2 \Phi_n}{dx^2} \tag{10.78}$$

and

$$\int_1^R y \frac{\partial}{\partial y}\left(\frac{1}{y}\frac{\partial \Psi}{\partial y}\right) E_{1n}(y)\,dy = \left[E_{1n}(y)\frac{\partial \Psi}{\partial y}\right]_1^R - \lambda_n \int_1^R \frac{\partial \Psi}{\partial y} E_{0n}\,dy$$

$$= \left[-\lambda_n E_{0n}(y)\Psi\right]_1^R - \lambda_n^2 \int_1^R \Psi E_{1n}(y)\,dy$$

$$= -\frac{\lambda_n}{2}\left[E_{0n}(1) - R^2 E_{0n}(R)\right] - \lambda_n^2 \Phi_n$$

With Eq. (10.75) this may be written

$$\int_1^R y \frac{\partial}{\partial y}\left(\frac{1}{y}\frac{\partial \Psi}{\partial y}\right) E_{1n}(y)\,dy = -\lambda_n^2 \Phi_n - \frac{\lambda_n^2}{2} A_n \qquad (10.79)$$

The last term in Eq. (10.54), $\alpha_i^2 \Psi$, transforms directly to $\alpha_i^2 \Phi_n$, so that by combining Eqs. (10.54) and (10.76–10.79) an equation for Φ_n is obtained,

$$\frac{d^2 \Phi_n}{dx^2} - \left(\lambda_n^2 - \alpha_i^2\right)\Phi_n = -\left(\lambda_n^2 - \alpha_i^2\right)T_n^{(i)} \qquad (10.80)$$

where

$$T_n^{(i)} \equiv \frac{1}{\lambda_n^2 - \alpha_i^2}\left[\left(\omega_i - \frac{\lambda_n^2}{2}\right)A_n - \alpha_i \beta_i B_n\right]$$

The solution to Eq. (10.80) consists of exponentials and is conveniently grouped in the forms

$$\Phi_n = \begin{cases} \dfrac{D_n^{(0)}}{\eta_n^{(0)}}\exp\left[\eta_n^{(0)}(x - x_1)\right] + T_n^{(0)} & x < x_1 \\[2em] \dfrac{D_n^{(i)}\cosh\left[\eta_n^{(i)}(x - x_i)\right] - C_n^{(i)}\cosh\left[\eta_n^{(i)}(x_{i+1} - x)\right]}{\eta_n^{(i)}\sinh\left[\eta_n^{(i)}(x_{i+1} - x_i)\right]} + T_n^{(i)} \\[2em] \qquad\qquad\qquad\qquad\qquad\qquad\qquad x_i < x < x_{i+1} \\[2em] -\dfrac{C_n^{(N)}}{\eta_n^{(N)}}\exp\left[\eta_n^{(N)}(x_N - x)\right] + T_n^{(N)} & x_N < x \end{cases}$$

$$(10.81)$$

where there has been introduced $\eta_n^{(i)}$ defined by $\eta_n^{(i)} = \sqrt{\lambda_n^2 - \alpha_i^2}$.

Note that the boundary conditions $\partial\Phi/\partial x \to 0$ as $x \to \pm\infty$ are satisfied. The coefficients $C_n^{(i)}$ and $D_n^{(i)}$ are to be determined from the matching conditions at the disks [Eq. (10.50)]. The condition $[\partial\Phi/\partial x]_j = 0$ gives immediately that $D_n^{(i-1)} = C_n^{(i)}$ and continuity of Φ at each disk then gives the set of equations

$$\frac{C_n^{(1)}}{\eta_n^{(0)}} + T_n^{(0)} = \frac{C_n^{(2)} - C_n^{(1)}\cosh\left[\eta_n^{(1)}(x_2 - x_1)\right]}{\eta_n^{(1)}\sinh\left[\eta_n^{(1)}(x_2 - x_1)\right]} + T_n^{(1)}$$

$$\frac{C_n^{(j)}\cosh\left[\left(\eta_n^{(j-1)}(x_j - x_{j-1})\right]\right] - C_n^{(j-1)}}{\eta_n^{(j-1)}\sinh\left[\eta_n^{(j-1)}(x_j - x_{j-1})\right]} + T_n^{(j-1)}$$

$$= \frac{C_n^{(j+1)} - C_n^{(j)}\cosh\left[\eta_n^{(j)}(x_{j+1} - x_j)\right]}{\eta_n^{(j)}\sinh\left[\eta_n^{(j)}(x_{j+1} - x_j)\right]} + T_n^{(j)}$$

$$\frac{C_n^{(N)}\cosh\left[\eta_n^{N-1}(x_N - x_{N-1})\right] - C_n^{(N-1)}}{\eta_n^{(N-1)}\sinh\left[\eta_n^{(N-1)}(x_N - x_{N-1})\right]} + T_n^{(N-1)} = -\frac{C_n^{(N)}}{\eta_n^{(N)}} + T_n^N$$

$$(10.82)$$

Although unwieldy in appearance, this coefficient matrix is diagonally dominant and tridiagonal, and is solved extremely rapidly on a computer. Once the coefficients have been determined, the values of Φ_n follow from Eq. (10.81) and the values of $\Psi(x, y)$ from Eq. (10.74). The tangential velocity then follows directly from

$$V = (1/y)(-\alpha_i\Psi + \beta_i) \qquad (10.83)$$

and the axial velocity from

$$W = -\frac{1}{y}\frac{\partial\Psi}{\partial y} = -\sum_{n=1}^{\infty} \lambda_n\Phi_n(x)E_{0n}(y) \qquad (10.84)$$

The Radial Equilibrium Limit of the Solutions

Very useful summations are obtained by noting that, because the $T_n^{(j)}$ are independent of x, the portions of the eigenfunction expansions with coefficients $T_n^{(j)}$ correspond to the radial equilibrium solutions already obtained in Sec. 9.4. Thus, with A of Eq. (9.77) replaced by $-\omega_i$ [see Eq. (10.52)], it follows that

$$\sum_{n=1}^{\infty} T_n^{(i)} y E_{1n}(y) = -\left(\frac{-\omega_i}{\alpha_i^2} + \frac{\beta_i}{\alpha_i} + \frac{1}{2} \right) \frac{y U_1(\alpha_i y)}{U_1(\alpha_i)}$$

$$-\left(\frac{-\omega_i R^2}{\alpha_i^2} + \frac{\beta_i}{\alpha_i} + \frac{R^2}{2} \right) \frac{y W_1(\alpha_i y)}{R U_1(\alpha_i)} - \frac{\omega_i}{\alpha_i^2} y^2 + \frac{\beta_i}{\alpha_i} \qquad (10.85)$$

In the case where $\alpha_i = 0$, Eq. (9.83) gives

$$\sum_{n=1}^{\infty} T_n^{(i)} y E_{1n}(y) = -\frac{\omega_i}{8} y^4 + \left[\frac{\omega_i}{4}(R^2+1) - 1 \right] \frac{y^2}{2} - \frac{\omega_i}{8} R^2 \quad (10.86)$$

The summations relate principal contributions of the summations to the determination of the stream function. Similar summations are obtained for the axial velocity from the relationship $W = -(1/y)(d\Psi/dy)$. Thus, it follows directly from Eqs. (10.85) and (10.86) that

$$-\frac{1}{y}\frac{d}{dy}\left[\sum_{n=1}^{\infty} T_n^{(i)} y E_{1n}(y) \right] = -\sum_{n=1}^{\infty} \lambda_n T_n^{(i)} E_{0n}(y) = \left(-\frac{\omega_i}{\alpha_i} + \beta_i + \frac{\alpha_i}{2} \right)$$

$$\times \frac{U_0(\alpha_i y)}{U_1(\alpha_i)} + \left(-\frac{\omega_i R^2}{\alpha_i} + \beta_i + \frac{\alpha_i R^2}{2} \right) \frac{W_0(\alpha_i y)}{R U_1(\alpha_i)} + \frac{2\omega_i}{\alpha_i^2} \qquad (10.87)$$

and when $\alpha_i = 0$

$$-\frac{1}{y}\frac{d}{dy}\left[\sum_{n=1}^{\infty} T_n^{(i)} y E_{1n}(y) \right] = -\sum_{n=1}^{\infty} \lambda_n T_n^{(i)} E_{0n}(y)$$

$$= \frac{\omega_i}{2} y^2 - \frac{\omega_i}{4}(R^2+1) + 1 \qquad (\alpha_i = 0) \qquad (10.88)$$

Expressions for the Axial Force and Torque

When the special forms of blade loading described by Eq. (10.51) are considered, particularly simple forms of the integrals of Eqs. (10.40) and (10.46) result. Thus, the dimensionless torque may be written as

$$\frac{\tau_{1-2}}{\rho_0 w_0^2 r_h^3} = -2\pi \int_{-1/2}^{-R^2/2} [(yV)_2 - (yV)_1] \, d\Psi$$

With Eq. (10.51), integration leads to

$$\frac{\tau_{1-2}}{\rho_0 w_0^2 r_h^3} = \pi(R^2 - 1)\left[\frac{R^2+1}{4}(\alpha_2 - \alpha_1) + (\beta_2 - \beta_1) \right] \qquad (10.89)$$

The dimensionless, incompressible, and perfect form of Eq. (10.46) may be written

$$
\frac{M}{r_h^2 w_0^2} = \int_1^R \left[y \left(H - \frac{\Psi}{2} \frac{dH}{d\Psi} \right) - \frac{yV}{2y} \left(yV - \Psi \frac{dyV}{d\Psi} \right) + \frac{1}{2} \frac{d}{dy} \left(\frac{\Psi}{y} \frac{d\Psi}{dy} \right) \right] dy
$$

(10.90)

Utilizing the dimensionless radial equilibrium form of Eq. (10.31) together with Eqs. (10.51), this expression may be manipulated to give

$$
\frac{M}{r_h^2 w_0^2} = \int_1^R \left\{ y \left(H - \frac{\omega_i \beta_i}{2\alpha_i} \right) + \frac{\omega_i}{2} y\Psi + \frac{1}{2} \frac{d}{dy} \left[\frac{(\Psi - \beta_i/\alpha_i)}{y} \frac{d\Psi}{dy} \right] \right\} dy
$$

Integration then gives

$$
\frac{M}{r_h^2 w_0^2} = \left(H - \frac{\omega_i \beta_i}{2\alpha_i} \right) \frac{R^2 - 1}{2} + \frac{1}{2} \left(\frac{R^2}{2} + \frac{\beta_i}{\alpha_i} \right) W_{eR}
$$

$$
- \frac{1}{2} \left(\frac{1}{2} + \frac{\beta_i}{\alpha_i} \right) W_{e1} + \frac{\omega_i}{2} \int_1^R y\Psi \, dy
$$

(10.91)

The integral in this expression is easily obtained in terms of elementary functions when Eq. (10.84) (for Ψ) is used. The terms W_{eR} and W_{e1} refer to the equilibrium axial velocity at tip and hub, respectively, and follow directly from Eq. (10.87). The dimensionless axial force across the blade row then follows from

$$
\frac{F_A}{\rho_0 r_h^2 w_0^2} = 2\pi \left[\frac{M}{r_h^2 w_0^2} \right]
$$

(10.92)

Solution for a Single Row

The equation set (10.82) allows rapid computer evaluation of the fluid velocities when a larger number of actuator disks exist in the annulus. For small numbers of actuator disks, however, there is some advantage to analytically inverting the matrix to obtain explicit forms for the desired quantities. The simplest imaginable case is that for which only one disk is present, and for this case the solution of Eq. (10.81) reduces to

$$
\Phi_n = \frac{C_n^{(1)}}{\eta_n^{(0)}} \exp\left[\eta_n^{(0)}(x - x_1) \right] + T_n^{(0)} \qquad x < x_1
$$

$$
= - \frac{C_n^{(1)}}{\eta_n^{(1)}} \exp\left[\eta_n^{(1)}(x_1 - x) \right] + T_n^{(1)} \qquad x_1 < x
$$

(10.93)

The matching condition equation (10.82) becomes simply

$$\frac{C_n^{(1)}}{\eta_n^{(0)}} + T_n^{(0)} = -\frac{C_n^{(1)}}{\eta_n^{(1)}} + T_n^{(1)} \qquad (10.94)$$

It is thus evident that the flow adjusts in an exponential fashion from its far upstream equilibrium condition as it approaches the actuator disk, and then exponentially relaxes toward its far downstream value as it departs the disk. It is apparent from Eq. (10.60) that λ_n is given approximately by $\lambda_n \approx n\pi/(R-1)$. Closer investigation of the terms[7] also indicates that α_i must be quite small compared to λ_1 if reverse flow is not to occur. Thus, it can be seen that the slowest decaying harmonic of the series decays approximately as $\exp[-\pi|x - x_1|/(R-1)]$ away from the row. This can be a useful approximation when attempting to estimate upstream effects.

The final forms of the solutions will be summarized below, but before doing so it will be of value to note that when conditions at the blade row are considered, the terms may be regrouped somewhat to ensure more rapid convergence of the summational terms. This is particularly useful because the disappearance of the exponential decay terms in the summations causes a much slower convergence of the summational terms. A familiar result from linearized theories is that the stream function at the disk would be one-half of the far upstream and far downstream values. Utilizing this result leads to

$$\left(\Phi_n\right)_{x=x_1} = \frac{T_n^{(0)} + T_n^{(1)}}{2} + \frac{T_n^{(0)} - T_n^{(1)}}{2} + \frac{C_n^{(1)}}{\eta_n^{(0)}}$$

with Eq. (10.94), this may be manipulated to give

$$\left(\Phi_n\right)_{x=x_1} = \frac{T_n^{(0)} + T_n^{(1)}}{2} + \frac{T_n^{(0)} - T_n^{(1)}}{2}\left[\frac{\left(\eta_n^{(0)}\right)^2 - \left(\eta_n^{(1)}\right)^2}{\left(\eta_n^{(0)} + \eta_n^{(1)}\right)^2}\right] \qquad (10.95)$$

The second group in this expression will always be found to be very small and, further, will be found to decrease with increasing n much faster than the terms $C_n/\eta_n^{(0)}$.

As a particular example consider a uniform flow approaching a stator that introduces a "solid-body-like" rotation. In terms of the above parameters, this means

$$\beta_0 = \omega_0 = \alpha_0 = 0, \qquad \beta_1 = \omega_1 = 0, \qquad \alpha_1 = \alpha$$

$$\eta_n^{(0)} = \lambda_n, \qquad \eta_n^{(1)} = \sqrt{\lambda_n^2 - \alpha^2} \equiv \eta_n$$

and

$$\sum_{n=1}^{\infty} T_n^{(0)} y E_{1n}(y) = -\frac{y^2}{2} \equiv \Psi_{-\infty}$$

$$\sum_{n=1}^{\infty} T_n^{(1)} y E_{1n}(y) = -\frac{y U_1(\alpha y)}{2 U_1(\alpha)} - \frac{R y W_1(\alpha y)}{2 U_1(\alpha)} \equiv \Psi_{\infty}$$

$$-\sum_{n=1}^{\infty} \lambda_n T_n^{(0)} E_{0n}(y) = 1 = W_{-\infty}$$

$$-\sum_{n=1}^{\infty} \lambda_n T_n^{(1)} E_{0n}(y) = \frac{\alpha U_0(\alpha y)}{2 U_1(\alpha)} + \frac{\alpha R W_0(\alpha y)}{2 U_1(\alpha)} \equiv W_{\infty}$$

Then with Eq. (10.80)

$$T_n^{(0)} = -\frac{A_n}{2}, \qquad T_n^{(1)} = \frac{-\lambda_n^2}{\lambda_n^2 - \alpha^2} \frac{A_n}{2}$$

so that

$$T_n^{(0)} - T_n^{(1)} = \frac{\alpha^2}{\lambda_n^2 - \alpha^2} \frac{A_n}{2} \equiv \frac{\alpha^2}{\eta_n^2} \frac{A_n}{2}$$

Thus

$$\Psi = \Psi_{-\infty} - \frac{\alpha^2}{2} \sum_{n=1}^{\infty} \frac{A_n y E_{1n}(y)}{\eta_n(\lambda_n + \eta_n)} \exp[\lambda_n(x - x_1)] \qquad x < x_1$$

$$\Psi = \Psi_{\infty} + \frac{\alpha^2}{2} \sum_{n=1}^{\infty} \frac{\lambda_n A_n y E_{1n}(y)}{\eta_n^2(\lambda_n + \eta_n)} \exp[\eta_n(x_1 - x)] \qquad x > x_1$$

$$\Psi = \frac{1}{2}(\Psi_{-\infty} + \Psi_{\infty}) + \frac{\alpha^4}{4} \sum_{n=1}^{\infty} \frac{A_n y E_{1n}(y)}{\eta_n^2(\lambda_n + \eta_n)^2} \qquad x = x_1$$

The tangential velocity follows from Eq. (10.83) to give

$$V = 0 \qquad x < x_1$$

$$V = -(\alpha/y)\Psi \qquad x > x_1$$

$$V = -\frac{\alpha}{2y}(\Psi_{-\infty} + \Psi_{\infty}) - \frac{\alpha^5}{4y} \sum_{n=1}^{\infty} \frac{A_n y E_{1n}(y)}{\eta_n^2(\lambda_n + \eta_n)^2} \qquad x = x_1^+$$

The axial velocities follow from Eqs. (10.84), (10.87), and (10.88) to give

$$W = 1 + \frac{\alpha^2}{2} \sum_{n=1}^{\infty} \frac{\lambda_n A_n E_{0n}(y)}{\eta_n(\lambda_n + \eta_n)} \exp[\lambda_n(x - x_1)] \qquad x < x_1$$

$$W = W_{\infty} - \frac{\alpha^2}{2} \sum_{n=1}^{\infty} \frac{\lambda_n^2 A_n E_{0n}(y)}{\eta_n^2(\lambda_n + \eta_n)} \exp[\eta_n(x_1 - x)] \qquad x > x_1$$

$$W = \frac{1}{2}(1 + W_{\infty}) - \frac{\alpha^4}{4} \sum_{n=1}^{\infty} \frac{\lambda_n A_n E_{0n}(y)}{\eta_n^2(\lambda_n + \eta_n)^2} \qquad x = x_1$$

Example values for the dimensionless velocities have been calculated for the case where $R = 3$ and $\alpha = 0.84$. The results are shown in Figs. 10.5 and 10.6.

The results indicate that, for this rather large swirl introduced by the stator, substantial perturbations to the axial velocity profile are introduced. The resulting stream surface shapes and ambient diffusion or acceleration can now be obtained to determine the overall flowfield in which the blade row is imbedded. These results in turn will allow the establishment of the correct cascade geometry to determine the cascade flowfield.

Examples with more blade rows are relatively easily obtained, although if more than two blade rows are considered, the analytic inversion of the coefficient matrix [Eq. (10.82)] becomes very messy algebraically. It is much easier to simply program the matrix and numerically invert it. More complicated example solutions are provided in Refs. 7 and 8.

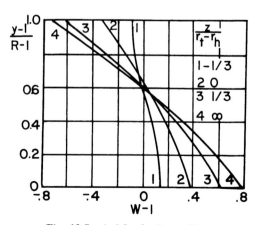

Fig. 10.5   Axial velocity profiles.

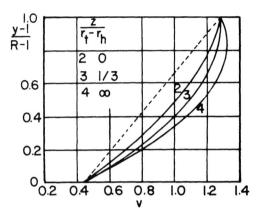

Fig. 10.6 Tangential velocity profiles.

10.6 Advanced Problems in Throughflow Theory

The example solutions illustrated in Sec. 10.5 allow rapid calculation of the flowfields existing in an annulus with many blade rows, but the solutions, although quite instructive, have incorporated into them many limitations. Thus, it was necessary to assume that the fluid was perfect and incompressible, that the blade loading (in terms of H and yV) was such as to lead to linear equations, and that the annulus had parallel walls. More advanced investigations allow some or all of these restrictions to be relaxed, so in the following sections a brief description of some of the methods available for solution of the more general forms of the throughflow equation will be outlined.

Effects of Compressibility

When the flow has substantial compressibility effects, the full coupled equations (10.30) and (10.31) together with the boundary conditions and matching conditions must be utilized. In virtually all calculation schemes, these equations are solved in an iterative manner, with Eq. (10.31) solved first (by finite difference or finite element techniques) with an assumed density distribution and then the density "updated" by utilizing the obtained values for ψ in Eq. (10.30). Equation (10.30) is then solved (iteratively) to determine a new set of values for the density, and the coupled iteration process continued until convergence is obtained. Simple Newtonian iteration is usually used to solve Eq. (10.30) for the density ratio. Thus, write

$$F \equiv \left(2h_t - v^2\right)\left(\frac{\rho}{\rho_0}\right)^2 - 2h_0 \exp\left[\frac{\gamma}{C_p}(s - s_0)\right]\left(\frac{\rho}{\rho_0}\right)^{\gamma+1}$$

$$-\left[\left(\frac{1}{r}\frac{\partial \psi}{\partial r}\right)^2 + \left(\frac{1}{r}\frac{\partial \psi}{\partial z}\right)^2\right] = 0 \qquad (10.96)$$

Then

$$\left(\frac{\rho}{\rho_0}\right)_{j+1} = \left(\frac{\rho}{\rho_0}\right)_j - \frac{F_j}{F_j'} \tag{10.97}$$

where

$$F' = 2\left[2h_t - v^2\right]\frac{\rho}{\rho_0} - 2(\gamma + 1)h_0\exp\left[\frac{\gamma}{C_p}(s - s_0)\right]\left(\frac{\rho}{\rho_0}\right)^\gamma \tag{10.98}$$

Note that

$$\left(2h_t - v^2\right) = \left(2h + u^2 + w^2\right) = 2h\left(1 + \frac{\gamma - 1}{2}M_m^2\right) \tag{10.99}$$

where M_m is the meridional Mach number.

Equation (10.99) may be combined with Eq. (10.98) to give

$$F' = 4h\frac{\rho}{\rho_0}\left\{1 + \frac{\gamma - 1}{2}M_m^2 - \frac{\gamma + 1}{2}\frac{h_0}{h}\left(\frac{\rho}{\rho_0}\right)^{\gamma-1}\exp\left[\frac{\gamma}{C_p}(s - s_0)\right]\right\}$$

It is evident from integration of the Gibbs equation (Sec. 2.9), however, that

$$\frac{h_0}{h}\left(\frac{\rho}{\rho_0}\right)^{\gamma-1}\exp\left[\frac{\gamma}{C_p}(s - s_0)\right] = 1$$

and the equation for F' hence becomes

$$F' = -2(\gamma - 1)h\frac{\rho}{\rho_0}\left(1 - M_m^2\right) \tag{10.100}$$

It is thus evident that it is the meridional Mach number (rather than the "total" Mach number) that plays the crucial role in determining the mathematical behavior of the solution. It is evident also that Newtonian iteration will fail in the vicinity of $M_m = 1$. As a result, most calculation schemes are limited to the description of flows with meridional Mach numbers less than unity, even though the total Mach number may be far in excess of unity. It is to be noted that because the density ratio itself is primarily dependent on the total Mach number, very large density changes can occur for flows with modest meridional Mach numbers. Thus, great care should be taken to properly calculate the resulting density distribution.

Hawthorne and Ringrose[9] considered the perfect flow of a calorically perfect gas through actuator disks contained within a parallel-walled annulus. The flow considered had a free vortex distribution, but was considered to be compressible. The analysis, which is one of the very few analytical treatments (as compared to numerical) of compressible flows in turbomachines, was carried out by linearizing the equations. Both the radial velocity and meridional Mach number, as well as the turning induced by the blades, were considered to be small. The results are very useful, because by restricting the study to free vortex flows in parallel-walled annuli, the perturbations in axial velocity that arise must be caused by the effect of compressibility alone. Example cases can be calculated with ease because of the analytical form of the results leading to easy physical interpretation of the results.

Fully numerical methods can usually be categorized into one of three methods: streamline curvature, finite difference, or finite element. The streamline curvature method utilizes an approximate method of solving the throughflow equations[10-13] and as such rests somewhat between the radial equilibrium method and methods that solve the full throughflow equations. The essence of the method is that approximate forms of the stream tube curvature are assumed, so that the describing equations become quite simplified. Various "curve fits" are assumed for the several example solutions detailed in Refs. 10-13.

Finite difference methods, as might be expected, incorporate finite difference approximations to the various derivatives appearing in the throughflow equation. A finite number of "nodal points" are selected, and as a result a finite set of algebraic equations results to be solved for the values of the stream function at each point. This leads to the requirement to invert a sizable matrix, and such methods are hence often referred to as matrix methods. Wu[1] in his pioneering work utilized a finite difference technique and many investigators have utilized the method since that time. Examples are given in Refs. 14 and 15.

In common with most techniques, the finite difference methods encounter computational difficulties when the meridional Mach numbers approach unity. A further difficulty arises when curved boundaries are encountered, because very complicated computational "stencils" are required to insure numerical stability. Davis[15] considers the flow in very highly curved channels and finds a 15-point stencil is required to adequately represent the Taylor series expansion of the derivatives of the stream function. A quasiorthogonal finite difference net is introduced to aid the computation. It is to be noted that the extreme curvature of the boundaries in the Davis study are such as to render most streamline curvature techniques incapable of describing the flowfield. This is because the curvature of the stream tube is very difficult to estimate in the approximate way required for streamline curvature techniques when such extreme boundary curvatures are present.

In the fairly recent past, finite element methods have been developed in the hopes of circumventing some of the difficulties found in applying finite difference techniques. Hirsch and Warzee[3] describe an investigation where the finite element method is applied to the description of flows in axial turbomachines. The compressible throughflow equations are derived in the (r, z) plane and a Galerkin finite element method is applied, leading to a system of equations for the unknown stream function. The curved hub and tip boundaries are well fit by utilizing high-order isoparametric quadrilateral elements. The method hence does not require approximation of the streamline curvature (with possible introduction of numerical instabilities or errors), nor does it require the use of complicated and extensive stencils with the resultant programming difficulty and possibly increased computational times.

Oates and Carey[16] and Oates, Knight, and Carey[17] present studies that also involve the use of a finite element approximation. In addition, a variational functional Γ is defined where

$$\Gamma = \int_{-\infty}^{\infty} \int_{\text{hub}}^{\text{tip}} \left[p - p_e + \frac{\rho}{\rho_0} (U^2 + W^2) \right] y \, dy \, dz \qquad (10.101)$$

where p_e is a (dimensionless) reference pressure.

It is shown that when the formal variation of Γ and $\delta\Gamma$ is taken and put equal to zero, the throughflow equation, boundary conditions, and all matching conditions are automatically satisfied. The method developed in Refs. 16 and 17 then involves putting the discrete approximation (using finite elements) of $\delta\Gamma$ equal to zero. A further useful manipulation is introduced in that, rather than solving for $\psi(y, z)$, the equations are transformed to solve for $y(\psi, z)$. By this artifice, the flow domain is mapped to a rectangular domain with $\psi = -\frac{1}{2}$ and $\psi = -R^2/2$ on the horizontal boundaries. These several manipulations lead to a very efficient computational scheme.

Reference 4 gives an extensive review and comparison of many throughflow calculation techniques.

All of the fully numerical calculation schemes described above can include the effects of variations in hub and tip radii. Several analytic studies to explore such effects have been carried out, however, and offer the advantage of relative simplicity in calculation of desired example cases. References 2 and 18 consider a study that includes the effect of variation in annulus radii when both the wall slopes and the annulus contraction or expansion are restricted to be small. A similar study is reported in Ref. 8, except that the restriction to small contraction or expansion is not required, although the small wall slope must be retained. Finally, Ref. 19 describes the passage of swirling flows through conical ducts. In all of the investigations into the effects of wall shape reported in Refs. 2, 8, 18, and 19, the flows are considered perfect and incompressible.

References

[1]Wu, C. H., "A General Theory of Three-Dimensional Flow in Subsonic and Supersonic Turbomachines of Axial, Radial and Mixed Flow Type," NACA TN 2604, 1952.

[2]Marble, F. E., "Three-Dimensional Flow in Turbomachines," *Aerodynamics of Turbines and Compressors, High Speed Aerodynamics and Jet Propulsion*, Vol. X, Sec. F, Princeton University Press, Princeton, N.J., 1954.

[3]Hirsch, Ch. and Warzee, G., "A Finite-Element Method for Through Flow Calculations in Turbomachines," *Transactions of ASME, Journal of Fluids Engineering*, Vol. 98, Ser. 1, Sept. 1976, pp. 403–421.

[4]*Throughflow Calculations in Axial Turbomachinery*, AGARD CP 195, Oct. 1976.

[5]Gray, A. and Mathews, G. B., *A Treatise on Bessel Functions and Their Applications to Physics*, Dover Publications, New York, 1966.

[6]Jahnke, E. and Emde, F., *Tables of Functions*, Dover Publications, New York, 1945.

[7]Oates, G. C., "Actuator Disc Theory for Incompressible Highly Rotating Flows," *Journal of Basic Engineering*, Vol. 94, Ser. D, Sept. 1972, pp. 613–621.

[8]Oates, G. C. and Knight, J., "Throughflow Theory for Turbomachines," AFAPL-TR-73-61, June 1973.

[9]Hawthorne, W. R. and Ringrose, J., "Actuator Disc Theory of the Compressible Flow in Free-Vortex Turbomachines," *Proceedings of the Institution of Mechanical Engineers*, Vol. 178, Pt. 31, April 1963–64, pp. 1–13.

[10]Smith, L. H., Traugott, S. C., and Wislicenus, G. F., "A Practical Solution of a Three-Dimensional Problem of Axial Flow Turbomachinery," *Transactions of ASME*, Vol. 75, July 1953, p. 789.

[11]Wu, C. H. and Wolfenstein, L., "Applications of Radial-Equilibrium Conditions to Axial Flow Compressor and Turbine Design," NACA Rept. 955, 1955.

[12]Seipell, C., "Three-Dimensional Flow in Multistage Turbines," *Brown Boveri Review*, Vol. 45, No. 3, 1958, p. 99.

[13]Wilkinson, D. H., "Stability, Convergence and Accuracy of Two-Dimensional Streamline Curvature Methods Using Quasi-Orthogonals," *Proceedings of the Institution of Mechanical Engineers*, Vol. 184, March 1970, p. 108.

[14]Marsh, H., "A Digital Computer Program for the Through-Flow Fluid Mechanics in an Arbitrary Turbomachine Using a Matrix Method," National Gas Turbine Establishment, Rept. R282, 1966.

[15]Davis, W. R., "A General Finite Difference Technique for the Compressible Flow in the Meridional Plane of a Centrifugal Turbomachine," ASME Paper 75 GT 121, International Gas Turbine Conference, Houston, Texas, March 2–6, 1975.

[16]Oates, G. C. and Carey, G. F., "A Variational Formulation of the Compressible Throughflow Problem," AFAPL-TR-74-78, Nov. 1974.

[17]Oates, G. C., Knight, C. J., and Carey, G. F., "A Variational Formulation of the

Compressible Throughflow Problem," *Transactions of ASME, Journal of Engineering for Power*, Ser. A, Vol. 98, Jan. 1976, pp. 1–8.

[18]Oates, G. C., "Theory of Throughflow in Axial Turbomachines with Variable Wall Geometry," California Institute of Technology, Jet Propulsion Center, Pasadena, Tech. Note 1, AFOSR TN 59-680, Aug. 1959.

[19]Horlock, J. H. and Lewis, R. I., "Non-Uniform Three-Dimensional and Swirling Flows Through Diverging Ducts and Turbomachines," *International Journal of Mechanical Science*, Vol. 3, 1961, pp. 170–196.

Problems

10.1   "Magnetofluid-dynamics" (MFD) involves the study of the interaction of magnetic and electric fields and fluids. When a plasma is considered (such that there is no significant net electrical charge in the fluid), the effect of the added interactions is to introduce the "Lorentz force" $\mathbf{j} \times \mathbf{B}$ into the momentum equation and the energy addition/second $\mathbf{j} \cdot \mathbf{E}$ into the first law of thermodynamics. Thus, if the viscous contributions are negligible, the momentum equation and equation for the stagnation enthalpy become

$$\rho \frac{D\mathbf{u}}{Dt} = \mathbf{j} \times \mathbf{B} - \nabla p$$

$$\rho \frac{Dh_t}{Dt} = \frac{\partial p}{\partial t} + \mathbf{j} \cdot \mathbf{E}$$

where $\mathbf{j}$ is the electrical current density, $\mathbf{B}$ the magnetic field, and $\mathbf{E}$ the electric field.

(a) Show that the equation for the variation in entropy may be written

$$\rho T \frac{DS}{Dt} = \mathbf{j} \cdot \mathbf{E}'$$

and find $\mathbf{E}'$ in terms of $\mathbf{u}$, $\mathbf{B}$, and $\mathbf{E}$.

(b) Noting the second Maxwell equation $\nabla \cdot \mathbf{B} = 0$ and the equation for current continuity $\nabla \cdot \mathbf{j} = 0$, show that an equation for the variation in vorticity can be written in the form

$$\rho \frac{D\boldsymbol{\omega}/\rho}{Dt} = (\boldsymbol{\omega} \cdot \nabla)\mathbf{u} + \nabla \frac{1}{\rho} \times [\mathbf{j} \times \mathbf{B} - \nabla p] + \frac{1}{\rho}[(\mathbf{B} \cdot \nabla)\mathbf{j} - (\mathbf{j} \cdot \nabla)\mathbf{B}]$$

10.2   Consider the ideal incompressible flow of a fluid through a stator represented as an actuator disk at $z = 0$. The flow approaches the disk from a uniform state far upstream in the parallel-walled annulus, and swirl is imparted such that in terms of the dimensionless variables,

$$yV = -\alpha\Psi + \beta$$

(a) Obtain analytical forms for Ψ, W, and V in terms of A_n, B_n, and other prescribed variables for both the upstream and downstream quantities.

(b) Obtain somewhat simplified forms for Ψ, W, and V at a location just downstream of the disk (at $z = 0^+$).

(c) For the case $R = 2.8$, $\alpha = 0.6$, and $\beta = 0.84$, calculate and plot $(W - 1)$ vs $(y - 1)/(R - 1)$ at $z/(r_t - r_h) = -\frac{1}{3}, 0, \frac{1}{3}$, and ∞.

(d) For the value of δ and R found in part (c), calculate and plot $(W - 1)$ vs $(y - 1)/(R - 1)$ at $z/(r_t - r_h) = -\frac{1}{3}, 0, +\frac{1}{3}$, and ∞.

10.3 For conditions as in Problem 10.2, except that the stator imparts a swirl such that

$$(yV)^2 = -\delta\Psi$$

(a) Obtain analytical forms for Ψ, W, and V in terms of y, δ, and R appropriate for each of the regions $z < 0$ and $z > 0$.

(b) Obtain a closed-form solution for $W_{z=0}$ in terms of y, δ, and R.

(c) For the case $R = 3$ find the value of δ that just leads to $W_\infty(R) = 0$.

(d) For the value of δ and R found in part (c), calculate and plot $(W - 1)$ vs $(y - 1)/(R - 1)$ at $z/(r_t - r_h) = -\frac{1}{3}, 0, +\frac{1}{3}$, and ∞.

(e) For the same values as part (c), plot V vs $(y - 1)/(R - 1)$ at $z/(r_t - r_h) = 0^+, \frac{1}{3}$, and ∞.

10.4 Consider the ideal flow of an incompressible fluid through a rotor at x_1 and then through a stator at x_2. The annulus radii are constant and the rotor introduces a (dimensionless) swirl given by

$$yV = -\alpha\Psi$$

The stator removes all the swirl.

(a) Obtain expressions for the swirl and stagnation enthalpy valid for each of the three regions 0, 1, and 2. Write the appropriate partial differential equation for the stream function in each region.

(b) Obtain the radial equilibrium form of the stream function in each of the three regions.

(c) Utilize the matching conditions to solve for any remaining unknowns so as to obtain analytical forms of the solutions in terms of y, α, R, and Ω (dimensionless rotor speed).

(d) Obtain simplified forms of the solutions valid at each of the actuator disks.

(e)   Calculate and plot $(W - 1)$ vs $(y - 1)/(R - 1)$ at both $x = x_1$ and $x = x_2$ for the case $\alpha = 0.7$, $\Omega = 1$, $R = 3$, and $x_1 = 0$, $x_2 = 2$.

10.5   Consider the ideal flow of an incompressible fluid through a stator row at x_1 and then through a rotor row at x_2. The rows introduce swirl such that in each region

$$yV = 0 \qquad\qquad x < x_1$$

$$yV = -\alpha_1\Psi + \beta_1 \qquad\qquad x_1 < x < x_2$$

$$yV = -\alpha_2\Psi + \beta_2 \qquad\qquad x_2 < x < \infty$$

(a)   Obtain analytical solutions for W valid in each of the three regions in terms of y, α_1, α_2, β_1, β_2, x_1, x_2, and Ω, the dimensionless rotor speed.

(b)   Indicate how you would obtain analytical forms for W valid at x_1 and x_2 that would hasten convergence of the series.

11. CASCADE FLOWS

11.1 Introduction

When the throughflow field (Chap. 10) has been determined, the blade profiles necessary to induce the desired fluid conditions can then be (approximately) determined by consideration of the cascade flowfield. As previously discussed (Sec. 9.1), a cascade flowfield is obtained by "unwrapping" the desired meridional surface that has been determined from the throughflow analysis. The required blade geometries necessary to give the desired flow turning efficiently for the particular "strip" considered are then obtained by experimental and/or theoretical consideration of a quasi-two-dimensional configuration such as that indicated in Fig. 11.1.

It should be noted that because the meridional surface will, in general, have a streamwise varying cross-sectional area (with its attendant imposed pressure gradients), the cascade wind tunnel should be constructed to impose this desired area variation. This is by no means a simple experimental task, because the upper and lower walls will have to be adjusted (in a curved fashion) to include not only the area variation actually occurring in the throughflow, but also the corrections that have to be made for the growth of the wind-tunnel sidewall boundary layers.

Even when the streamwise variation of the cross-sectional area is well approximated, a further problem of considerable difficulty remains. It is apparent that the flow is (very nearly) periodic in the actual (annular) blade row, so it is important that enough blades be incorporated in the two-dimensional cascade to ensure that the required periodicity occurs over the middle (test) blades. The results of considerable discussion on this matter are reported in Ref. 1. The specialists' estimates of the minimum number of blades required were 5–15 for subsonic cascades and 3–9 for supersonic cascades. The required minimum number of blades was considered to depend somewhat on the purpose of the tests, so that, for example, if only the surface pressure distribution of a blade was to be determined, relatively few blades would be required. In contrast, if an accurate estimate of the cascade losses was desired, a large number of blades would be required.

It is apparent that the approximation to exact periodicity can be aided somewhat by shaping the sidewalls to approximate the expected approach streamline shape. In practice, this is rarely attempted because the shape of the approaching streamline will, of course, vary with the loading on the

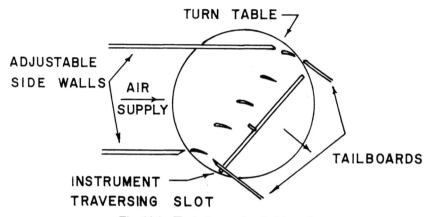

Fig. 11.1 Typical cascade wind tunnel.

cascade. As indicated in Fig. 11.1, "tailboards" are sometimes used to reduce the effects of the external flow interaction at the sidewalls.

Usually, the number of blades utilized in a given facility cannot be greatly increased because of the resulting small size of the blades (if the solidity is to be maintained at the desired value). Excessively small blades create difficulties in that it becomes hard to maintain the appropriate range of Reynolds numbers. Further, when detailed flow information such as the surface pressure distribution is desired, small blade sizes lead to great instrumentation difficulties.

In spite of these many difficulties, the results of careful cascade tests remain a most important source of information for evaluating the performance of and determining the detailed flow behavior of the many candidate blade profiles considered for use in turbomachinery. It should be recognized, however, that even "routine" cascade tests should be carried out with a great deal of care.

11.2 Cascade Losses

When a cascade has been successfully constructed to minimize the problems discussed in the preceding section, the performance can be determined by traversing stagnation pressure probes and yaw meters across the exit plane.

The detailed information obtained from the instrument traverses can be presented in the form indicated in Fig. 11.2. Each setting of the inlet incidence angle and Mach number will have identified with it one such graph.

Customarily, the detailed information contained in Fig. 11.2 is averaged in one of several ways so that the effect of variation in the angle of attack can be presented in a single graph such as that shown in Fig. 11.3.

A common form of averaging is that of mass flow averaging. Thus, for example, the gas exit angle β_2 would be obtained in terms of the mass-flow-

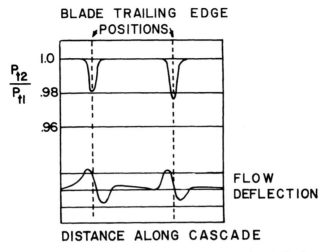

Fig. 11.2   Variation in stagnation pressure and flow deflection.

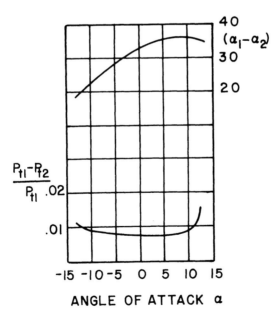

Fig. 11.3   Variation of average stagnation pressure loss and turning angle with angle of attack.

averaged tangential momentum and mass-averaged axial momentum. Thus,

$$\tan \beta_2 = \int_0^S \rho u v \, d\ell \bigg/ \int_0^S \rho u^2 \, d\ell \qquad (11.1)$$

Similarly, the mass-flow-averaged pressure loss would be obtained from

$$\Delta p_t = \int_0^S \rho u (p_{t_1} - p_{t_2}) \, d\ell \bigg/ \int_0^S \rho u \, d\ell \qquad (11.2)$$

It is evident, particularly in the case of the mass-flow-averaged pressure loss that considerable difficulty will arise when the results of different cascade measurements are compared. This is because ΔP_t as defined by Eq. (11.2) will change (for the same cascade test) with the distance from the cascade exit at which the measurements are taken. This problem is present even if the cascade wall losses do not intrude into the measurement region, because the wake mixing process itself introduces further entropy gains.

It will be recalled from Sec. 6.3 that great care must be taken when employing average values to describe component performance. The warnings and examples of that section can again be referenced with regard to the problem of depicting cascade performance.

As a result of the difficulty of interpreting mass-flow-averaged pressure losses, it has been suggested (for example, Ref. 2) that "mixed-out" values of the pressure loss be used to compare the performance as determined from different cascades. The mixed-out value corresponds to that value of the pressure loss that would exist if the fluid were allowed to fully mix in an ideal (no sidewall friction) constant-area mixer. It is also pointed out in Ref. 2 that the results will have to be carefully interpreted no matter what

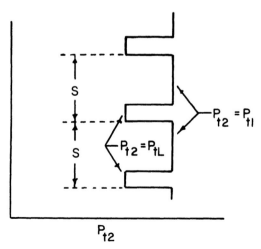

Fig. 11.4 Simplified stagnation pressure profile.

averaging method is used, because the blade wakes will have a substantial interaction with the following blade rows.

A measure of the expected size of the effect of averaging techniques can be obtained from a simple example. Thus, assume a simple abrupt wake to be imbedded in a flow that otherwise has experienced no stagnation pressure decrease, Fig. 11.4. Denoting the ratio of the mass flow with stagnation pressure decrement to the mass flow without stagnation pressure decrement as α, it follows that

$$\frac{\Delta p_{t \text{ mass av}}}{p_{t_1}} = \frac{1}{p_{t_1}} \frac{\int_0^S \rho u (p_{t_1} - p_{t_2}) \, d\ell}{\int_0^S \rho u \, d\ell} = \frac{\alpha}{1 + \alpha} \left(1 - \frac{p_{tL}}{p_{t_1}} \right) \quad (11.3)$$

The mixed-out stagnation pressure losses for this simple flow may be calculated by utilizing the results of Sec. 6.3 directly. As an example consider the case with a Mach number of the high-pressure stream M_1 equal to 0.6, $\alpha = 0.2$, $\gamma = 1.4$, and the stagnation temperature ratio of the two streams equal to unity. The result is shown in Table 11.1.

These simple calculations indicate that it is very important to be aware of which averaging technique has been utilized for a given data set and, in addition, at which location the measurement traverse was taken. Note that an alternate way of looking at this problem is to realize that if two data sets were compared where one set of measurements were taken far downstream (where conditions approach mixed-out) and $\Delta p_{t \text{ mass av}}/p_{t_1}$ were found to be 0.0333, and then a second set of data taken in close proximity of the cascade exit, which also had $\Delta p_{t \text{ mass av}}/p_{t_1} = 0.0333$, the second cascade would in fact have almost double the losses of the first.

In an actual turbomachine, the various loss mechanisms are quite interactive. Thus, the flow profiles departing one blade row affect the losses produced in the following row. In addition, losses occurring on the annulus walls and at the blade tips affect the total losses produced within the blade row and are themselves influenced by the blade losses. It is the hope in representing the entire flowfield as a compilation of three two-dimensional fields that knowledge of the "pure" cascade flowfield will allow accurate estimates to be made of other losses. Thus, for example, when determining the performance of a cascade, every effort is made to remove the influence

Table 11.1   Stagnation Pressure Losses

p_{tL}/p_{t_1}	$\Delta p_{t \text{ mixed out}}/p_{t_1}$	$\Delta p_{t \text{ mass av}}/p_{t_1}$
0.95	0.0090	0.0083
0.9	0.0197	0.0167
0.85	0.0336	0.0250
0.8	0.0625	0.0333

of the cascade wall losses from the data. Later, however, the losses experienced in the turbomachine at the annulus walls will be estimated with the help of the cascade results.

Several major flow interactions have been classified and attempts made to analyze them. An important example of such flows is termed the secondary flow. Secondary flows have been thoroughly reviewed in Ref. 3, where many example applications of secondary flow analyses are described. With regard to their applicability for use in the analysis of flows in turbomachines, secondary flow analyses are particularly useful in describing the flow through the compressor rows. Figure 11.5 depicts a distorted flow profile such as could be produced by the annuli walls approaching a stator row. When fluid with such a profile enters the pressure field of the blades, the low-momentum air is turned more ("overturned") than the high-momentum air, leading to the secondary flow patterns indicated.

Secondary flow in compressors is particularly amenable to analysis because to a good approximation the viscous stresses may be ignored in the region of the blades. This is an allowable approximation because the adverse pressure gradient imposed by the blades moves the flow toward separation with consequent low viscous stresses. The effects of viscosity are, of course, implicitly present in the distorted entry profile that has been developed by the long approach flow over the annulus walls. It is to be noted that the favorable high-pressure gradient found over much of the turbine blade profile leads to thin boundary layers that must be analyzed using (three-dimensional) boundary-layer techniques.

Cascade tests are necessarily conducted with single rows of blades, so the effects of upstream blade rows (particularly moving blade rows) cannot be simulated. The effects of shed blade wakes can be substantial, however, and several extensive studies to describe the flow interactions have been conducted. Reference 4 extends the analytical models of the earlier studies of Refs. 5 and 6 and provides experimental verification of the analytical

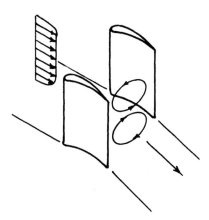

Fig. 11.5   Secondary flow patterns.

predictions. A more recent study is reported in Ref. 7, where a method of directly estimating the stagnation temperature profile following a stator in terms of the rotor blade loss factor is presented. The latter study has particular utility in the description of flow in high Mach number stages.

In summary, it should be noted that the various losses occurring in turbomachines, other than two-dimensional profile losses described above, are not those to be found in a cascade wind-tunnel investigation. The cascade flowfield, however, appears as the "parent" flowfield for all these secondary flows, just as the throughflow field appeared as the parent for the cascade flowfield. Hence the results of cascade studies are of utility not only in determining the blade losses, but also in establishing the cascade flowfield so that the secondary losses can be estimated.

11.3 Cascade Notation

Typical cascade notation is illustrated in Fig. 11.6 where

$$
\begin{aligned}
\text{subscript } 1 &= \text{inlet condition} \\
\text{subscript } 2 &= \text{outlet condition} \\
\beta &= \text{flow angle} \\
\gamma &= \text{angle of blade camber line} \\
\gamma_1 &= \text{stagger angle} \\
\theta^* &= \text{blade camber angle} = \gamma_1 - \gamma_2 \\
\phi &= \text{flow turning angle} = \beta_1 - \beta_2 \\
\delta &= \text{deviation angle} = \beta_2 - \gamma_2 \\
\alpha &= \text{angle of attack} = \beta_1 - \gamma_1 \\
w &= \text{"axial" } (x) \text{ velocity} \\
v &= \text{magnitude of the velocity} \\
S &= \text{spacing} \\
C &= \text{chord} \\
\sigma &= \text{solidity} = C/S
\end{aligned}
$$

When presenting the data, as already depicted in Fig. 11.3, it is usual to depict the stagnation pressure loss in a dimensionless manner, such as

$$ \Delta p_t / \tfrac{1}{2}\rho_1 V_1^2 $$

It is customary to describe an airfoil shape in terms of its thickness distribution about a prescribed camber line. The camber line is often taken to be of parabolic shape (which reduces to a circular arc profile as a special case). Many attempts have been made to relate the blade camber angle and other geometric properties of the blade to the deviation angle. (See particularly Chap. 6 of Ref. 8.) A quite convenient approximate form has also been suggested[9,10] to relate the value of the deviation angle at design conditions δ^* to the blade geometry and flow exit angle at design β_2^*. The suggested

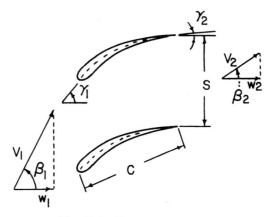

Fig. 11.6 Cascade notation.

relationship is

$$\delta^* = \left(m/\sqrt{\sigma} \right)\theta^* \tag{11.4}$$

where

$$m = 0.23(2a/C)^2 + 0.1(\beta_2^*/50)$$

In this expression a is the distance from the blade leading edge to the point of maximum camber and β_2^* is to be measured in degrees.

By utilizing this approximate relation, the blade turning angle necessary to introduce the desired flow turning can be easily estimated.

11.4 Calculational Methods

Now consider the quasi-two-dimensional flow in the cascade plane. Figure 11.7 illustrates the coordinate system. The flow will be assumed to be of depth b, where b is a function of x alone.

It is apparent that if vorticity is to be present in the flowfield it must be directed perpendicular to the cascade plane. With ideal (isentropic) flow assumed, with uniform properties far upstream, Eq. (10.25) then gives directly

$$\boldsymbol{\omega} \times \mathbf{u} = 0 \tag{11.5}$$

but because $\boldsymbol{\omega}$ must be perpendicular to $\mathbf{u}$, $\boldsymbol{\omega}$ itself must be zero. It is thus possible to define a potential ϕ, such that

$$\mathbf{u} = \nabla\phi \tag{11.6}$$

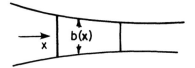

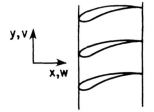

Fig. 11.7 Cascade coordinate system.

The quasi-two-dimensional form of the continuity equation may be written

$$\frac{\partial(\rho u b)}{\partial x} + \frac{\partial(\rho v b)}{\partial y} = 0 \qquad (11.7)$$

Utilizing this equation, a stream function may be defined by the relationships

$$u = \frac{1}{\rho b}\frac{\partial \psi}{\partial y}, \qquad v = \frac{-1}{\rho b}\frac{\partial \psi}{\partial x} \qquad (11.8)$$

The condition that the vorticity is zero,

$$\frac{\partial u}{\partial y} - \frac{\partial v}{\partial x} = 0 \qquad (11.9)$$

then gives an equation for the stream function that may be written

$$\nabla^2 \psi = \frac{1}{b}\frac{db}{dx}\frac{\partial \psi}{\partial x} + \frac{1}{\rho}\frac{\partial \rho}{\partial y}\frac{\partial \psi}{\partial y} + \frac{1}{\rho}\frac{\partial \rho}{\partial x}\frac{\partial \psi}{\partial x} \qquad (11.10)$$

As in Sec. 10.2, a subsidiary equation for the density may be obtained by utilizing the isentropic condition. Thus

$$\frac{\rho}{\rho_1} = \left(\frac{T}{T_1}\right)^{1/(\gamma-1)} = \left[\frac{h_t - \frac{1}{2}(u^2 + v^2)}{h_t - \frac{1}{2}(u_1^2 + v_1^2)}\right]^{1/(\gamma-1)}$$

and hence with Eq. (11.8)

$$h_t \rho^2 - \left[h_t - \tfrac{1}{2}\left(u_1^2 + v_1^2 \right) \right] \rho_1^2 \left(\frac{\rho}{\rho_1} \right)^{\gamma+1} - \frac{1}{2b^2} \left[\left(\frac{\partial \psi}{\partial y} \right)^2 + \left(\frac{\partial \psi}{\partial x} \right)^2 \right] = 0$$

(11.11)

Equations (11.10) and (11.11) are two coupled nonlinear equations for ψ and ρ in terms of the far upstream conditions and the prescribed area variations. When the flow can be approximated as incompressible, the equation for the stream function reduces to

$$\nabla^2 \psi = \frac{1}{b} \frac{\mathrm{d}b}{\mathrm{d}x} \frac{\partial \psi}{\partial x}$$

(11.12)

The solution to the above equations is usually dependent upon the application of fully numerical techniques, although analytical methods have been applied in special cases. References 11 and 12 report on a study of incompressible flow with area variation. A perturbation analysis is carried out in which it is necessary to restrict the wall slope to be small.

A method that applies conformal mapping techniques to two-dimensional ($\mathrm{d}b/\mathrm{d}x = 0$) transonic flows is described in Ref. 13. This paper represents an application of a highly developed analytical technique. In conformal mapping techniques, the cascade geometry is transformed to a geometry that is much more simply analyzed (either analytically or numerically) and the results of the analysis in the simple plane are then transformed back to the more complex plane. Because of the utility of such techniques, the next section considers a relatively simple example problem that leads to an exact solution for the flowfield.

Two-Dimensional, Inviscid, Incompressible Flow

When the case of flow of an incompressible, inviscid fluid in a strictly two-dimensional channel is considered, the equations describing the flowfield simplify greatly and, in fact, both the velocity potential and the stream function satisfy Laplace's equation. As a result, the very powerful techniques of complex variable theory can be used, including the technique of conformal mapping. Before embarking upon a formal analysis of the problem, however, it is instructive to first consider the force relationships and circulation relationships for a two-dimensional cascade. Figure 11.8 depicts the flow through such a cascade.

Noting that the forces on the two streamlines depicted are equal and that continuity ensures that the axial (x) velocity remains unchanged, the momentum equation gives

$$F_x = (p_1 - p_2) S \Delta b$$

(11.13)

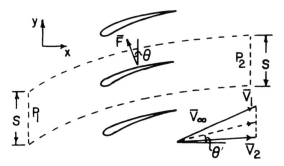

Fig. 11.8 Flow through a two-dimensional cascade.

and

$$F_y = \rho u(v_1 - v_2) S\Delta b \tag{11.14}$$

Utilizing Bernoulli's equation to relate the pressure difference to velocity differences, the vector force may then be written

$$\mathbf{F} = \rho S\Delta b(v_1 - v_2)\left[-\tfrac{1}{2}(v_2 + v_1)\mathbf{e}_x + u\mathbf{e}_y\right] \tag{11.15}$$

Thus the magnitude of the force F is given by

$$F = \rho S\Delta b(v_1 - v_2)V_\infty$$

where

$$V_\infty = \left[\left(\frac{v_2 + v_1}{2}\right)^2 + u^2\right]^{\frac{1}{2}} \tag{11.16}$$

Noting that

$$\tan\theta = (v_2 + v_1)/2u = \tan\theta'$$

it can be seen that the force $\mathbf{F}$ is perpendicular to the velocity $\mathbf{V}_\infty$, where

$$\mathbf{V}_\infty = \tfrac{1}{2}(\mathbf{V}_1 + \mathbf{V}_2) \tag{11.17}$$

The circulation about the contour follows immediately to give

$$\Gamma = S(v_1 - v_2) \tag{11.18}$$

so that just as with an isolated airfoil, there is obtained

$$F = \rho\Gamma V_\infty \Delta b \tag{11.19}$$

The Complex Potential

When, as in the case of frictionless two-dimensional flow through cascades, the flow may be considered irrotational, a velocity potential may be defined as in Eq. (11.6). The continuity equation (10.1) then gives for incompressible flow,

$$\nabla^2 \phi = 0 \qquad (11.20)$$

The stream function also satisfies Laplace's equation, as may be seen directly from Eq. (11.10) for the case of incompressible flow with constant-flow depth b. Thus, define a complex potential W by

$$W = \phi + i\psi \qquad (11.21)$$

and note that

$$\nabla^2 W = 0 \qquad (11.22)$$

If the complex potential is to satisfy Laplace's equation, it must be a function of $z \equiv x + iy$ only. Thus,

$$\frac{\partial W}{\partial x} = \frac{dW}{dz}\frac{\partial z}{\partial x} = \frac{dW}{dz} \qquad (11.23)$$

but from Eqs. (11.6) and (11.8)

$$u = \frac{\partial \phi}{\partial x} = \frac{\partial \psi}{\partial y}, \qquad v = \frac{\partial \phi}{\partial y} = -\frac{\partial \psi}{\partial x} \qquad (11.24)$$

and hence

$$\frac{dW}{dz} = u - iv \qquad (11.25)$$

In the following sections, a transformation function is introduced to allow solution for the complex potential W (for special cases of the cascade geometry). Once W is obtained, the velocity components may be obtained directly from Eq. (11.25).

The Cascade Transformation

In order to illustrate the use of conformal mapping techniques, a transformation is considered that maps a cascade of straight-line airfoils onto a circle. The resulting flow in the "circle plane" is relatively easy to analyze, and the cascade mapping function allows the flow in the circle plane to be mapped back into the "physical" plane. This is a relatively simple example of a highly developed analytical technique that has been extensively reported elsewhere. (For example, see Refs. 14 and 15.)

The transformation function to be investigated is

$$z = \frac{S}{2\pi} \left(e^{-i\beta} \ell n \frac{e^{\psi} + \zeta}{e^{\psi} - \zeta} + e^{i\beta} \ell n \frac{e^{\psi} + 1/\zeta}{e^{\psi} - 1/\zeta} \right) \qquad (11.26)$$

It will be shown that this function takes a unit circle in the ζ plane and maps it into a straight-line cascade in the z plane, with geometry as indicated in Fig. 11.9. Note that for notational convenience, the geometry has been rotated so that the flat-plate airfoils are horizontal in the z plane.

To verify that Eq. (11.26) has the desired transformation properties, first note that on the unit circle (where $\zeta = e^{i\phi}$) the second term in the brackets of Eq. (11.26) is the complex conjugate of the first. Thus, the imaginary part of the expression must be zero (or a constant) and hence the circle is mapped onto the x axis in the z plane.

A further property of the transformation is evident in that the points $\zeta = \pm e^{\psi}$ map into $\pm \infty$ in the z plane. It can be seen also that the mapping is multiple valued by considering the mapping in the neighborhood of the point $\zeta = e^{\psi}$. Thus, consider the behavior of the related point z when proceeding around the point $\zeta = e^{\psi}$ in a small circle of radius r. That is, consider

$$\zeta = e^{\psi} + re^{i\theta}$$

and change θ from 0 to 2π rad. It is evident from Eq. (11.26) that all of the logarithmic terms return to their original value except the term containing $e^{\psi} - \zeta$. Thus, denoting the difference of the final and initial values of z as $z_2 - z_1$, there is obtained

$$z_2 - z_1 = -\frac{S}{2\pi} e^{-i\beta} \left[\ell n(-re^{i2\pi}) - \ell n(-re^{i0}) \right]$$

$$= -\frac{S}{2\pi} e^{-i\beta} \ell n \, e^{i2\pi}$$

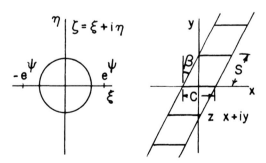

Fig. 11.9 The cascade transformation.

hence

$$z_2 - z_1 = -iSe^{-i\beta} = -S\sin\beta - iS\cos\beta \qquad (11.27)$$

Thus, the transformation from the ζ to the z plane is not unique, but rather a given point in the ζ plane can be reproduced n times in the z plane simply by going around the point

$$\zeta = e^{\psi} \qquad \left(\text{or } \zeta = -e^{\psi}\right)$$

n times. Each point is removed from the previous by the vector $-S\sin\beta - iS\cos\beta$, which leads to the geometrical relationship indicated in Fig. 11.9. The ζ plane may be considered to consist of an infinite number of Riemann sheets, each sheet mapping to a given strip in the z plane. The z plane itself can be considered a two-sheet plane because both the inside and outside of the unit circle map into the same strip in the z plane. Only the mapping of the exterior of the circle will be considered here.

It has been shown that Eq. (11.26) transforms the unit circle to a series of straight-line airfoils in the z plane. The airfoils are staggered at an angle of β and have a spacing of magnitude S. In order to determine the airfoil chord in terms of the properties of the mapping function, locate the transformation singularities that occur at $dz/d\zeta = 0$. (It is at these singularities that the transformation is not conformal, and hence the angle of the circular profile is not conserved in transforming to the z plane.) Thus, note

$$\frac{dz}{d\zeta} = 0 = e^{-i\beta}\left(\frac{1}{e^{\psi} + \zeta} + \frac{1}{e^{\psi} - \zeta}\right) + e^{i\beta}\left(\frac{1}{\zeta + e^{-\psi}} - \frac{1}{\zeta - e^{-\psi}}\right)$$

After some manipulation it follows that

$$\zeta^2 = \zeta_0^2 = \frac{e^{i\beta + \psi} + e^{-(i\beta + \psi)}}{e^{-i\beta + \psi} + e^{i\beta - \psi}} \qquad (11.28)$$

It can be noted that the numerator and denominator are complex conjugates and, hence, of course, the singularities exist on the unit circle in the ζ plane. A more convenient form of Eq. (11.28) is obtained in terms of the angle to the location of the (rear) singularity ϕ_0 where

$$\zeta_0 = e^{i\phi_0}$$

There is thus obtained

$$\frac{\zeta_0^2 - 1}{\zeta_0^2 + 1} = \frac{e^{i\beta} - e^{-i\beta}}{e^{i\beta} + e^{-i\beta}}\frac{e^{\psi} - e^{-\psi}}{e^{\psi} + e^{-\psi}}$$

hence

$$i \tan \phi_0 = i \tan \beta \tanh \psi$$

or

$$\phi_0 = \tan^{-1}(\tan \beta \tanh \psi) \qquad (11.29)$$

The location of the blade leading edge in the ζ plane follows immediately by noting $\tan(\phi_0 + \pi) = \tan \phi_0$. Hence, the leading-edge singularity occurs at $\phi_0 + \pi$.

A location x on the blade in the ζ plane may be determined by writing $\zeta = e^{i\phi}$. Upon substituting into Eq. (11.26) and after some manipulation, there is then obtained

$$\frac{2\pi x}{S} = \cos \beta \, \ell n\left(\frac{\cosh \psi + \cos \phi}{\cosh \psi - \cos \phi}\right) + 2 \sin \beta \tan^{-1}\left(\frac{\sin \phi}{\sinh \psi}\right) \qquad (11.30)$$

The location of the blade trailing edge is obtained by inserting the value of ϕ_0 from Eq. (11.29) into Eq. (11.30) and the leading edge follows with $\phi = \phi_0 + \pi$ to give $x_\ell = -x_t$. Then, with $C = 2x_t$,

$$\frac{\pi C}{2S} = \cos \beta \, \ell n\left(\frac{\sqrt{\sinh^2\psi + \cos^2\beta} + \cos \beta}{\sinh \psi}\right)$$

$$+ \sin \beta \tan^{-1}\left(\frac{\sin \beta}{\sqrt{\sinh^2\psi + \cos^2\beta}}\right) \qquad (11.31)$$

This expression relates the parameter ψ to the solidity $\sigma = C/S$. Thus, when the solidity is prescribed, the equation may be solved (iteratively) for ψ and the appropriate transformation function determined. Before considering the behavior of the flowfield, one last characteristic of the transformation will be observed. Thus, it is noted that the point $\zeta \Rightarrow \infty$ transforms to

$$z \Rightarrow \frac{S}{2\pi} e^{-i\beta} \ell n(-1) = \frac{S}{2\pi} e^{-i\beta} \ell n[e^{i(2n+1)\pi}]$$

hence

$$z \Rightarrow S(n + \tfrac{1}{2})(\sin \beta + i \cos \beta)$$

Thus, $\zeta \Rightarrow \infty$ transforms into points midway between the blades. See Fig. 11.10.

This concludes the investigation of the properties of the transformation. The behavior of the flowfield will be considered in the next section. Note, however, that the effect of the transformation has been to bring upstream and downstream infinity in the x plane and into the proximity of the circle in the ζ plane. When circulation about the circle exists, the angle of flow in the proximity of $\zeta = \pm e^{\psi}$ ($z = \pm \infty$) can be affected. Thus, unlike the case for an isolated two-dimensional airfoil, the angle of turning of the fluid can be made other than zero.

The Cascade Flowfield

Figure 11.11 depicts aspects of the flowfield in the coordinate system being presently considered.

The mean velocity V_{∞} was previously defined in Eq. (11.16), so the mean complex velocity may be designated

$$(u - iv)_{\infty} = V_{\infty} e^{-i\alpha}$$

Equation (11.18) gave the circulation in terms of the blade spacing change and tangential velocities, which here may be written

$$\Delta v = \Gamma / 2S \qquad (11.32)$$

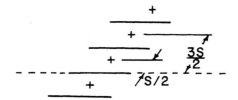

Fig. 11.10 Location of $\zeta \to \infty$ in the Z plane.

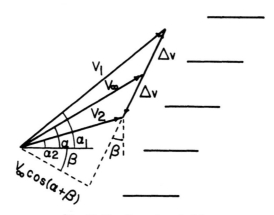

Fig. 11.11 Cascade velocities.

Thus, the desired far upstream and far downstream complex velocities may be written

$$\left.\frac{dW}{dz}\right|_{-\infty} = V_{\infty}e^{-i\alpha} - i\frac{\Gamma}{2S}e^{i\beta} \tag{11.33}$$

$$\left.\frac{dW}{dz}\right|_{+\infty} = V_{\infty}e^{-i\alpha} + i\frac{\Gamma}{2S}e^{i\beta} \tag{11.34}$$

Figure 11.12 indicates these incremental velocity relationships.

In order to create the desired velocities at upstream and downstream infinity, as prescribed in Eqs. (11.33) and (11.34), the flow behavior in the circle plane in the vicinity of the points $\pm e^{\psi}$ is investigated. Thus, by placing (complex) sources at these points, the flow at $\pm \infty$ will be affected in the physical plane. It will then be necessary to adjust conditions in the vicinity of the circle in the ζ plane in order to satisfy both the boundary condition of no flow through the circle and the Kutta condition at the trailing edge of the blades in the z plane. It is to be noted that the adjustments to match the boundary and Kutta conditions can be carried out without further affecting the flow properties at $z = \pm \infty$ because $dz/d\zeta$ approaches infinity as ζ approaches $\pm e^{\psi}$ (i.e., as z approaches infinity) so the only terms that will contribute to dW/dz at $z = \pm \infty$ are those causing $dW/d\zeta$ to also approach infinity. The only terms that lead to $dW/d\zeta$ approaching infinity as ζ approaches $\pm e^{\psi}$ arise from the complex sources located at $\zeta = \pm e^{\psi}$.

Now place a complex source of strength A at $\zeta = -e^{\psi}$, so that the complex potential in the vicinity of the source W_{s-} may be written

$$W_{s-} = (A/2\pi)\ell n(\zeta + e^{\psi}) \tag{11.35}$$

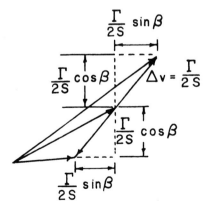

Fig. 11.12   Incremental velocity relationships.

To investigate the behavior near the source write $\zeta = -e^{\psi} + \delta$, where δ is very small, and obtain

$$\frac{dW_{s-}}{d\zeta} = \frac{A}{2\pi}\frac{1}{\zeta + e^{\psi}} = \frac{A}{2\pi\delta} \tag{11.36}$$

From Eq. (11.26) note

$$\frac{dz}{d\zeta} = \frac{S}{\pi}\left(e^{-i\beta}\frac{e^{\psi}}{e^{2\psi} - \zeta^2} - e^{i\beta}\frac{e^{-\psi}}{\zeta^2 - e^{-2\psi}}\right) \tag{11.37}$$

In the vicinity of $\zeta = -e^{\psi}$ (where $\delta \Rightarrow 0$) this approaches

$$\frac{dz}{d\zeta} \Rightarrow \frac{Se^{-i\beta}}{2\pi\delta} \tag{11.38}$$

Combination of Eqs. (11.36) and (11.38) then gives

$$\left.\frac{dW}{dz}\right|_{-\infty} = \left(\left.\frac{dW_{s-}}{d\zeta}\right|_{-e^{\psi}}\right)\left(\left.\frac{dz}{d\zeta}\right|_{-e^{\psi}}\right)^{-1} = \frac{A}{S}e^{i\beta}$$

with Eq. (11.33),

$$A = V_{\infty}Se^{-i(\alpha+\beta)} - i\frac{\Gamma}{2} \tag{11.39}$$

In a similar manner, placing a source of strength B at the location $\zeta = e^{\psi}$ and satisfying Eq. (11.34) leads to

$$B = -V_{\infty}Se^{-i(\alpha+\beta)} - i\frac{\Gamma}{2} \tag{11.40}$$

With the two sources at locations $\zeta = \pm e^{\psi}$, the complex potential $F(\zeta)$ in the ζ plane is given by

$$F(\zeta) = \frac{SV_{\infty}}{2\pi}e^{-i(\alpha+\beta)}\ell n\frac{\zeta + e^{\psi}}{\zeta - e^{\psi}} - i\frac{\Gamma}{4\pi}\ell n(\zeta^2 - e^{2\psi}) \tag{11.41}$$

A complex potential satisfying the boundary condition for flow around the circle can now be obtained by adding a complex function of the form of Eq. (11.41), but with i replaced by $-i$ (where i appears explicitly) and with ζ replaced by $1/\zeta$. This result (known as the circle theorem[16]) follows because the added function becomes the complex conjugate of $F(\zeta)$ when ζ is on the unit circle; hence, ψ is zero on the circle. (And, of course, the circle is thus a streamline.) Finally, a circulation Γ^* is added at $\zeta = 0$ (which does not violate the boundary conditions) to give the complex potential W in the ζ plane. Thus,

$$W = \frac{SV_\infty}{2\pi} e^{-i(\alpha+\beta)} \ell n \frac{e^\psi + \zeta}{e^\psi - \zeta} - i\frac{\Gamma}{4\pi} \ell n\left(e^{2\psi} - \zeta^2\right)$$

$$+ \frac{SV_\infty}{2\pi} e^{i(\alpha+\beta)} \ell n \frac{e^\psi + 1/\zeta}{e^\psi - 1/\zeta} + i\frac{\Gamma}{4\pi} \ell n\left(e^{2\psi} - \frac{1}{\zeta^2}\right) + i\frac{\Gamma^*}{2\pi} \ell n\zeta \quad (11.42)$$

Investigation of this expression reveals that application of the circle theorem has introduced further singularities at the conjugate points of $\pm e^\psi$, that is at $\zeta = \pm e^{-\psi}$. Figure 11.13 indicates the location and strength of all the singularities in the circle plane.

It is evident that, as must be the case, the total source strength is zero. The net circulation about the circle is Γ^* and the net circulation about the entire field is $\Gamma^* - \Gamma$. It will be recalled that when the ζ plane is entirely traversed at infinity, the image point in the z plane simply traverses a point midway between the blades. Obviously, the circulation about such a point must be zero, so $\Gamma^* = \Gamma$. The expression for the velocity potential in the circle plane may hence be written

$$W = \frac{SV_\infty}{2\pi}\left(e^{-i(\alpha+\beta)} \ell n \frac{e^\psi + \zeta}{e^\psi - \zeta} + e^{i(\alpha+\beta)} \ell n \frac{\zeta + e^{-\psi}}{\zeta - e^{-\psi}}\right) - \frac{i\Gamma}{4\pi} \ell n \frac{e^{2\psi} - \zeta^2}{e^{2\psi}\zeta^2 - 1}$$

$$(11.43)$$

The complex potential given by Eq. (11.43) describes a flowfield that, when transformed to the z plane, matches the upstream and downstream conditions as well as the boundary conditions for flow around the blades.

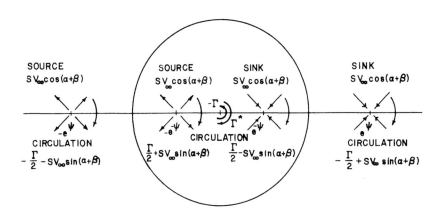

SOURCE
$SV_\infty \cos(\alpha+\beta)$

SOURCE
$SV_\infty \cos(\alpha+\beta)$

SINK
$SV_\infty \cos(\alpha+\beta)$

SINK
$SV_\infty \cos(\alpha+\beta)$

$-\Gamma$

Γ^*

CIRCULATION
$-\frac{\Gamma}{2} - SV_\infty \sin(\alpha+\beta)$

CIRCULATION
$\frac{\Gamma}{2} + SV_\infty \sin(\alpha+\beta)$

$\frac{\Gamma}{2} - SV_\infty \sin(\alpha+\beta)$

CIRCULATION
$-\frac{\Gamma}{2} + SV_\infty \sin(\alpha+\beta)$

Fig. 11.13   Singularities in the circle plane.

To complete the description of the flowfield it remains to determine an appropriate value for the circulation Γ. Such a condition is obtained by applying the Kutta condition that the velocity not be infinite at the trailing edge of the blades. At the location of the trailing edge [where $\phi = \phi_0$ as in Eq. (11.29)] the derivative $dz/d\zeta$ is zero. Thus, to prevent an infinite velocity at the trailing edge in the z plane, Γ must be chosen to put $dW/d\zeta = 0$ at $\zeta = e^{i\phi_0}$.

Taking $dW/d\zeta$ from Eq. (11.43) and putting $\zeta = e^{i\phi}$, the complex velocity components for the flow on the circle are obtained, to give after some manipulation.

$$\left.\frac{dW}{d\zeta}\right|_{\zeta = e^{i\phi}} = e^{-i\phi}\left(u_r - iu_\phi\right)$$

$$= \frac{i}{2\pi}\frac{e^{-i\phi}}{\cosh 2\psi - \cos 2\phi}\left\{4SV_\infty\left[\sin\phi\cosh\psi\cos(\alpha + \beta)\right.\right.$$

$$\left.\left. - \cos\phi\sinh\psi\sin(\alpha + \beta)\right] + \Gamma\sinh 2\psi\right\} \qquad (11.44)$$

It can be seen that as required, $u_r = 0$. The Kutta condition requires $u_\phi = 0$ when $\phi = \phi_0$. Thus, from Eq. (11.29)

$$\sin\phi_0 = \frac{\sinh\psi\sin\beta}{\sqrt{\sinh^2\psi + \cos^2\beta}}, \qquad \cos\phi_0 = \frac{\cosh\psi\cos\beta}{\sqrt{\sinh^2\psi + \cos^2\beta}} \qquad (11.45)$$

The value of the circulation necessary to satisfy the Kutta condition Γ_K then follows to give

$$\Gamma_K = \frac{2SV_\infty\sin\alpha}{\sqrt{\sinh^2\psi + \cos^2\beta}} \qquad (11.46)$$

When this value of the circulation is substituted into Eq. (11.44), it follows after some manipulation that

$$\left.\frac{dW}{d\zeta}\right|_{\zeta = e^{i\phi}} = -\frac{i2SV_\infty}{\pi}\frac{e^{-i\phi}}{\sin 2\phi_0}\frac{\cosh\psi\sin\psi}{\sqrt{\sinh^2\psi + \cos^2\beta}}\frac{\sin(\phi_0 - \phi)}{\sinh^2\psi + \sin^2\phi}$$

$$\times\left\{\cos\beta\sin(\alpha + \beta) - \sin\alpha\cos\phi_0\frac{\cos[(\phi_0 + \phi)/2]}{\cos[(\phi_0 - \phi)/2]}\right\} \qquad (11.47)$$

To obtain the complex velocity on the blade surface, note that on the blade $\zeta = e^{i\phi}$ and $z = x$, so that

$$\frac{dz}{d\zeta} = -ie^{-i\phi}\frac{dx}{d\phi} \qquad (11.48)$$

$dx/d\phi$ may be obtained directly from Eq. (11.30). It follows that

$$\frac{dz}{d\zeta} = -\frac{i2S}{\pi}\frac{e^{-i\phi}}{\sin 2\phi_0}\frac{\cosh\psi\sinh\psi}{\sqrt{\sinh^2\psi + \cos^2\beta}}\frac{\sin(\phi_0 - \phi)}{(\sinh^2\psi + \sin^2\phi)}\sin\beta\cos\beta$$

$$(11.49)$$

The velocity on the blade surface follows directly from Eqs. (11.47) and (11.49) to give

$$\left.\frac{dW}{dz}\right|_{blade} = \frac{dW}{d\zeta}\left(\frac{dz}{d\zeta}\right)^{-1}$$

$$= \frac{V_\infty}{\sin\beta\cos\beta}\left\{\cos\beta\sin(\alpha + \beta) - \sin\alpha\cos\phi_0\frac{\cos[(\phi_0 + \phi)/2]}{\cos[(\phi_0 - \phi)/2]}\right\}$$

$$(11.50)$$

To complete the description of the flow in the cascade plane, the various flow angles will be obtained. With reference to Fig. 11.11, note the relationships

$$\frac{\Delta v}{V_\infty\cos(\alpha + \beta)} = \tan(\alpha + \beta) - \tan(\alpha_2 + \beta)$$

$$= \tan(\alpha_1 + \beta) - \tan(\alpha + \beta) \qquad (11.51)$$

Equations (11.32), (11.46), and (11.51) then lead to

$$\alpha_1 = \arctan\left[\sin\alpha\frac{\sqrt{\sinh^2\psi + \cos^2\beta} + \cos\beta}{\sqrt{\sinh^2\psi + \cos^2\beta}\,(\cos\alpha) + \sin\beta\sin\alpha}\right]$$

$$\alpha_2 = \arctan\left[\sin\alpha\frac{\sqrt{\sinh^2\psi + \cos^2\beta} - \cos\beta}{\sqrt{\sinh^2\psi + \cos^2\beta}\,(\cos\alpha) - \sin\beta\sin\alpha}\right] \qquad (11.52)$$

These equations complete the desired description of the flowfield. They are summarized in a form suitable for calculation in the next section.

Summary of the Equations—Cascade Transformation

Inputs: $\qquad\qquad\qquad\qquad \beta,\ C/S,\ \alpha,\ \phi$

Outputs: $\qquad\qquad\qquad\qquad x/C,\ u/V_\infty,\ \alpha_1,\ \alpha_2$

Equations:

$$\frac{\pi C}{2S} = \cos \beta \, \ell n \frac{\sqrt{\sinh^2 \psi + \cos^2 \beta} + \cos \beta}{\sinh \psi} + \sin \beta \tan^{-1} \frac{\sin \beta}{\sqrt{\sinh^2 \psi + \cos^2 \beta}}$$

$$\phi_0 = \arctan(\tan \beta \tanh \psi)$$

$$\frac{x}{C} = \frac{1}{2\pi C/S} \left[\cos \beta \, \ell n \frac{\cosh \psi + \cos \phi}{\cosh \psi - \cos \phi} + 2 \sin \beta \tan^{-1} \frac{\sin \phi}{\sinh \psi} \right]$$

$$\frac{u}{V_\infty} = \frac{1}{\sin \beta \cos \beta} \left\{ \cos \beta \sin(\alpha + \beta) - \sin \alpha \cos \phi_0 \frac{\cos[(\phi_0 + \phi)/2]}{\cos[(\phi_0 - \phi)/2]} \right\}$$

$$\alpha_1 = \arctan \left[\sin \alpha \frac{\sqrt{\sinh^2 \psi + \cos^2 \beta} + \cos \beta}{\sqrt{\sinh^2 \psi + \cos^2 \beta} (\cos \alpha) + \sin \beta \sin \alpha} \right]$$

$$\alpha_2 = \arctan \left[\sin \alpha \frac{\sqrt{\sinh^2 \psi + \cos^2 \beta} - \cos \beta}{\sqrt{\sinh^2 \psi + \cos^2 \beta} (\cos \alpha) - \sin \beta \sin \alpha} \right]$$

Example Results—Two-Dimensional Straight-Line Cascade

The equations summarized in the preceding section lead to rapid computation of the performance variables of this ideal straight-line cascade. The results are useful for detecting the design trends of real cascades and, in fact, techniques to relate the performance of more complicated (and more realistic) geometries to an equivalent straight-line cascade have been developed (for example, Ref. 15). The singularity in fluid velocity remains at the blade leading edges, but the fluid velocities at blade locations away from the leading edges exhibit the tendencies of the flowfields existing on real geometries.

As an example calculation Fig. 11.14 shows the deviation angle (α_2) vs angle of attack (α_1) for three values of blade stagger angle and two values of the solidity.

It is apparent that the reduced blade loading existing for the case of high solidity greatly reduces the tendency of the flow to depart from the angle of the blade trailing edge. The deviation angle also varies strongly with the stagger angle, with large stagger angles causing increases in the deviation angle. It is evident from Fig. 11.15 that increasing the stagger angle leads to geometrical separation of the blades, so that the pressure fields of the blades do not interact as much at higher stagger angles.

It is of interest to note that this effect of increasing stagger angle is not as limiting as it might at first appear. Thus, the example calculations of Secs. 9.3 and 9.4 indicated that the required turning angles near the blade tips are

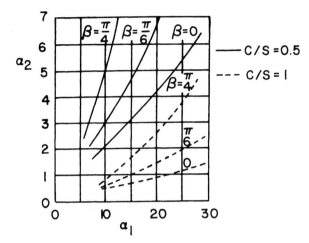

Fig. 11.14   Deviation angle vs angle of attack.

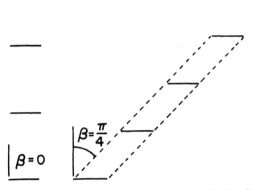

Fig. 11.15   Cascades at two stagger angles $C/S = 0.5$.

very much reduced, compared to those at the hub, for a compressor or turbine with a large tip-to-hub ratio. Thus, even though the geometry of a compressor or turbine is such that solidity decreases and the blade stagger angle increases with increasing radius, the effects of these geometry changes are much mitigated by the large reduction in required turning angle of the cascade.

As a final calculational example, the deviation angle vs solidity has been calculated for the case of constant angle of attack for $\beta = \pi/4$. (Note that this calculation required iteration of the input variable α to obtain the desired angle of attack $\alpha_1 = 15$ deg.) The results shown in Fig. 11.16 indicate once again the sensitivity of the deviation angle to solidity.

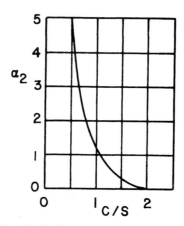

Fig. 11.16 Deviation angle vs solidity.

These relatively simple results provide a method of quickly estimating the effects of design choices upon the overall cascade performance. Extensions of the method, such as those described in Refs. 13–15 lead to rapid computer solution for much of the detailed information of realistic (ideal) flowfields. As has hopefully been evident throughout this book, however, the analytical cascade results contribute only a portion of the information desired in the very complicated process of designing the optimum compressor. Careful experimental studies, in the form of cascade tests, have contributed invaluable information for researchers and designers. Finally, the effects of interactive loss mechanisms, such as wake shedding and thence wake chopping by the following blade row, annulus wall boundary-layer buildup, and interaction with the blade pressure fields, blade boundary-layer buildup, and interaction with the centrifugal field, etc., must all be included in a careful program to develop the best possible compressor or turbine.

References

[1]Pianko, M. (ed.), *Modern Methods of Testing Rotating Components of Turbomachines*, AGARD-A6-167, Sept. 1972.

[2]Taylor, E. S., "Boundary Layers, Wakes and Losses in Turbomachines," Massachusetts Institute of Technology, Cambridge, GTL Rept. 105, April 1971.

[3]Hawthorne, W. R., "The Applicability of Secondary Flow Analysis to the Solution of Internal Flow Problems," *Fluid Mechanics of Internal Flow*, edited by G. Sovran, Elsevier, Amsterdam, 1967, pp. 238–269.

[4]Lefcort, M. D., "Wake Induced Unsteady Blade Forces in Turbo-Machines," Massachusetts Institute of Technology, Cambridge, GTL Rept. 70, Nov. 1962.

[5]Kemp, N. H and Sears, W. R., "The Unsteady Forces Due to Viscous Wakes in

Turbomachines," *Journal of Aeronautical Sciences*, Vol. 22, July 1955, pp. 478–483.

[6]Meyer, R. X., "Interference Due to Viscous Wakes Between Stationary and Rotating Blades in Turbomachines," Dr. of Engr. Thesis, Johns Hopkins University, Baltimore, Md., 1955.

[7]Kerrebrock, J. L. and Mikolajczak, A. A., "Intra-Stator Transport of Rotor Wakes and Its Effect on Compressor Performance," *Transactions of ASME, Journal of Engineering for Power*, Ser. A, Vol. 92, Oct. 1970, pp. 359–368.

[8]Johnsen, A. I. and Bullock, R. O. (eds.) *Aerodynamic Design of Axial-Flow Compressors*, NASA SP-36, 1965.

[9]Vincent, E. T., *Theory and Design of Gas Turbines and Jet Engines*, McGraw-Hill Book Co., New York, 1950.

[10]Cohen, H., Rogers, G. F. C., and Saravanamutto, H. J. H., *Gas Turbine Theory*, John Wiley & Sons, New York, 1973.

[11]Mani, R. and Acosta, A. J., "Quasi-Two-Dimensional Flows Through a Cascade," *Transactions of ASME, Journal of Power*, Vol. 90, No. 2, April 1968.

[12]Wilson, M. B., Mani, R., and Acosta, A. J., "A Note on the Influence of Axial Velocity Ratio on Cascade Performance," *Fluid Mechanics, Acoustics and Design of Turbomachinery*, Pt. I, NASA SP-304, 1974, pp. 101–133.

[13]Ives, D. C., "A Modern Look at Conformal Mapping Including Multiple Connected Regions," *AIAA Journal*, Vol. 14, Aug. 1976, pp. 1006–1011.

[14]Lighthill, M. J., "A Mathematical Method of Cascade Design," British Aeronautical Research Council, R & M No. 2104, 1945.

[15]Weinig, F. S., "Theory of Two-Dimensional Flow Through Cascades," *Aerodynamics of Turbines and Compressors, High Speed Aerodynamics and Jet Propulsion*, Vol. 10, Princeton University Press, Princeton, N.J., 1964.

[16]Milne-Thomson, L. M., *Theoretical Hydrodynamics*, The Macmillan Co., New York, 1961.

Problems

11.1   Investigate the behavior of a flow with complex potential

$$W = \frac{A}{2\pi} \ell n \frac{z}{a}$$

(a) Find expressions for u_r and u_θ.

(b) Evaluate the volume flow rate Q through a contour encircling the origin.

(c) Evaluate the divergence of the velocity.

(d) Check that the divergence theorem is consistent with the results of parts (b) and (c).

11.2   Investigate the behavior of a flow with complex potential

$$W = -\frac{iB}{2\pi}\ell n\frac{z}{a}$$

(a) Find expressions for u_r and u_θ.

(b) Evaluate the circulation Γ about a contour encircling the origin.

(c) Evaluate the vorticity.

(d) Check that Stokes' theorem is consistent with the results of parts (b) and (c).

11.3   Show that the complex potential

$$W = Ua(z/a)^{\pi/\alpha}$$

corresponds to flow in a corner of α radians.

11.4   Using the results of Problem 11.3 and the circle theorem, show that if a circular segment of radius a is placed at the origin in a corner of $\pi/2$ rad, the velocities in the flowfield are given by

$$u_r = 2U\left(\frac{r}{a}\right)\cos(2\theta)\left[1-\left(\frac{a}{r}\right)^4\right]$$

$$u_\theta = -2U\left(\frac{r}{a}\right)\sin(2\theta)\left[1+\left(\frac{a}{r}\right)^4\right]$$

11.5   (a) Use the circle theorem to obtain the complex potential for uniform flow past a circle of radius a, centered at the origin.

(b) Find expressions for ψ, ϕ, u_r, and u_θ in terms of the reference flow velocity U and a, r, and θ.

(c) Show that the pressure coefficient on the body

$$C_p = \frac{p-p_\infty}{\frac{1}{2}\rho U^2}$$

is given by $C_p = 1 - 4\sin^2\theta$. Sketch C_p vs θ in the range $\pi/2 \le \theta \le \pi$.

(d) Show that the dimensionless axial force acting on the cylinder from the "nose" back to the angle β is given by

$$\frac{F}{\frac{1}{2}\rho U^2(2a)} = \sin\beta - \frac{4}{3}\sin^3\beta$$

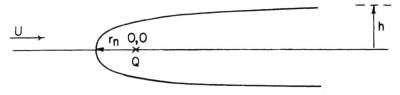

Fig. A

11.6   Consider a uniform stream to flow past a "source" of strength Q located at the origin (Fig. A). The complex potential of such a flow is

$$W = Uz + (Q/2\pi)\ell n\, z$$

   (a)   Show that the radial distance to the nose of the body from the origin r_n is given by $r_n = Q/2\pi U$.
   (b)   Show that the source strength Q is given in terms of the eventual height of the body h by $Q = 2Uh$ and hence also $r_n = h/\pi$.
   (c)   Show that the equation of the body may be written

$$y = h[1 - (\theta/\pi)]$$

   (d)   Show that the pressure coefficient $C_p = (p - p_\infty)/\tfrac{1}{2}\rho U^2$ may be written in terms of the angle from the nose β as

$$C_p = -\frac{\sin^2\beta}{\beta^2} + \frac{2\sin\beta\cos\beta}{\beta}$$

   (e)   Show that the dimensionless axial force on the upper half body

$$\frac{F}{\tfrac{1}{2}\rho U^2 h} = \int_{\beta=0}^{\beta} C_p\, d\left(\frac{y}{h}\right)$$

may be written $F/\tfrac{1}{2}\rho U^2 h = \dfrac{\sin^2\beta}{\pi\beta}$.

11.7   When axisymmetric flow past a point source of strength M is considered (Fig. B), solution to the equations yields

$$\psi = \frac{1}{2}Ur^2 - \frac{Mx}{R}$$

where

$$u = -\frac{1}{r}\frac{\partial\psi}{\partial x}, \qquad w = \frac{1}{r}\frac{\partial\psi}{\partial r}$$

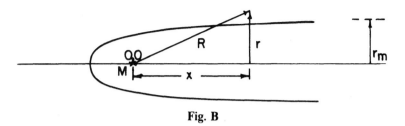

Fig. B

(a) Show that

$$\frac{\partial 1/R}{\partial x} = \frac{-x}{R^3}, \qquad \frac{\partial 1/R}{\partial r} = \frac{-r}{R^3}$$

(b) Show that $U = Mr/R^3$ and $w = U + Mx/R^3$.

(c) Show that the equation for the body may be written

$$Mx = \left(\tfrac{1}{2}Ur^2 - M\right)R$$

and that r at $x \to \infty \equiv r_m = \sqrt{4M/U}$.

(d) Defining y by $y = r/r_m$, show that the pressure coefficient on the body is given by

$$\frac{p - p_\infty}{\tfrac{1}{2}\rho U^2} = 1 - 4y^2 + 3y^4$$

(e) Show by integrating the axial force that the dimensionless force (from the nose to y)

$$\frac{F}{\tfrac{1}{2}\rho v^2\left(\pi r_m^2\right)} = \int_0^r \frac{p - p_\infty}{\tfrac{1}{2}\rho U^2}\,\frac{2\pi r\,dr}{\left(\pi r_m^2\right)}$$

is given by

$$\frac{F}{\tfrac{1}{2}\rho U^2\left(\pi r_m^2\right)} = y^2\left(1 - 2y^2 + y^4\right)$$

11.8   Provide the detailed manipulations leading to Eq. (11.29).

11.9   Provide the detailed manipulations leading to Eq. (11.31).

11.10   (a) Show that the numerical value of the transformation variable ψ may be obtained from the Newtonian iteration expression

$$x_{j+1} = x_j - F/F'$$

where

$$x = \sinh \psi$$

$$F = \frac{\pi c}{2S} - \cos \beta \, \ell n \left(\frac{\sqrt{x^2 + \cos^2 \beta} + \cos \beta}{x} \right) - \sin \beta \tan^{-1} \left(\frac{\sin \beta}{\sqrt{x^2 + \cos^2 \beta}} \right)$$

$$F' = \frac{\sqrt{x^2 + \cos^2 \beta}}{x(1 + x^2)}$$

(b) Show also $\psi = \ell n(x + \sqrt{x^2 + 1})$.

(c) Show that for the special case of $\beta = 0$,

$$\psi = \ell n \left[\frac{(\exp \pi c/2S) + 1}{(\exp \pi c/2S) - 1} \right]$$

and hence

$$\sinh \psi = \frac{2 \exp \pi c/2S}{(\exp \pi c/S) - 1} = \left[\sinh \left(\frac{\pi c}{2S} \right) \right]^{-1}$$

11.11   Provide the detailed manipulations leading to Eq. (11.46).

11.12   Provide the detailed manipulations leading to Eq. (11.50).

11.13   Verify that $dz/d\zeta$, as given by Eq. (11.49), does not go to zero as β goes to zero.

11.14   Show (using the results of Problem 10c, if you wish) that in the special case of zero stagger angle β the deviation angle α_2 is given in terms of the angle of attack α_1 by

$$\alpha_2 = \arctan \left(\exp - \pi \frac{C}{S} \tan \alpha_1 \right)$$

APPENDIX A.   CHARACTERISTICS OF THE STANDARD ATMOSPHERE

h ($ft/10^3$)	T/T_{SL}	P/P_{SL}	ρ/ρ_{SL}	a/a_{SL}	h ($ft/10^3$)
0	1.0000	1.0000	1.0000	1.0000	0
1	0.9931	0.9644	0.9711	0.9366	1
2	0.9863	0.9298	0.9428	0.9966	2
3	0.9794	0.8963	0.9151	0.9395	3
4	0.9725	0.8637	0.8881	0.9662	4
5	0.9656	0.8321	0.8617	0.9827	5
6	0.9583	0.8014	0.8359	0.9792	6
7	0.9519	0.7717	0.8107	0.9757	7
8	0.9450	0.7429	0.7861	0.9722	8
9	0.9381	0.7149	0.7621	0.9636	9
10	0.9313	0.6878	0.7386	0.9650	10
11	0.9244	0.6616	0.7157	0.9615	11
12	0.9175	0.6362	0.6933	0.9579	12
13	0.9107	0.6115	0.6715	0.9543	13
14	0.9038	0.5877	0.6502	0.9507	14
15	0.8969	0.5646	0.6295	0.9471	15
16	0.8901	0.5422	0.6092	0.9434	16
17	0.8832	0.5206	0.5895	0.9382	17
18	0.8764	0.4997	0.5702	0.9361	18
19	0.8695	0.4795	0.5514	0.9325	19
20	0.8626	0.4599	0.5332	0.9238	20
21	0.8558	0.4410	0.5153	0.9251	21
22	0.8489	0.4227	0.4980	0.9214	22
23	0.8420	0.4051	0.4811	0.9176	23
24	0.8352	0.3880	0.4646	0.9139	24
25	0.8233	0.3716	0.4486	0.9101	25
26	0.8215	0.3557	0.4330	0.9063	26
27	0.8146	0.3404	0.4178	0.9026	27
28	0.8077	0.3256	0.4030	0.8987	28
29	0.8009	0.3113	0.3887	0.8949	29
30	0.7940	0.2975	0.3747	0.8911	30

h (ft/10^3)	T/T_{SL}	P/P_{SL}	ρ/ρ_{SL}	a/a_{SL}	h (ft/10^3)
31	0.7872	0.2843	0.3611	0.8872	31
32	0.7803	0.2715	0.3479	0.8834	32
33	0.7735	0.2592	0.3351	0.8795	33
34	0.7666	0.2474	0.3227	0.8758	34
35	0.7598	0.2360	0.3106	0.8717	35
36	0.7529	0.2250	0.2988	0.8677	36
37	0.7519	0.2145	0.2852	0.8671	37
38	0.7519	0.2044	0.2719	0.8671	38
39	0.7519	0.1949	0.2592	0.8671	39
40	0.7519	0.1858	0.2471	0.8671	40
41	0.7519	0.1771	0.2355	0.8671	41
42	0.7519	0.1688	0.2245	0.8671	42
43	0.7519	0.1609	0.2140	0.8671	43
44	0.7519	0.1534	0.2040	0.8671	44
45	0.7519	0.1462	0.1945	0.8671	45
46	0.7519	0.1046	0.1391	0.8671	46
47	0.7519	0.1329	0.1767	0.8671	47
48	0.7519	0.1267	0.1685	0.8671	48
49	0.7519	0.1208	0.1606	0.8671	49
50	0.7519	0.1151	0.1531	0.8671	50
52	0.7519	0.1046	0.1391	0.8671	51
54	0.7519	0.9507 − 1	0.1264	0.8671	54
56	0.7519	0.8640	0.1149	56	
58	0.7519	0.7852	0.1044	58	
60	0.7519	0.7137	0.9492 − 1	0.8671	60
62	0.7519	0.6486 − 1	0.8627 − 1	0.8671	62
64	0.7519	0.5895	0.7841	0.8671	64
66	0.7519	0.5358	0.7126	0.8671	66
68	0.7519	0.4870	0.6477	0.8671	68
70	0.7519	0.4426	0.5887	0.8671	70
72	0.7519	0.4023 − 1	0.5351 − 1	0.8671	72
74	0.7519	0.3657	0.4864	0.8671	74
76	0.7519	0.3324	0.4421	0.8671	76
78	0.7519	0.3022	0.4019	0.8671	78
80	0.7519	0.2747	0.3653	0.8671	80
85	0.7602	0.2166 − 1	0.2849 − 1	0.8719	85
90	0.7760	0.1715	0.2210	0.8809	90
95	0.7917	0.1365	0.1724	0.8893	95
100	0.8074	0.1091	0.1351	0.8986	100
110	0.8388	0.7063 − 2	0.8420 − 2	0.9159	110
120	0.8702	0.4649	0.5342	0.9329	120
130	0.9016	0.3106	0.3445	0.9495	130

h $(\text{ft}/10^3)$	T/T_{SL}	P/P_{SL}	ρ/ρ_{SL}	a/a_{SL}	h $(\text{ft}/10^3)$
140	0.9329	0.2105	0.2257	0.9659	140
150	0.9642	0.1446	0.1500	0.9819	150
160	0.9809	0.1004 – 2	0.1024 – 2	0.9904	160
170	0.9809	0.6986 – 3	0.7122 – 3	0.9904	170
180	0.9620	0.4855	0.5047	0.9808	180
190	0.9215	0.3330	0.3614	0.9600	190
200	0.8810	0.2246	0.2550	0.9386	200

Notation: Single digit preceded by a minus sign indicates power of 10 by which associated and following tabulated values should be multiplied, e.g., 0.2468 – 2 = 0.002468.

Notes: (1) Data from "*U.S. Extension of the ICAO Standard Atmosphere*," 1958. (2) Sea Level Values: $T_{SL} = 518.69°\,\text{R}$, $P_{SL} = 2116.2$ psf, $\rho_{SL} = 0.0023769$ slug/ft^3, and $a_{SL} = 1116.4$ fps.

APPENDIX B.   SAE GAS TURBINE ENGINE NOTATION

1. Purpose

1.1   This Aerospace Recommended Practice (ARP 755A) provides performance station identification and nomenclature systems for gas turbine engines.

1.2   The systems presented herein are for use in all communications concerning engine performance such as computer programs, data reductions, design activities, and published documents.

2. Station Identification

The following station numbering system will be used to identify the points in the gas flow path that are significant to engine performance definition.

2.1   Basis of System   The system provides for the consistent definition of the process being undergone by the gas, regardless of the type of engine cycle. The five main processes that are isolated are: air intake, compression in engine compressors, heat addition, expansion in turbines, and expansion in nozzles.

2.2   Primary Stream   The station numbers required to identify the processes for the primary gas flow are:

0 Freestream air conditions	5 Last turbine discharge
1 Inlet/engine interface	6 Available for mixer, afterburner, etc.
2 First compressor front face	7 Engine/exhaust nozzle interface
3 Last compressor discharge	8 Exhaust nozzle throat
4 Burner discharge	9 Exhaust nozzle discharge

2.3   Multiple Streams   Extension of the primary flow numbering scheme to multiple streams (e.g., the bypass flow of a turbofan engine) is obtained by prefixing a digit to the numbers in Sec. 2.2.

2.3.1   Unity (1) will be used for the innermost bypass duct.

Examples: 12 First compressor front face tip section (if different from Station 2)

13 End of compression of bypass flow

17 Bypass duct/exhaust nozzle interface

18 Bypass exhaust nozzle throat

2.3.2   To avoid conflict with two-digit primary stream intermediate stations (see Sec. 2.4), the prefixing of bypass duct streams with the digits 2 through 8 should be avoided where possible. The digit 9 will be used to identify ejector nozzle flow or for a second bypass duct.

Example: 98 Ejector exhaust nozzle throat

2.3.3   If, however, two or more flow paths are mixed, succeeding numbers will be consistent with the innermost stream. For example, primary flow numbers are to be used when primary flow is mixed with a bypass flow.

2.3.4   The first digit of the primary stream, and the first two digits of the innermost and second bypass ducts, will be numeric only.

2.3.5   Property values (or flow rates) for individual streams are always average (or total) quantities. Where primary and bypass streams are differentiated by separate stations and there is a need to describe the average (or total) properties at a plane including both streams, an alphanumeric station will be created. This station will be coplanar with the primary and bypass stations and formed by appending a letter to the hub station identification.

Appendage of the letter A (e.g., 1A, 6A) is reserved to describe the combined properties of all the streams in that plane. For example, when Stations 1 and 11 define the primary and innermost bypass streams at the inlet/engine interface, Station 1A is defined as encompassing both Stations 1 and 11. More than two streams can be handed in a similar manner.

2.4   Intermediate Stations   For identification of intermediate stations, numeric or, if necessary, alphabetic subdivision will be used for the appended symbols. The numbering of stations intermediate to those indicated in Secs. 2.2 and 2.3 should, where possible, be limited to two digits that will be chosen to prevent duplication, and will be assigned in an ascending or alphabetic sequence that corresponds to the direction of flow. For example, a primary intermediate station between Stations 1 and 2 for a bypass engine may be identified as 1B to avoid conflict with the innermost bypass duct first compressor front face tip section, Station 12.

2.5   Figures   Figures B1 and B2 are examples of the applications of this system to several typical engine configurations.

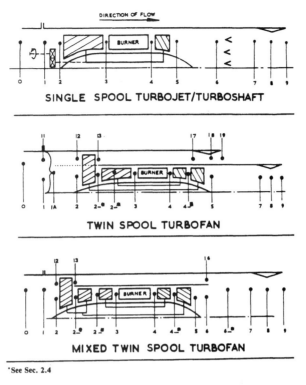

SINGLE SPOOL TURBOJET/TURBOSHAFT

TWIN SPOOL TURBOFAN

MIXED TWIN SPOOL TURBOFAN

*See Sec. 2.4

Fig. B1   Example of station identification—1.

3.   Nomenclature

This nomenclature has been compiled to provide a uniform method of naming variables associated with gas turbine engines. Its use is encouraged for all communications involving engine performance including computer programs. There are two columns of symbols. The first column presents the recommended symbols for general use and is restricted to upper case letters to be compatible with the computer. The second column presents alternate symbols that are retained because of their widespread use. Lower case letters, subscripts and superscripts, Greek letters, and other specialized characters have been avoided in the recommended symbols. It is hoped that a single system will soon evolve, on one hand through changes in common usage resulting from greater familiarity with the computer, and on the other hand through development of computer practices permitting a wider range of symbols than is now possible.

3.1   Basic Symbols   This section includes the symbols used to derive basic parameters and will normally form the leading letter, or letters, in compound groups. Most of these symbols will be expanded by the addition of a

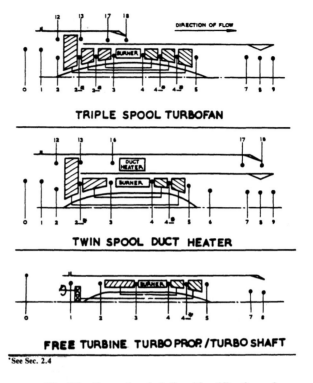

TRIPLE SPOOL TURBOFAN

TWIN SPOOL DUCT HEATER

FREE TURBINE TURBO PROP./TURBO SHAFT

*See Sec. 2.4

Fig. B2   Example of station identification—2.

station number, component symbol, or stream identification as contained in later sections. Examples of some resulting compound groups are contained in Sec. 4.

3.1.1   *Properties and Fundamental Parameters*

	Recommended	Alternate
Area, geometric	A	
Altitude (Geopotential pressure)	ALT	
Angle	ANG	α, β, γ, etc.
Density	RHØ	ρ
Efficiency, adiabatic	E	η
Enthalpy—total per unit mass	H	
Entropy—total per unit mass	S	
Force, thrust	F	
Frequency	FY	f
Heat transfer rate	QU	Q
Inertia—polar moment (see Sec. 3.4.3)	XJ	J
Length	XL	L
Mass	GM	m
Mass flow rate	W	

Power	PW	
Pressure—total	P	
Rotational speed	XN	N
Temperature—total	T	
Time	TIME	t
Torque	TRQ	
Velocity	V	
Viscosity	VIS	
Volume	VØL	v
Weight	WT	w

3.1.2 *Commonly Used Ratios, Functions, etc.* This section contains symbol groupings which, although they are exceptions to the general system, have been retained because of their widespread use in industry.

	Recommended	Alternate
Blow-out margin	BØM	
Bypass ratio	BPR	
Coefficient or constant	C	
Delta (pressure/standard SLS pressure)	DEL	δ
Discharge coefficient	CD	
Drag	FD	
Entropy function	PHI	ϕ
Error	Y	
Fuel/air ratio	FAR	
Fuel lower heating value	FHV	
Fuel specific gravity	FSG	
Gas constant (per unit mass)	R	
Light-off margin	XLØM	LØM
Mach number	XM	M
Mechanical equivalent of heat	CJ	
Molecular weight	XMW	MW
Power lever angle	PLA	
Ratio of specific heats	GAM	γ
Relative humidity	RH	
Reynolds number	RE	
Reynolds number index	RNI	
Rotor blade angular position	RØP	
Specific fuel consumption	SFC	
Specific gravity	SG	
Specific heat at constant pressure	CP	
Surge margin	SM	
Stator blade angular position	STP	
Tangential wheel speed	U	
Theta (temperature/standard SLS temperature)	TH	θ
Velocity dynamic head	VH	q
Velocity of sound	VS	a

Water (liquid)/air ratio	WARL
Water (vapor)/air ratio	WAR

3.2   Operating Symbols   The letters in this section describe operations and will normally be embedded in compound groups.

	Recommended	Alternate
Derivative with respect to time	U	d/dt
Derivative with respect to following symbol	U_	d/d_
Difference (see Sec. 3.4.2)	D	– or Δ
Quotient, ratio (when not followed by U)	Q	/
Square root	R	

3.3   Descriptive Symbols   This section includes recommended symbols that describe the basic parameters and will usually be the trailing letter, or letters, in compound groups. It is subdivided into a part describing the fluid, another containing symbols describing parts of the engine, and a descriptive symbols part. No alternate symbols are recognized in this section.

3.3.1   Fluid Description   Some properties and fundamental parameters (e.g., pressure, flow rate) that refer to the fluid may require additional description to indicate the composition and use of the fluid. The following letters should be appended directly after the basic symbols of Sec. 3.1 (see Sec. 3.4.1 for additional notes on fluid description):

Air	A
Bleed	B
Boundary layer	BL
Coolant	CL
Fuel	F
Leakage	LK
Water	W

3.3.2   Engine Description   Some parameters that refer to engine components or rotors (e.g., efficiency, rotor speed, surge margin, torque) require more specific description. This should be provided by appending the station number (see Sec. 2) at inlet to the relevant component or rotor after the basic symbols of Sec. 3.1. (An alternate method, included because of its widespread use in industry, is to append the following symbols. The use of this alternate method is not encouraged because of possible confusion with other descriptive symbols.)

Afterburner	AB
Boattail	BT
Burner	B
Compressor	C

Engine	E
Heat exchanger	EX
High-pressure component or rotor	H
Intermediate pressure component or rotor	I
Low-pressure component or rotor	L
Power turbine or rotor	PT
Turbine	T

3.3.3 *General Description* The following general descriptive symbols should be appended after the basic symbols of Sec. 3.1:

Average	AV
Ambient	AMB
Conductivity	K
Controlled variable	C
Diameter	DI
Distortion	DIST
Effective	E
Extraction	X
Gross	G
High (maximum)	H
Ideal	I
Installed	IN
Low (minimum)	L
Map value	M
Net	N
Parasitic	PAR
Polytropic	P
Radius	RAD
Ram	RAM
Referred (corrected)	R
Relative	REL
Sea level	SL
Sensed parameter	SE
Shaft delivery (output)	SD
Standard	STD
Static	S
Swirl	SW
Tip	TIP
Total	T
True air speed	TAS

3.4 Additional Notes

3.4.1 To describe the position within the engine of parameters associated with a fluid, the numbers detailed in the station identification system of Sec. 2 should be appended. The letters of Sec. 3.3.1 should precede these station numbers if both are required.

3.4.2   The embedded D, which identifies a difference (see Sec. 3.2), should be used wherever the compound group of of symbols is of an acceptable length. However, D may also be used as a leading symbol when contraction of the compound group of symbols is necessary.

3.4.3   The symbols XJ (polar moment of inertia) should be appended by a component identification symbol (Sec. 3.3.2). The component should be that to which all associated inertias are algebraically referred.

3.4.4   A gas property followed by S denotes a static quantity; otherwise a stagnation condition is implied.

3.4.5   The symbol X was prefixed to leading symbols I, J, K, L, M, and N for computer purposes.

3.4.6   It is recognized that it may be required to limit the number of characters per parameter name. When this limitation is not compatible with the recommended nomenclature of this ARP, the parameter name may be shortened.

3.4.7   Throughout this document: 0 denotes the numeric symbol and Ø denotes the alphabetic symbol.

4. Examples

Some examples of compound groups formed from recommended symbols are contained in this section.

4.1   Groups formed by basic symbols together with one or more descriptive symbols:

AE	Effective area
ANGBT	Boattail angle
ANGSW	Swirl angle
CFG	Gross thrust coefficient
CPSTD	Standard SLS pressure
CQU	Overall heat-transfer coefficient
CQUBL	Heat-transfer film (boundary-layer) coefficient
CQUK	Thermal conductivity
CQUL	Coefficient of linear thermal expansion
CR	Universal gas constant
CTSTD	Standard SLS temperature
CV	Nozzle velocity coefficient
DTAMB	Ambient temperature minus standard day ambient temperature
DPW	Unbalanced power
DTRQ	Unbalanced torque
EP	Polytropic efficiency
ERAM	Ram pressure recovery

FG	Gross thrust
FGI	Ideal gross thrust
FN	Net thrust
FNIN	Installed net thrust
FRAM	Ram drag
HF	Enthalpy of fuel
HS	Static enthalpy
PAMB	Ambient pressure
PB	Bleed flow total pressure
PREL	Relative pressure
PS	Static pressure
PWPAR	Parasitic power
PWSD	Delivered shaft power
PWX	Power extraction
SFCIN	Installed specific fuel consumption
TAMB	Ambient temperature
TLK	Total temperature of leakage gas
TRQSD	Delivered shaft torque
TS	Static temperature
UTIP	Tangential wheel tip speed
VANG	Angular velocity
VTAS	Aircraft velocity (true air speed)
WA	Airflow rate
WF	Fuel flow rate
WFT	Total fuel flow rate
WW	Water flow rate
XNSD	Delivered shaft speed

4.2 Groups formed by basic symbols together with descriptive symbols and station numbers:

CD8	Primary nozzle flow discharge coefficient
CV8	Primary nozzle velocity coefficient
DT1	Temperature to be added to T1
FAR4	Fuel/air ratio at Station 4
FG19	Bypass nozzle gross thrust
F7	Stream thrust at Station 7
HA3	Total enthalpy of air at Station 3
PB3	Bleed flow total pressure at Station 3
PS4QS3	Static pressure ratio; Station 4 divided by Station 3
PW4	High-pressure turbine power
P1QAMB	Ram pressure ratio
P3	Total pressure at Station 3
P3U	Time rate of change of total pressure at Station 3
P4Q3	Total pressure ratio; Station 4 divided by Station 3
P6D7	Total pressure change from Station 6 to Station 7
TRQ2	Low-pressure compressor torque
TS0STD	Standard atmospheric temperature
T2UN2	Rate of change of T2 with respect to N2

WA2	Airflow rate at Station 2
WB3	Bleed flow rate at Station 3
WLK3	Leakage flow rate at Station 3
W1R	Referred engine inlet flow rate
W3R2	Flow rate at Station 3 referred to Station 2
XJ2	Polar moment of inertia of spool containing low-pressure compressor with all inertias referred to that component
XN2	Low-pressure compressor rotor speed
XN2H	Maximum low-pressure compressor rotor speed
XN21L	Minimum speed of rotor whose compressor inlet is at Station 21

APPENDIX C. OATES COMPANION SOFTWARE

Daniel H. Daley
William H. Heiser
Jack D. Mattingly
David T. Pratt

The enclosed software is intended for use with *Aerothermodynamics of Gas Turbine and Rocket Propulsion,* by Gordon C. Oates. This comprehensive set of programs may be used with the problems and design analyses discussed in the book. The information below details the necessary system requirements, and describes installation and operation procedures. The software opening screen menus, their program titles, and associated topics in the textbook are also given.

1. Getting Started

1.1 System Requirements

IBM PC 386/486/586/Pentium or compatible computer with at least 640 Kb RAM, a hard drive with 1.0 Mb of available disk space for storing **OATES,** and EGA or better video capability. **OATES** is designed to load and run on any DOS-based operating system.

1.2 Installation

Insert the distribution disk into the A: or B: drive as appropriate (the A: drive is assumed below), and log onto that drive. Assuming that you want to install **OATES** on the C: hard drive, then at the A:\ DOS prompt, type INSTALL A: C:, and press the <Enter> key. (To install **OATES** on a different hard drive, for example the D: drive, type INSTALL A: D: and press <Enter>.) The installation program will create a new directory OATES on the target hard drive, and will then transfer all the OATES executable files, as well as the required files ECAP.DEF and EOPP.DEF, to the new directory.

1.3 Operation

With all of the files from the distribution disk installed in the directory c:\OATES, log on to that directory (type CD OATES and press <Enter>), and at the prompt c:\OATES, type OATES and press <Enter>. Respond to the menu selections and prompts with keyboard entries, arrow keys, mouse clicks, or a mixture of all modes. (Note that although there are many ".EXE" files present, only OATES.EXE can be launched by the user. All of the other ".EXE" files are executed by **OATES** in response to the menu selections.)

1.4 General Information

After **OATES** has been installed on your hard disk, you can browse the menu on the opening screen to read the one-line descriptions of each program that appears at the bottom of the screen. Authoring credits and software information can be read from options on the opening screen menu as well.

1.5 Writing Screens to File

Input and output screens, in either graphics or text mode, can be saved to file as follows. 1) Run **OATES** from the Microsoft Windows MS-DOS prompt. 2) Press the <Print Screen> key to copy the screen contents to the Windows Clipboard. 3) Use the Windows Clipboard Viewer, or a text editor or word processor that can access the Windows Clipboard to save either graphics or text mode screens to a file.

2. Opening Screen Menu/Program	Textbook Site
2.1 Atmosphere	
Atmosphere	Appendix A, p. 435
2.2 Quasi–1D Flows	
Ideal Constant-Area Interaction	Chapter 2, p. 47
Adiabatic Constant-Area Flow with Friction	Chapter 2, p. 50
Nozzle Flow Equations	Chapter 2, p. 52
Rocket Nozzle Performance	Chapter 3, p. 70
Normal Shock Waves	Chapter 6, p. 206
Oblique Shock Waves	Chapter 6, p. 211
2.3 Gas Turbine	
Engine Cycle Analysis, Ideal	Chapter 5, p. 121
Engine Cycle Analysis, Nonideal	Chapter 7, p. 231
Engine Off-Design Performance	Chapter 8, p. 277

2.4 Rocket Combustion

T_c and p_c given	Chapter 3, p. 84
h_c and p_c given	Chapter 3, p. 84
Isentropic Expansion	Chapter 3, p. 85

All programs can be run in either the SI or English unit system.

3. Program Descriptions

3.1 Atmosphere

The **Atmosphere** program gives the variation of atmospheric properties with altitude from sea level to 86 km (282.15 kft), taken from U.S. Government Printing Office, "U.S. Standard Atmosphere, 1976," Washington, D.C., 1976. Please note that the atmospheric information in Appendix A is from 1958.

3.2 Quasi–1D Flows

The **Quasi–1D Flows** menu contains a collection of utility programs for evaluating traditional compressible flow functions and rocket nozzle performance for calorically perfect gases and values of Cp/Cv prescribed by the user. The compressible flow functions take their structures from the venerable "Gas Tables" of Keenen and Kaye, published in 1948, and from the 1953 NACA report 1135, "Equations, Tables and Charts for Compressible Flow." The program collection consists of **Ideal Constant-Area Heat Interaction,** including Eq. (2.92) from the textbook; **Adiabatic Constant-Area Flow with Friction,** including Eq. (2.98) from the textbook, **Nozzle Flow Equations** (Isentropic Flows); **Rocket Nozzle Performance; Normal Shock Waves;** and **Oblique Shock Waves.** The **Rocket Nozzle Performance** program is included to facilitate the calculation of the performance of a nozzle under a wide variety of conditions.

With few exceptions, all of the compressible flow function programs can be entered with any property ratio listed, just as one would enter the Keenan and Kaye or NACA 1135 gas tables. When an entered variable is out of range, the appropriate range of the variable is displayed and the user is prompted to enter another value. Whenever an output function is double valued (usually subsonic or supersonic) for the input value, the user is queried for the output value of interest.

3.3 Gas Turbine

The **Gas Turbine** menu contains the **Engine Cycle Analysis Program (ECAP)** and the **Engine Off-Design Performance Program (EOPP).** Choosing one of these programs brings up that program's home screen, which has the following pull-down menus: **File, Cycle, Data, Variable, Units, Output,** and **Help.**

ECAP in the **Gas Turbine** menu provides a means for determining the variation in gas turbine engine performance with cycle design variables such as compressor ratio. The program is based on the engine models contained in Chapter 5, Ideal Cycle Analysis, and Chapter 7, Nonideal Cycle Analysis. ECAP is very useful for examining the trends of an engine's specific thrust and specific fuel consumption with changes in applicable design variables. The following actions are required to run the ECAP program: with the **Cycle** menu, choose one of seven engine cycles and either an ideal or non-ideal analysis model; with the **Data** menu, enter the engine operating conditions in the edit fields of the data screen and/or select the variable to be optimized; with the **Variable** menu, choose one of eight iteration variables along with its applicable range and increment; with the **Units** and **Output** menus, choose the desired units and output devices; and with the **File** menu, choose run. The seven engine cycles contained in the program are 1) ramjet, 2) turbojet, 3) turbojet with afterburner, 4) turbofan, 5) turbofan with afterburners, 6) mixed turbofan, and 7) turboprop.

EOPP of the **Gas Turbine** menu is based on the engine models contained in Chapter 8, Engine Off-Design Performance. EOPP is very useful for examining the variation of a given engine's performance with changes in flight conditions and throttle setting. The following actions are required to run the EOPP program: with the **Cycle** menu, choose one of four engine cycles; with the **Data** menu, enter the reference point data, calculate the reference point performance, enter off-design data, and/or select the variable to be optimized; with the **Variable** menu, choose one of nine iteration variables along with its applicable range and increment; with the **Units** and **Output** menus, choose the desired units and output devices; and with the **File** menu, choose run. The four engine cycles contained in the program are 1) fixed area turbine turbojet, 2) fixed area turbine turbofan, 3) fixed area turbine turboprop, and 4) variable area turbine turbojet.

3.4 Rocket Combustion

The **Rocket Combustion** menu contains three programs that calculate chemical equilibrium properties and composition of products of combustion for cryogenic and storable liquid bipropellants. As opposed to the method of the textbook, the very robust ZGM (Zeleznik–Gordon–McBride) algorithm for Gibbs function minimization is used to calculate equilibrium states of combustion products in each of the programs. The T_c **and** P_c **given and** H_c **and** P_c **given** programs calculate the product's properties and composition for assigned rocket combustion chamber temperature/pressure and enthalpy/pressure, respectively. The **Isentropic Expansion** program calculates the properties and composition for isentropic expansion or compression from the rocket combustion chamber pressure to a prescribed pressure, for both equilibrium and frozen flows.

Index

Texts Published in the AIAA Education Series

(Continued on the next page)

Texts Published in the AIAA Education Series (continued)

Published by
American Institute of Aeronautics
and Astronautics, Inc.
Reston, Virginia

To download your software and any software updates, please go to http://www.aiaa.org/publications/supportmaterials. Select your title, follow the instructions provided, and enter the following password: **ramjet.**

Many of the topics introduced in this book are discussed in more detail in other AIAA publications. For a complete listing of titles in the Education Series as well as other AIAA publications, please visit http://www.aiaa.org.